Ecological Networks:
Linking Structure to Dynamics
in Food Webs

Santa Fe Institute
Studies in the Sciences of Complexity

Lecture Notes Volume

Author	*Title*
Eric Bonabeau, Marco Dorigo, and Guy Theraulaz	Swarm Intelligence: From Natural to Artificial Systems
Mark E. J. Newman and Richard Palmer	Modeling Extinction

Proceedings Volumes

Editor	*Title*
James H. Brown and Geoffrey B. West	Scaling in Biology
Timothy A. Kohler and George J. Gumerman	Dynamics in Human and Primate Societies
Lee A. Segel and Irun Cohen	Design Principles for the Immune System and Other Distributed Autonomous Systems
H. Randy Gimblett	Integrating Geographic Information Systems and Agent-Based Modeling Techniques
James P. Crutchfield and Peter Schuster	Evolutionary Dynamics: Exploring the Interplay of Selection, Accident, Neutrality, and Function
David Griffeath and Cristopher Moore	New Constructions in Cellular Automata
Murray Gell-Mann and Constantino Tsallis	Nonextensive Entropy—Interdisciplinary Applications
Lashon Booker, Stephanie Forrest, Melanie Mitchell, and Rick Riolo	Perspectives on Adaptation in Natural and Artificial Systems
Erica Jen	Robust Design: A Repertoire of Biological, Ecological, and Engineering Case Studies
Kihong Park and Walter Willinger	The Internet as a Large-Scale, Complex System
Lawrence E. Blume and Steven N. Durlauf	The Economy as an Evolving Complex System, III
Mercedes Pascual and Jennifer A. Dunne	Ecological Networks: Linking Structure to Dynamics in Food Webs

Ecological Networks: Linking Structure to Dynamics in Food Webs

Editors

Mercedes Pascual
University of Michigan

Jennifer A. Dunne
Pacific Ecoinformatics and Computational Ecology Lab
Santa Fe Institute

Santa Fe Institute
Studies in the Sciences of Complexity

UNIVERSITY PRESS

2006

OXFORD
UNIVERSITY PRESS

Oxford University Press, Inc., publishes works that further
Oxford University's objective of excellence
in research, scholarship, and education.

Oxford   New York
Auckland   Cape Town   Dar es Salaam   Hong Kong   Karachi
Kuala Lumpur   Madrid   Melbourne   Mexico City   Nairobi
New Delhi   Shanghai   Taipei   Toronto

With offices in
Argentina   Austria   Brazil   Chile   Czech Republic   France   Greece
Guatemala   Hungary   Italy   Japan   Poland   Portugal   Singapore
South Korea   Switzerland   Thailand   Turkey   Ukraine   Vietnam

Published by Oxford University Press, Inc.
198 Madison Avenue, New York, New York 10016

www.oup.com

Oxford is a registered trademark of Oxford University Press

Library of Congress Cataloging-in-Publication Data
Ecological networks : linking structure to dynamics in food webs /
edited by Mercedes Pascual, Jennifer A. Dunne.
p. cm. — (Santa Fe Institute studies in the sciences of complexity
(Oxford University Press))
Includes bibliographical references (p.).
ISBN-13 978-0-19-518816-5
ISBN 0-19-518816-0
1. Food chains (Ecology) I. Pascual, Mercedes. II. Dunne, Jennifer A.
III. Proceedings volume in the Santa Fe Institute studies in the sciences
of complexity.
QH541.F648 2005
577'.16—dc22   2005047788

9 8 7 6 5 4 3 2 1

Printed in the United States of America
on acid-free paper

About the Santa Fe Institute

The *Santa Fe Institute* (SFI) is a private, independent, multidisciplinary research and education center, founded in 1984. Since its founding, SFI has devoted itself to creating a new kind of scientific research community, pursuing emerging science. Operating as a small, visiting institution, SFI seeks to catalyze new collaborative, multidisciplinary projects that break down the barriers between the traditional disciplines, to spread its ideas and methodologies to other individuals, and to encourage the practical applications of its results.

All titles from the *Santa Fe Institute Studies in the Sciences of Complexity* series carry this imprint which is based on a Mimbres pottery design (circa A.D. 950–1150), drawn by Betsy Jones. The design was selected because the radiating feathers are evocative of the outreach of the Santa Fe Institute Program to many disciplines and institutions.

Contributors List

Andrew P. Allen, *National Center for Ecological Analysis and Synthesis, 735 State Street, Suite 300, Santa Barbara, CA 93101; e-mail: drewa@nceas.ucsb.edu*

David Alonso, *University of Michigan, Department of Ecology and Evolutinary Biology, 830 N. University Avenue, Ann Arbor, MI 48109-1048; e-mail: dalonso@umich.edu*

Richard Baltensperger, *University of Fribourg, Department of Mathematics, Pérolles, Chemin du Musée 23, CH-1700 Fribourg, Switzerland; e-mail: richard.baltensperger@unifr.ch*

Carolin Banašek-Richter, *EcoNet Lab, Darmstadt University of Technology, Schnittspahnstr. 10, D-64287 Darmstadt, Germany; e-mail: richter@bio.tu-darmstadt.de*

Jordi Bascompte, *Integrative Ecology Group, Estación Biológica de Doñana, Apdo. 1056, Sevilla, 41080 Spain; e-mail: bascompte@ebd.csic.es*

Eric L. Berlow, *University of California, San Diego, White Mountain Research Station, 3000 E. Line St., Bishop, CA 93514, and Pacific Ecoinformatics and Computational Ecology Lab, 1604 McGee Ave., Berkeley, CA 94703; e-mail: eric@wmrs.edu*

Louis-Felix Bersier, *Unit of Ecology and Evolution, Department of Biology, University of Fribourg, Ch. du Musée 10, CH-1700 Fribourg, Switzerland; e-mail: louis-felix.bersier@unine.ch*

James H. Brown, *University of New Mexico, Department of Biology, Albuquerque, NM 87131; e-mail: jhbrown@unm.edu*

Guido Caldarelli, *La Sapienza University, Department of Physics, Piazzale. A. Moro 5, Rome 00185 Italy; e-mail: guido.caldarelli@roma1.infn.it*

Cecile Caretta Cartozo, *Ecole Polytechnique Federale de Lausanne (BSP), Laboratoire de Biophysique, Statistique-ITP-FSB, Lausanne 1015, Switzerland; e-mail: cecile.carettacartozo@epfl.ch*

Marie-France Cattin, *Zoological Institute, Rue Emile-Argand 11, C.P. 2, CH-2007, Neuchâtel, Switzerland; e-mail: marie-france.cattin@ne.ch*

Stephanie S. Chow, *Computation and Neural Systems 136-93, California Institute of Technology, Pasadena, CA 91125; e-mail: steph@caltech.edu*

Sarah Cobey, *University of Michigan, Department of Ecology and Evolutionary Biology, 830 N. University Avenue, Ann Arbor, MI 48109-1048; e-mail: cobey@umich.edu*

Laura J. Cushing, *Pacific Ecoinformatics and Computational Ecology Lab, 1604 McGee Ave., Berkeley, CA 94703; e-mail: laracushing@hotmail.com*

Andy P. Dobson, *Princeton University, Department of Ecology and Evolutionary Biology Eno Hall, Princeton, NJ 08544-1003; e-mail: andy@eno.princeton.edu*

Barbara Drossel, *Darmstadt University of Technology, Hochschulstrasse 6, Darmstadt 64289 Darmstadt; e-mail: barbara.drossel@physik.tu-darmstadt.de*

Jennifer A. Dunne, *Pacific Ecoinformatics and Computational Ecology Lab, 1604 McGee Ave., Berkeley, CA 94703, and Santa Fe Institute, 1399 Hyde Park Road, Santa Fe, NM 87501; e-mail: jdunne@santafe.edu*

Jean-Pierre Gabriel, *Département de mathématiques, Ch. du Musée 23, CH-1700 Fribourg, Switzerland; e-mail: jean-pierre.gabriel@unifr.ch*

Diego Garlaschelli, *INFM UdR Siena, Department of Physics, University of Siena, Via Roma 56, 53100, Siena, Italy; e-mail: garlaschelli@csc.unisi.it*

James F. Gillooly, *University of New Mexico, Department of Biology, 77 Castetter Hall, Albuquerque, NM 87106; e-mail: gillooly@unm.edu*

Pedro Jordano, *Integrative Ecology Group, Estación Biológica de Doñana, Apdo. 1056, Sevilla, 41080 Spain; e-mail: jordano@ebd.csic.es*

Armand M. Kuris, *Western Ecological Research Center, USGS, Marine Science Institute, University of California, Santa Barbara, CA 93106; e-mail: kuris@lifesci.ucsb.edu*

Kevin D. Lafferty, *Ecology, Evolution and Marine Biology, University of California, Santa Barbara, CA 93106; e-mail: lafferty@lifesci.ucsb.edu*

Simon Levin, *Princeton University, Department of Ecology and Evolutionary Biology, 203 Eno Hall/Guyot Hall, Princeton, NJ 08544-1003; e-mail: slevin@eno.princeton.edu*

Alan McKane, *University of Manchester, Theory Group, School of Physics and Astronomy, Manchester M13 9PL, England; e-mail: alan.mckane@manchester.ac.uk*

Neo D. Martinez, *Pacific Ecoinformatics and Computational Ecology Lab, 1604 McGee Ave., Berkeley, CA 94703, and Rocky Mountain Biological Laboratory, P.O. Box 519, Crested Butte, CO 81224; e-mail: neo@sfsu.edu*

Jane Memmott, *University of Bristol, School of Biological Sciences, Bristol BS8 1UG Great Britain; e-mail: jane.memmott@bris.ac.uk*

José M. Montoya, *Universitat Pompeu Fabra, Dr. Aiguader 80, Barcelona 08003 Spain; e-mail: jose.montoya@upf.edu*

Scott D. Peacor, *Department of Fisheries and Wildlife, Michigan State University, East Lansing, MI 48824, and Great Lakes Environmental Research Laboratory (NOAA), 2205 Commonwealth Blvd., Ann Arbor, MI 48105; e-mail: peacor@msu.edu*

Mercedes Pascual, *University of Michigan, Department of Ecology and Evolutinary Biology, 830 N. University Avenue, Ann Arbor, MI 48109-1048; e-mail: pascual@umich.edu*

Rick Riolo, *University of Michigan, Center for the Study of Complex Systems, 2071 Randall Lab 1120, Ann Arbor, MI 48103; e-mail: rlr@merit.edu*

Diego Ruiz-Moreno, *University of Michigan, Department of Ecology and Evolutinary Biology, 830 N. University Avenue, Ann Arbor, MI 48109-1048; e-mail: drmoreno@umich.edu*

Ricard Solé, *Universitat Pompeu Fabra, Complex Systems Lab-GRIB, Dr Aiguader 80, Barcelona 08003, Spain; e-mail: ricard.sole@cexs.upf.es*

Joshua Weitz, *Princeton University, Department of Ecology and Evolutionary Biology, Eno Hall, Princeton, NJ 08544; e-mail: jsweitz@princeton.edu*

Claus Wilke, *Keck Graduate Institute, 535 Watson Drive, Claremont, CA 91711; e-mail: wilke@kgi.edu*

Rich J. Williams, *Pacific Ecoinformatics and Computational Ecology Lab, 1604 McGee Ave., Berkeley, CA 94703, and San Francisco State University, Computer Science Department, Thornton Hall 906, 1600 Holloway Avenue, San Francisco, CA 94132; e-mail: rich@sfsu.edu*

Workshop Participants

David Alonso, *University of Michigan*

Jordi Bascompte, *Estación Biológica de Doñana*

Eric Berlow, *University of California, San Diego, White Mountain Research Station, and Pacific Ecoinformatics and Computational Ecology Lab*

Ulrich Brose, *Darmstadt University of Technology*

Claudia Codeço, *Oswaldo Cruz Foundation (Fiocruz)*

Joel Cohen, *Rockefeller University*

Lara Cushing, *Pacific Ecoinformatics and Computational Ecology Lab*

Andy Dobson, *Princeton University*

Barbara Drossel, *Darmstadt University of Technology*

Jennifer Dunne, *Pacific Ecoinformatics and Computational Ecology Lab and Santa Fe Institute*

Doug Erwin, *National Museum of Natural History, Smithsonian Institution and Santa Fe Institute*

Gregor Fussmann, *McGill University*

Jamie Gillooly, *University of New Mexico*

Michelle Girvan, *Santa Fe Institute*

Alan Hastings, *University of California, Davis*

Alex Herman, *Santa Fe Institute*

Sanjay Jain, *Indian Institute of Science*

Michio Kondoh, *Netherlands Institute of Ecology*

Simon Levin, *Princeton University*

Per Lundberg, *Lund University*

Neo Martinez, *Pacific Ecoinformatics and Computational Ecology Lab and Rocky Mountain Biological Laboratory*

Alan McKane, *University of Manchester*

Carlos Melian, *National Center for Ecological Analysis and Synthesis*

Jane Memmott, *University of Bristol*

José Montoya, *Universitat Pompeau Fabra and Universidad de Alcalá*

Diego Moreno, *University of Michigan*

Mercedes Pascual, *University of Michigan*

Scott Peacor, *Michigan State University*

Tamara Romanuk, *University of Quebec, Montreal and Pacific Ecoinformatics and Computational Ecology Lab*

Ricard Solé, *Universitat Pompeau Fabra*

Chris Warren, *University of Michigan*

Joshua Weitz, *Princeton University*

Claus Wilke, *Keck Graduate Institute*

Rich Williams, *Pacific Ecoinformatics and Computational Ecology Lab and San Francisco State University*

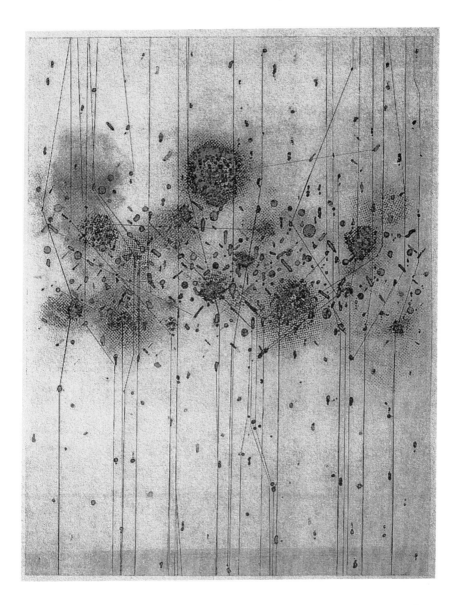

Nodes and links (by Roberto Pascual 2005, printed with permission).

Dedication

To:

Maria Spada Pascual

Ann Louise Dunne

Agnes Edmunds Cohen

Contents

Preface

This book arose from a recent workshop "From Structure to Dynamics in Complex Ecological Networks" held in February, 2004, at the Santa Fe Institute (SFI). The workshop brought together ecologists working on food web theory, structure, and dynamics with researchers from other fields in ecology and complex systems who are interested in large adaptive networks. The main goals of the workshop were to explore the boundaries of what is known about the relationship between structure and dynamics in large ecological networks and to define directions for future progress in this field. Chapters in this book developed out of the many interesting talks presented at the meeting and from lively discussions following those talks, as well as separate smaller group discussions on specific topics. We thank all of the participants for their input, ideas, and hard work, and for being a part of the stimulating discourse that makes science exciting, challenging, and fun.

The workshop was made possible by the generous support of the James S. McDonnell Foundation through a Centennial Fellowship to Mercedes Pascual. We thank the foundation for supporting not just this meeting but Complex Systems research in general and for promoting the freedom to explore ideas. This workshop was also funded through grants from the National Science Foundation, DBI-0234980 and ITR-0326460, that support Jennifer A. Dunne and other members of the Pacific Ecoinformatics and Computational Ecology Lab (PEaCE Lab). Thanks go to the PEaCE Lab for its participation and support, particularly for adding *joie de vivre* to the process.

Additional support was provided by the Santa Fe Institute. We could not think of a better place than SFI to hold this workshop. The setting is magical and the intellectual environment highly stimulating. The organizational support was excellent. We particularly thank Sarah Hayes, the Event and Meeting Manager, for making the logistical arrangements and helping the workshop to run smoothly, and we also appreciate the support given by the rest of the staff at SFI during the workshop. We give many thanks to Ronda Butler-Villa, the Director of Publications, Facilities, and Personnel, for shepherding this book along the winding path from proposal to contract, to Della Ulibarri, the Production Manager, who oversaw the transformation of the book from a hodgepodge of final drafts to a polished set of final proofs, and to Laura Ware, Publications Assistant, who copy edited the book.

We take this opportunity to remember Peter Yodzis who could not attend the workshop and passed away during the preparation of this volume—his contributions to ecology were extensive and fundamental. In particular, he made great strides in integrating theory and mathematical approaches with empirical data and applied concerns within the area of food-web research. He is cited extensively within this volume, and his research will continue to shape future work in this subject. He will be deeply missed.

We hope that some of the excitement of the workshop comes across in the chapters and in the many open areas for future research sketched in the concluding chapter on "Challenges for the Future," and that readers enjoy reading this book as much as we enjoyed the time at SFI and assembling these pages.

Mercedes Pascual
University of Michigan

Jennifer A. Dunne
Pacific Ecoinformatics and
   Computational Ecology Lab
Santa Fe Institute

Introduction

"It is a very long haul from handling a small group of four species like the lemon tree, the nightshade, the black scale, and a chalcid parasite, to the contemplation of the almost inconceivable and profuse richness of a tropical rain forest, or even to the several thousand species living in Wytham Woods, Berkshire. It is a question for future research, but an urgent one, how far one has to carry complexity in order to achieve any sort of equilibrium."

— Charles Elton,
*Ecology of Invasions by Animals
and Plants*, 1958

From Small to Large Ecological Networks in a Dynamic World

Mercedes Pascual
Jennifer A. Dunne

.

Food webs are one of the most useful, and challenging, objects of study in ecology. These networks of predator-prey interactions, conjured in Darwin's image of a "tangled bank," provide a paradigmatic example of complex adaptive systems. While it is deceptively easy to throw together simplified caricatures of feeding relationships among a few taxa as can be seen in many basic ecology text books, it is much harder to create detailed descriptions that portray a full range of diversity of species in an ecosystem and the complexity of interactions among them (fig. 1). Difficult to sample, difficult to describe, and difficult to model, food webs are nevertheless of central practical and theoretical importance. The interactions between species on different trophic (feeding) levels underlie the flow of energy and biomass in ecosystems and mediate species' responses to natural and unnatural perturbations such as habitat loss. Understanding the ecology and mathematics of food webs, and more broadly, ecological networks, is central to understanding the fate of biodiversity and ecosystems in response to perturbations.

Ecological Networks: Linking Structure to Dynamics in Food Webs, edited by Mercedes Pascual and Jennifer A. Dunne, Oxford University Press.

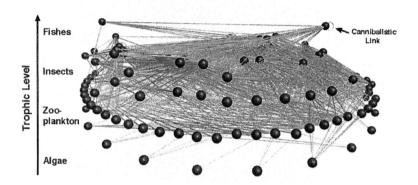

FIGURE 1   A detailed food web of Little Rock Lake, WI, with 997 feedings links among 92 taxa (Martinez 1991). Ecosystems are complex networks consisting of many species that interact in many different ways with each other and their environment. Food webs focus on the feeding relationships among co-occurring species in a particular habitat. In this image, each node represents a "trophic species" that may be a biological species, a group of species, a life-history stage of a species, or organic matter such as detritus (Briand and Cohen 1984). Each trophic species in a food web is functionally distinct; i.e., has a set of predators and prey that differ, even if only by one predator or prey item, from those of other trophic species. This image was produced using FoodWeb3D software written by R. J. Williams and provided by the Pacific Ecoinformatics and Computational Ecology Lab ⟨www.foodwebs.org⟩.

Research on ecological networks is also important for understanding the consequences of biodiversity itself for ecosystem function. Much theoretical and empirical food-web research, as well as other ecological research, has oriented itself around various notions of stability (Box 1). Ultimately, stability properties matter to the functioning of ecosystems and to the all-to-often unacknowledged services they provide to humans (Box 2). While a large body of research addresses the relationship between biodiversity and ecosystem functions such as primary productivity, the ecological networks considered in those studies are generally restricted to one or two trophic levels, and usually focus on interactions among competitors for one or a few resources (Kinzig et al. 2002; but see Montoya et al. 2003). The importance of extending studies of biodiversity-ecosystem function to include multiple trophic levels and complex predator-prey interactions has been recently emphasized (Dobson et al. in press; Worm and Duffy 2003). A better understanding of dynamics in large, complex networks is an important intermediate step in addressing how the functioning of ecosystems is influenced by structural properties of the underlying network.

Box 1: Different definitions and concepts of "stability" used in ecology

- **Local Asymptotic Stability (LAS):** An equilibrium is said to be locally stable if arbitrarily small perturbations away from this steady state always decay. Close to equilibrium, nonlinear systems can be approximated by a linear set of equations, $d\mathbf{x}/dt = \mathbf{A}\mathbf{x}$, that govern the dynamics of perturbations (with the vector $\mathbf{x}$ specifying the deviations from equilibrium). Written in this form, the system is specified by a matrix $\mathbf{A}$, the so-called community matrix in the case of ecological networks, whose dominant eigenvalue rules the exponential decay or growth of perturbations in the long term (May 1973; Pimm 1982). Local stability technically refers to asymptotic or long-term behavior, thus providing no information on the short-term or transient response to perturbations (Neubert and Caswell 1997; Chen and Cohen 2001b). For nonlinear systems, it is well known that the possible coexistence of multiple equilibria (and other attractors) further limits the relevance of this form of stability.

- **Resilience:** Resilience is closely related to the concept of local asymptotic stability and measures how fast a stable system returns to equilibrium following a perturbation away from it. Resilience is quantified as the absolute magnitude of the largest real part of any of the eigenvalues of the community matrix (Pimm and Lawton 1977; Pimm 1982). Ecosystem resilience has also been defined as the magnitude of perturbation that can be tolerated before a change in system control and structure (Holling and Gunderson 2002), but this definition is better described by the terms resistance or robustness.

- **Reactivity:** Reactivity is one of the measures characterizing the short-term transient response of a locally stable system to a perturbation away from equilibrium. It specifies whether perturbations can initially grow and be amplified, in spite of the eventual return of the system to equilibrium. Reactivity is calculated as the maximum instantaneous rate at which perturbations away from equilibrium can be amplified (Neubert and Caswell 1997; Chen and Cohen 2001b; Ruiz-Moreno et al. Chapter 7). Other relevant quantities characterizing the "stability" of transients are the size and time of the maximum amplification of perturbations (Neubert and Caswell 1997). The following measures move away from an emphasis on small perturbations in an arbitrarily small neighborhood of equilibria, and in some cases, away from the focus on equilibrium behavior altogether, allowing for stability concepts related to more complex nonlinear dynamics, such as cycles and chaos.

continued on next page

In the rest of this introduction we provide a brief sketch of the historically central problem in trophic ecology, the relationship between species diversity and community stability. This points the reader to a few of many key books and papers in food-web research (see also Box 3), and sets the stage for outlining how the chapters in this book further develop issues of complexity and stability and

---

Box 1 continued

• **Qualitative Global (Asymptotic) Stability (QGAS):** Qualitative stability refers to the tendency of a system to return to equilibrium when interaction coefficients are specified only by their sign and not by their magnitude. It is global in the sense that this return is also independent of initial conditions. Thus, QGAS can evaluate responses to large perturbations. The qualitative theory of nonlinear differential equations has been applied to the QGAS of food webs (Cohen et al. 1990; Chen and Cohen 2001a).

• **Permanence and persistence:** These measures do not rely on the existence of any single specific type of attractor but focus instead on whether species remain in the system. Thus, permanence measures whether the system's variables remain bounded and positive. More technically, a system is said to be permanent if the boundary of the positive quadrant of state space is a repellor (Hofbauer and Sigmund 1998; Chen and Cohen 2001a). One way to determine permanence is by numerical invasibility analysis, in which the species are deleted one at a time to examine if they can reinvade from arbitrarily small numbers. For more technical criteria (i.e., sufficient conditions for permanence) applied to Lotka-Volterra systems, see Jansen (1987), Law and Morton (1993), and Chen and Cohen (2001a). Persistence is in turn determined via numerical simulation, by examining the trajectories of the dynamical system for a large number of initial conditions and for a prescribed window of time (e.g., Martinez et al. Chapter 6). For the possible discrepancies between permanence and persistence, see Law and Morton (1993) or Chen and Cohen (2001).

• **Invasibility:** Invasibility measures the likelihood that new species are able to invade an established community of interacting species. It is generally evaluated in assembly models of ecological networks (e.g., Kokkoris et al. 1999).

• **Variability:** Variability measures the magnitude of fluctuations in species' numbers. For a given species, it is computed as the coefficient of variation (standard deviation divided by the mean) of its abundance over time (Pimm 1984).

• **Robustness:** Robustness focuses on the persistence of features of interest in a system's response to perturbations, particularly those the system does not normally experience in its development or history (Jen 2003). It has been measured in a variety of ways depending on the system being studied, for example the likelihood of cascading secondary extinctions resulting from primary biodiversity loss in ecological networks (e.g., Dunne et al. 2002b).

---

expand into new areas of structure and dynamics, especially within the context of broader theory related to networks and complex systems.

The nodes of Darwin's "tangled bank" are species, whose identity, abundances, and biomasses provide means of measuring biological diversity. The links between the nodes (fig. 1) represent feeding relationships, and potentially other interactions, that account for the connectivity of ecological networks. Species richness and connectance, the proportion of possible interactions that actually occur, have often been used in food-web research as basic measures of complexity. These simple measures are highlighted in the influential work of May (1972, 1973) on the relationship between complexity and ecosystem stability. In this and other early theoretical food-web research, dynamical stability was equated with the mathematical concept of local stability, measured as the tendency of an arbitrarily small perturbation to grow or contract in the proximity of an equilibrium point. Much has been said about the relevance of this concept in ecology (e.g., McCann 2000; Dunne et al. 2005), and other interpretations of ecosystem stability have been used, including resilience, invasibility, persistence, permanence, and robustness (e.g., Holling 1973; Pimm 1984; McCann, 1998; Kokkoris et al. 1999; Chen and Cohen 2001a; see also Box 1).

Connectance and species richness provide simple characterizations of ecological network structure by describing global properties of an entire network of interacting species. A much larger set of properties that describe other aspects of global structure as well as more local characteristics of species and links quickly accumulated to address the existence of regularities in the structure of food webs (Lawton 1989; Cohen et al. 1990a; Pimm et al. 1991). More recently, the explosion of research on network topology as a fundamental aspect of the study of complex systems (Strogatz 2001; Albert and Barabási 2002) has stimulated new efforts to unravel regularities of ecological network structure (e.g., Camacho et al. 2002; Dunne et al. 2002a; Garlaschelli et al. 2003; Jordano et al. 2003).

Species richness and connectance thus provided an early, and continuing, point of contact and integration between quantitative assessments of both dynamics and structure. Such integration has even deeper roots, since it emerged out of May's challenge to prior claims that complexity should enhance stability of ecosystems (e.g., Odum 1953; MacArthur 1955), which in turn emerged out of the very early recognition of the fundamental connection between ecological structure and function (Elton 1927). This core ecological notion continues to inform much current research. Interestingly, one of the most challenging problems across many scientific fields today is the relationship between the structure of complex networks and their nonlinear dynamics (Strogatz 2001). Complex networks are characterized not only by numerous components that interact, but by interactions that are nonlinear, are distributed non-randomly, and are adaptive, i.e., changing continuously in response to the state of the system itself. These characteristics obviously apply to food webs, and more broadly, to ecosystems. Ways that ecologists come to understand the interplay between structure and dynamics of ecosystems can, and should, influence research in other fields, as well as the reverse.

Within ecology, even though complex webs of many interacting species are widely observed in nature, understanding their persistence remains puzzling. This is largely due to methodological limitations. It is basically impossible to manipulate diverse assemblages of species in natural systems in a controlled, replicated, ethical fashion; experimentation using microcosms is limited in the diversity it can embrace; simple models may not be scalable to more complex systems or may oversimplify biology; and conducting nonlinear dynamical simulations of "virtual ecosystems" that are diverse and incorporate plausible biological processes is only now becoming computationally tractable, but presents its own challenges such as how to sensibly explore giant parameter spaces. Earlier complexity-stability research has generally focused on simple dynamical models such as linear approximations close to equilibrium and Lotka-Volterra equations; these early models often suggested that complexity gets in the way of stability (e.g., May 1972; Pimm and Lawton 1977, 1978). Since then, the inclusion of plausible structural elements, nonlinearities, and/or variable interaction strengths (De Angelis 1975; Yodzis 1981; Pimm 1982; de Ruiter et al. 1995; McCann and Hastings 1997; McCann et al. 1998) in simple models provide some glimpses at constraints that might lead to stability or persistence of diverse, complex systems. Understanding stability and persistence, in turn, can help us understand how and why ecosystems are robust or sensitive to human-related perturbations, the central concern of conservation biology.

The interplay between ecological structure and dynamics has many nuances. For example, the notion of incorporating not only the presence or absence of links between species but the strength of those interactions is clearly important for dynamical modeling (starting with De Angelis 1975), and there may be general patterns of how interaction strengths are distributed in ecological networks that relate to stability (Neutel et al. 2002). A different issue concerns how ecological network structure is studied. Most structural studies focus on "snapshots" of ecological networks, by compiling a master list of species and their observed interactions integrated over some sufficiently inclusive, albeit somewhat arbitrary, temporal and spatial scale. This strategy allows researchers to go beyond the contingent details of local dynamics of a few interacting species to look for a bigger, simpler picture where coarse-grained patterns, ideally with governing processes, may (or may not!) emerge (Brown 1995; Lawton 1999). However, structure is clearly dynamic at many scales, changing in time and space as the result of the plasticity of ecological interactions, environmental variation, and assembly processes that occur at both ecological and evolutionary time scales. Dynamic aspects of structure may also prove crucial for our understanding of stability and persistence.

While most observations and theory pertain to webs of trophic interactions, recent efforts that focus on parasites and mutualistic relationships underscore the importance of extending the scope of ecological network studies to embrace other kinds of interactions. Furthermore, the large and increasing body of ecological literature on higher-order interactions such as phenotypic plasticity (see review by Bolker et al. 2003) demonstrates that the strength of interactions between species is not constant but varies in response to various indirect effects that extend beyond the densities of directly interacting species (Peacor and Werner 2001). Consequences of such plasticity and other types of indirect effects, which can play as important a role as direct effects (Menge 1995), need to be explored

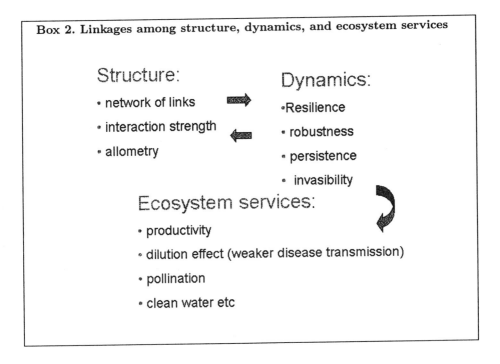

Box 2. Linkages among structure, dynamics, and ecosystem services

systematically with regard to their impact on the dynamics and structure of high-dimensional systems.

The first section of this book focuses on the history of and recent advances in uncovering regularities and universal patterns in the structure of complex ecological networks, with a focus initially on food webs that is followed by work on other types of interactions (parasitism, mutualism). The second section moves to the explicit coupling of structure and dynamics, and is followed by a set of chapters that address various aspects of adaptation at ecological and evolutionary time scales. The importance of an ecological network approach to conservation and restoration is emphasized in the final section, which includes an outline of a series of open empirical and theoretical questions. More challenges for the future are raised in the concluding chapter. A brief roadmap of these sections follows below.

STRUCTURE OF COMPLEX ECOLOGICAL NETWORKS

Within the study of complex systems, one pervasive subject is the characterization of the structure of biotic and abiotic networks that include dozens to millions of nodes. From studies of the structure of the internet and the World Wide Web, to those of the social links that underlie the transmission of innovation as well as infectious disease, network research is becoming omnipresent across disciplines, spawning thousands of papers and a number of popular books (e.g., Watts 1999, 2003; Barabási 2002; Buchanan 2002; Strogatz 2003). Well before the current re-popularization of research on the non-random, non-regular structure of "real-world" networks (Watts and Strogatz 1998), as informed by earlier graph theory (Erdös and Rényi 1960), ecologists were starting to characterize apparent generalities in the non-random structure of food-web networks (Cohen 1978) and to develop models inspired by graph theory to explain observed generalities (Cohen and Newman 1985). There is a rich and stormy history of empiricism and modeling related to complex food-web structure, with important ecological and graph-theory contexts that extend to the present day (Dunne Chapter 2; Cartozo et al. Chapter 3). Just as other kinds of network structure research have undergone a renaissance, there has been a recent renewed interest in potential generalities and simple models relating to the structure of food webs. This interest has resulted in a number of novel studies that are building on a new generation of improved food-web data (Dunne Chapter 2; Cartozo et al. Chapter 3).

For the most part, the links typically considered in the literature are conventional "predator-prey" interactions. Recent work has begun to expand this view to include other kinds of interactions. One important example is parasitism (Dobson et al. Chapter 4). The unintended introduction of parasites into ecosystems, such as rinderspest in the Serengeti and lampreys into the Great Lakes, provide some dramatic examples of the important role parasites can play in the dynamics of food webs (Dobson Chapter 4, Box C). While a few excellent studies have explicitly considered or focused on the role of parasites and parasitoids in ecological network structure (Huxham et al. 1996; Martinez et al. 1999; Memmott et al. 2000; Leaper and Huxhman 2002), more are needed. A recent study on coastal salt marshes (Lafferty et al. in press) illustrates the enormous field efforts required to obtain the data on both trophic and parasitic interactions. More importantly, it demonstrates the critical role played by parasites in both the structure and the biomass flow of ecosystems. Recent analysis of these data demonstrates that parasitic links are not distributed randomly upon the underlying trophic network (Warren et al. in prep). Instead, a clustering structure is apparent with measures that extend the concept of clustering coefficient (Dunne Chapter 2; Cartozo et al. Chapter 3) to networks with two types of links, trophic and parasitic ones. Because parasites are typically smaller than their hosts, their consideration is also likely to illuminate the role of size in determining network structure.

Research on ecological network structure is also being expanded to include plant-animal mutualisms (Bascompte and Jordano Chapter 5). While these are also a type of consumer-resource interaction, their consequences include positive effects that go far beyond classic "negative" feeding effects. For example, an animal feeds on the fruit of a plant, and then benefits that plant by acting as a dispersal vector for its seeds. These types of not-strictly-feeding interactions are revealing fascinating nonrandom network patterns that reflect coevolutionary dynamics likely to have important consequences for conservation. Those consequences (e.g., Memmott et al. 2004) are largely unexplored, and represent an important future area of research (Bascompte and Jordano Chapter 5; Memmott et al. Chapter 14).

INTEGRATING ECOLOGICAL STRUCTURE AND DYNAMICS

Two challenges are immediately obvious in addressing the dynamics of large networks. The first one concerns the need to specify appropriate non-random structures consistent with empirical patterns. The second one is that predator-prey and therefore food-web models are prototypical examples of nonlinear systems, capable as such of generating a variety of non-equilibrium behaviors including cycles and chaos (Hastings and Powell 1991; Fussman and Heber 2002). While mathematical analysis is difficult for low-dimensional nonlinear systems and usually impossible for large ones, the alternative of numerical exploration of parameter space through computer simulation is quickly limited by the size of the system being explored.

Static (probabilistic) models to generate the non-random trophic structure of ecological communities (Dunne Chapter 2) provide a starting point to investigate the dynamics of more realistic network models. For example, the niche model of Williams and Martinez (2000) successfully generates a network of links that shares many properties with empirical food webs. A dynamical model for trophic interactions can then be built upon the resulting structure (Martinez et al. Chapter 6). This set of differential equations for the nonlinear population dynamics of the interacting species is based on a bioenergetic model (Yodzis and Innes 1992) describing realistic biology with relatively simple parameters. The resulting hybrid model, which couples structure and dynamics, allows for the persistence of a surprisingly large number of species, as demonstrated with numerical explorations of parameter space and comparisons to the corresponding models for less realistic structures, including random ones. Such hybrid models provide a promising way to address feedbacks between structure and dynamics. An earlier example is found in the Lotka-Volterra cascade model (LVCM) which generates network structure via the cascade model, a predecessor of the niche model (Cohen et al. 1990b). Interestingly, the qualitative global stability (Box 1) of this hybrid model was studied analytically in the limit of a large number of species and long time sequences. Analytical results on these high-dimensional nonlinear systems are rare, if almost non-existent, and focus on ecological networks in which the signs and not the numerical values of the interaction coefficients are considered. They provide stability conditions that can be evaluated numerically (e.g., Chen and Cohen 2001a). As an alternative, numerical simulations of the dynamical system itself must be performed, but these still require a number of simplifying assumptions, such as similar values for specific parameters across species, to reduce the enormous size of parameter space (Martinez et al. Chapter 6).

Genetic algorithms (GAs; Holland 1975; Mitchell 1996) provide one possible numerical approach to explore the large space of parameters specifying the dynamics of networks (e.g., Sporns and Tononi 2002). They further allow the exploration of the large "network" space of possible structures, where the links between species are not predefined. Instead of specifying a structure and asking what are its dynamic consequences, one can address the opposite question of which structures exhibit particular dynamical properties of interest. Network space is explored by an "evolutionary" process that selects for those networks with the desired dynamical properties. Ruiz-Moreno et al. (Chapter 7) provides an example of an application to the local stability of food webs, and explores the relationship between network modularity and the response of the system to perturbations. Although this first application to food webs considers only linear dynamics close to equilibrium, the nonlinear case can also be explored but is computationally more taxing (see Sporns and Tononi [2002] for an example in neurobiology). While earlier work had considered the effect of compartments on long-term stability, results in this chapter indicate that modularity may be critical in short-term transient responses. The importance of considering short-term transient responses when evaluating the stability of food webs has also been recently emphasized by Chen and Cohen (2001b).

Another approach to reduce the size of network and parameter space consists of constraining parameters or structure based on empirical findings (e.g., Yodzis 1982; Emmerson and Raffaelli 2004). One potential avenue for realistic parameterizations is given by allometric scalings describing how specific parameters vary as a power-law function of the size of organisms (e.g., Raffaelli and Hall 2004). This approach is not new to biological oceanographers and plankton ecologists, but a substantial part of their work in this area concerned ecosystem models, and targeted steady-state dynamics and explanations for observed power-law distributions of biomass in pelagic ecosystems (e.g., Platt and Denman 1978). The few studies of the nonlinear behavior of plankton food-web models constructed with allometric scaling showed a striking propensity for unstable equilibria and pronounced fluctuations, raising questions about their applicability to nature (e.g., Moloney and Fields 1991). In the last decade, through a separate route not directly connected to trophic interactions, terrestrial ecologists have become increasingly fascinated by allometric scalings in both physiological and community patterns (e.g., Enquist and Niklas 2001; West et al. 1997; Williams 1997). While much elegant theory has focused on explanations of these patterns, steady-state (often linear) considerations have been the norm. The use of the patterns themselves in the formulation of dynamical models for ecological networks has not been pursued. Gillooly et al. (Chapter 8) begin to explore the connection of allometry to food web theory, with regard to the use of the two different ecological currencies of numbers and energy. The final chapter on challenges for the future (Pascual et al. Chapter 15) returns to allometry and emphasizes the importance of considering deviations from both steady-state conditions and simple power laws in community patterns, to understand responses to perturbations.

Finally, the daunting size of network space can be tackled with theoretical approaches on the assembly of communities (Yodzis 1982; Post and Pimm 1983; Law and Blackford 1992; Solé et al. 2002; see also the section on "Community Assembly" in the review by Hall and Rafaelli 1993). Species are allowed to sequentially invade a previously established local community from a predefined regional pool; parameter space is thus restricted by the trajectory of the assembly process itself for persistent local networks. This is a powerful approach but one in which properties of the regional pool can exert important constraints on the local assembly process. One avenue to define those constraints within the model itself is to incorporate explicitly the evolutionary time scale.

ECOLOGICAL NETWORKS AS EVOLVING, ADAPTIVE SYSTEMS

A number of theoretical studies are beginning to address how co-evolutionary processes shape the structure of the resulting food webs (McKane and Drossel Chapter 9), and through this structure, their response to perturbations. Evolutionary processes can also influence dynamical properties by modifying interactions over ecological time scales if sufficiently fast, as has been shown with a predator-prey model formulated at the individual level (Hargvigsen and Levin 1997). The central issue of "how evolution shapes ecosystem properties, and whether ecosystems become buffered to changes (more resilient) over their ecological and evolutionary development or proceed to critical states and the edge of chaos" (Levin 1998; see also Kauffman 1993) could have been explicitly written for the ecological networks that underlie ecosystems. In fact, the view of ecosystems as "prototypical examples of complex adaptive systems" (Levin 1998) is very much based on properties of ecological networks, namely the nonrandom distribution and nonlinear nature of interactions between species, and levels of selection that are below that of the whole structure. Different chapters in this book (Ruiz-Moreno et al. Chapter 7; Peacor et al. Chapter 10; Wilke and Chow Chapter 11) illustrate approaches from CAS. In addition, computational approaches centered on the concept of digital organisms begin to provide the tools to explore the assembly and dynamics of ecological networks *in silico* (Peacor et al. Chapter 10; Wilke and Chow Chapter 11). Several chapters (McKane and Drossel Chapter 9; Martinez Chapter 12; Pascual et al. Chapter 15) outline a series of open questions at the interface of ecology and evolution. The concept of instability boundaries, and the related idea of "self-organized instabilities" (Solé et al. 2002) are discussed, emphasizing the need to better understand under what conditions ecological networks converge to these boundaries (Pascual et al. Chapter 15).

Besides evolutionary and ecological time scales, the dynamics of ecological networks are influenced by changes that are behavioral and modify the structure of interactions. Phenotypic plasticity and other forms of higher-order interactions underscore the adaptive nature of food webs at shorter, behavioral, time scales (Peacor et al. Chapter 10). The structure itself must now be viewed as dynamic, continuously changing as a function of the state of the system. It is increasingly recognized that interaction strength varies not just as a function of the two species directly involved in the interaction, but with the abundance of other species in the network. Well-known examples of these so-called higher-order interactions are given by predator switching, in which predation on any one prey is also a function of the relative abundances of the other prey, and by predator (or pathogen) regulation of herbivores (Packer et al. 2003). For example, predator switching was incorporated in plankton ecosystem models, where its stabilizing effect on equilibria has been well known for a long time (e.g., Fasham et al. 1990). It was also considered in a recent theoretical study of large food webs to reexamine the question of complexity and stability (Kondoh 2003). Another suite of examples is provided by phenotypic plasticity. Changes in behavioral or physiological traits of a prey, driven by the presence of a predator, affect in turn its own ability to forage and therefore its own predation rate (e.g., Peacor et al. Chapter 10; Peacor and Werner 2001). Indirect interactions of this sort are not mediated by density but by traits, and contrast with the lethal effects of predators typically represented in graphs of food webs. On the dynamic front, experiments and models have shown that phenotypic plasticity can have significant effects, although only low numbers of species have been considered so far (see review by Bolker et al. 2003). These studies have also represented phenotypic plasticity with a prescribed functional form at the population level (e.g., Ives and Dobson 1987). Because these functional forms are not really known, there is significant scope for formulating models at the individual level and using the models themselves to examine representations at the more aggregated population level (Bolker et al. 2003; discussion in Peacor et al. Chapter 10).

STABILITY AND ROBUSTNESS OF ECOLOGICAL NETWORKS

Although the relationship between complexity and "stability" is a fascinating theoretical question that continues to prove mathematically and computationally challenging after many decades, it is important to recognize that a main motivation behind the early thoughts on this subject was to address applied issues in the conservation and management of ecological systems. For example, in his still cited masterwork "The Ecology of Invasions by Animals and Plants," Elton (1958) wrote that "the enormous problem still is to manage, control, and where necessary alter the pattern of food-chains in the world, without upsetting the balance of their populations. It is this last problem that has not by any means been solved, and which is exacerbated every year by the spread of species to new lands." On a more, if not overly, optimistic tone he added, "Once the notion is grasped that complexity of populations is a property of the community, to be studied and used in conservation, there is hardly any limit to the ways in which it could be introduced."

The last section of this book treats the subject of ecological networks and their robustness to perturbations within the realm of conservation and restoration. The effects of habitat destruction and fragmentation are addressed, as well as the collapse of ecological networks following spatial perturbations and species' extinctions. One way of thinking about these issues is in the context of ecosystem "robustness," where robustness is a type of stability that focuses on the persistence of features of interest in a system's response to perturbations, particularly those it does not normally experience in its development or history (Jen 2003). Thus, robustness is a useful way to think about ecosystem response to perturbations such as species loss. Memmott et al. (Chapter 14) outline a series of open questions for future theory but also empirical work. The case is convincingly made for the importance of a network, multi-trophic, approach to conservation and restoration. Solé et al. (Chapter 13) emphasize spatial considerations from a network perspective.

Ultimately, a significant proportion of the quality and possibly even the persistence of human life is dependent upon the conservation of natural ecosystems. The conservation of functioning ecosystems will only benefit from a deeper understanding of the robustness and resilience of ecological networks, and of the structural features that underlie their dynamical responses to perturbations. It is one of the defining scientific problems of the twenty-first century as only now do we have the computing power to examine problems which involve evolving nonlinear interactions among large numbers of different components whose sizes, rates of birth-death, interaction strengths, and other factors vary over many orders of magnitude. The problems have daunting scales of complexity, yet their solution will not only contribute to the development of the mathematics of complexity, they may also enhance our ability to preserve the invaluable diversity of nature.

Box 3. Food-web books and reviews

Selected books and reviews focused on food-web research are listed, with citations provided in chronological order. Several of the edited volumes listed under (A) contain useful review chapters, but they are not cited separately in the reviews listed under (B).

(A) Food-web books

May, R. M. 1973. *Stability and Complexity in Model Ecosystems.* Princeton University Press. Reprinted in 2001 as a "Princeton Landmarks in Biology" edition.

Cohen, J. E. 1978. *Food Webs and Niche Space.* Princeton University Press.

Pimm, S. L. 1982. *Food Webs. Chapman and Hall.* Reprinted in 2002 as a 2nd edition by University of Chicago Press.

DeAngelis, D. L., W. M. Post, and G. Sugihara, eds. 1983. "Current Trends in Food Web Theory." ORNL-5983, Oak Ridge National Laboratory.

Cohen, J. E., F. Briand, and C. M. Newman. 1990. *Community Food Webs: Data and Theory.* Springer-Verlag.

Christensen, V., and D. Pauly. 1993. "Trophic Models of Aquatic Ecosystems." ICLARM.

Polis, G. A., and K. O. Winemiller, eds. 1996. *Food Webs: Integration of Patterns and Dynamics.* Chapman and Hall.

Polis, G. A., M. E. Power, and G. R. Huxel, eds. 2003. *Food Webs at the Landscape Level.* University of Chicago Press.

Belgrano, A., U. Scharler, J. A. Dunne, and R. E. Ulanowicz, eds. 2005. *Aquatic Food Webs: An Ecosystem Approach.* Oxford University Press.

Moore, J., P. de Ruiter, and V. Wolters. In press. *Dynamic Food Webs: Multispecies Assemblages, Ecosystem Development, and Environmental Change.* Academic Press.

(B) Food-web reviews

May, R. M. 1983. "The Structure of Food Webs." *Nature* 301:566–568.

May, R. M. 1986. "The Search for Patterns in the Balance of Nature: Advances and Retreats." *Ecology* 67:1115–1126.

Strong, D. R., ed. 1988. "Food Web Theory: A Ladder for Picking Strawberries." Special Feature. *Ecology* 69:1647–1676. Includes five articles by R. T. Paine, J. E. Cohen and C. M. Newman, A. Hastings, S. L. Pimm and R. L. Kitching, and R. H. Peters.

continued on next page

Box 3 continued

Lawton, J. H., and P. H. Warren. 1988. "Static and Dynamic Explanations for Patterns in Food Webs." *Trends Ecol. & Evol.* 3:242–245.

Lawton, J. H. 1989. "Food Webs." In *Ecological Concepts*, ed. J. M. Cherett, 43–78. Oxford: Blackwell Scientific.

Schoener, T. W. 1989. "Food Webs from the Small to the Large." *Ecology* 70:1559–1589.

Pimm, S. L., J. H. Lawton, and J. E. Cohen. 1991. "Food Web Patterns and Their Consequences." *Nature* 350:669–674.

Cohen, J. E., R. A. Beaver, S. H. Cousins, D. L. De Angelis, and 20 others. 1993. "Improving Food Webs." *Ecology* 74:252–258.

Hall, S. J., and D. G. Raffaelli. 1993. "Food Webs: Theory and Reality." *Adv. Ecol. Res.* 24:187–239.

Warren, P. H. 1994. "Making Connections in Food Webs." *Trends Ecol. & Evol.* 9:136–141.

Martinez, N. D. 1995. "Unifying Ecological Subdisciplines with Ecosystem Food Webs." In *Linking Species and Ecosystems*, ed. C. G. Jones and J. H. Lawton, 166–175. New York: Chapman and Hall.

Hall, S. J., and D. G. Raffaelli. 1997. "Food-Web Patterns: What do We Really Know." In *Multi-Trophic Interactions in Terrestrial Systems*, ed. A. C. Gange, and V. K. Brown, 395–417. Oxford: Blackwell Scientific.

Martinez, N. D., and J. A. Dunne. 1998. "Time, Space, and Beyond: Scale Issues in Food-Web Research." In *Ecological Scale: Theory and Applications*, ed. D. Peterson, and V. T. Parker, 207–226. New York: Columbia University Press.

McCann, K. S. 2000. "The Diversity-Stability Debate." *Nature* 405:228–233.

Borer, E. T., K. Anderson, C. A. Blanchette, B. Broitman, S. D. Cooper, and B. S. Halpern. 2002. "Topological Approaches to Food Web Analyses: A Few Modifications May Improve our Insights." *Oikos* 99:397–401.

Jordán, F. 2002. "Searching for Keystones in Ecological Networks." *Oikos* 99:607–612.

Post, D. M. 2002. "The Long and Short of Food-Chain Length." *Trends in Ecol. & Evol.* 17:269–277.

Drossel, B., and A. J. McKane. 2003. "Modeling Food Webs." In *Handbook of Graphs and Networks: From the Genome to the Internet*, ed. S. Bornholt, and H. G. Schuster, 218–247. Berlin, Wiley-VCH.

continued on next page

Box 3 continued

Berlow, E. L., A.-M. Neutel, J. E. Cohen, P. De Ruiter, B. Ebenman, M. Emmerson, J. W. Fox, V. A. A. Jansen, J. I. Jones, G. D. Kokkoris, D. O. Logofet, A. J. McKane, J. Montoya, and O. L. Petchey. 2004. "Interaction Strengths in Food Webs: Issues and Opportunities." *J. Animal Ecol.* 73:585–598.

Jordán, F., and I. Scheuring. 2004. "Network Ecology: Topological Constraints on Ecosystem Dynamics." *Phys. Life Rev.* 1:139–229.

Moore, J. C., E. L. Berlow, D. C. Coleman, P. C. de Ruiter, Q. Dong, A. Hastings, N. Collins Johnson, K. S. McCann, K. Melville, P. J. Morin, K. Nadelhoffer, A. D. Rosemond, D. M. Post, J. L. Sabo, K. M. Scow, M. J. Vanni, and D. H. Wall. 2004. "Detritus, Trophic Dynamics and Biodiversity." *Ecol. Lett.* 7:584–600.

REFERENCES

Albert, R., and A.-L. Barabási. 2002. "Statistical Mechanics of Complex Networks." *Rev. Mod. Phys.* 74:47–97.

Barabási, A.-L. 2002. *Linked: The New Science of Networks.* Cambridge, MA: Perseus Publishing.

Bascompte, J., and P. Jordano. "The Structure of Plant-Animal Mutualistic Networks." This volume.

Bolker, B, M. Holyoak, V. Krivan, L. Rowe, and O. Schmitz. 2003. "Connecting Theoretical and Empirical Studies of Trait-Mediated Interactions." *Ecology* 84:1101–1114.

Borgmann, U. 1987. "Models on the Slope of, and Biomass Flow Up, the Biomass Size Spectrum." *Can. J. Fish. Aquat. Sci.* 44:136–140.

Briand, F., and J. E. Cohen. 1984. "Community Food Webs have Scale-Invariant Structure." *Nature* 398:330–334.

Brown, J. H. 1995. *Macroecology.* Chicago, IL: University of Chicago Press.

Buchanan, M. 2002. *Nexus: Small Worlds and the Groundbreaking Science of Networks.* New York: W.W. Norton and Company.

Camacho, J., R. Guimeà, and L. A. N. Amaral. 2002. "Robust Patterns in Food Web Structure." *Phys. Rev. Lett.* 88:228102-1.

Cartozo, C. C., G. Garlaschelli, and G. Caldarelli. "Graph Theory and Food Webs." This volume.

Chen, X., and J. E. Cohen. 2001a. "Global Stability, Local Stability and Permanence in Model Food Webs." *J. Theoret. Biol.* 212:223–235.

Chen, X., and J. E. Cohen. 2001b. "Transient Dynamics and Food-Web Complexity in the Lotka-Volterra Cascade Model." *Proc. Roy. Soc. Lond. Ser. B* 268:869–877.

Cohen, J. E. 1978. *Food Webs and Niche Space.* Princeton, NJ: Princeton University Press.

Cohen, J. E., F. Briand, and C. M. Newman. 1990a. *Community Food Webs: Data and Theory.* New York: Springer-Verlag.

Cohen, J. E., T. Jonsson, and S. R. Carpenter. 2003. "Ecological Community Description using the Food Web, Species Abundance, and Body Size." *PNAS* 100:1781–1786.

Cohen, J. E., T. Luczak, C. M. Newman, and Z.-M. Zhou. 1990b. "Stochastic Structure and Nonlinear Dynamics of Food Webs: Qualitative Stability in a Lotka-Volterra Cascade Model." *Proc. Roy. Soc. Lond. B.* 240:607–627.

Cohen, J. E., and C. M. Newman. 1985. "A Stochastic Theory of Community Food Webs: I. Models and Aggregated Data." *Proc. Roy. Soc. Lond. B* 224:421–448

De Angelis, D. L. 1975. "Stability and Connectance in Food Web Models." *Ecology* 56:238–243.

de Ruiter, P. C., A.-M. Neutel, and J. C. Moore. 1995. "Energetics, Patterns of Interaction Strengths, and Stability in Real Ecosystems." *Science* 269:1257–1260.

Dobson, A., and K. Lafferty. "Parasites and Food Webs." This volume.

Dobson, A. P., D. M. Lodge, J. Alder, G. Cumming, J. E. Keymer, H. A. Mooney, J. A. Rusak, O. E. Sala, D. H. Wall, R. Winfree, V. Wolters, and M. A. Xenopoulos. In Prep. "Habitat Loss, Trophic Collapse and the Decline of Ecosystem Services." *Ecology.*

Dunne, J. A. "The Network Structure of Food Webs." This volume.

Dunne, J. A., U. Brose, R. J. Williams, and N. D. Martinez. 2005. "Modeling Food-Web Dynamics: Complexity-Stability Implications." In *Aquatic Food Webs: An Ecosystem Approach,* ed. A. Belgrano, U. Scharler, J. A. Dunne, and R. E. Ulanowicz. New York: Oxford University Press.

Dunne, J. A., R. J. Williams, and N. D. Martinez. 2002a. "Food-Web Structure and Network Theory: The Role of Connectance and Size." *Proc. Natl. Acad. Sci. USA* 99:12917–12922.

Dunne, J. A., R. J. Williams, and N. D. Martinez. 2002b. "Network Structure and Biodiversity Loss in Food Webs: Robustness Increases with Connectance." *Ecol. Lett.* 5:558–567.

Durrett, R., and S. A. Levin. 1994. "On the Importance of Being Discrete (and Spatial)." *Theoret. Pop. Biol.* 46:363–394.

Elton, C. S. 1927. *Animal Ecology.* London: Sidgwick and Jacksons.

Emmerson, M. C., and D. Rafaelli. 2004. "Predator-Prey Body Size, Interaction Strength and the Stability of a Real Food Web." *J. Animal Ecol.* 73:399–409.

Enquist, B. J. and K. J. Niklas. 2001. "Invariant Scaling Relations across Tree-Dominated Communities." *Nature* 410:655–660.

Erdös, P, and A. Rényi. 1960. "On the Evolution of Random Graphs." *Publ. Math. Inst. Hung. Acad. Sci.* 5:17–61.

Fasham, M. J. R., H. W. Ducklow, and S. M. McKelvie. 1990. "A Nitrogen-Based Model of Plankton Dynamics in the Oceanic Mixed Layer." *J. Mar. Res.* 48:591–639.

Fussmann, G. F., and G. Heber. 2002. "Food Web Complexity and Chaotic Population Dynamics." *Ecol. Lett.* 5:394–401.

Gardner, M. R., and W. R. Ashby. 1970. "Connectance of Large Dynamic (Cybernetic) Systems: Critical Values for Stability." *Nature* 228:784.

Garlaschelli, D., G. Caldarelli, and L. Pietronero. 2003. "Universal Scaling Relations in Food Webs." *Nature* 423:165–168.

Gillooly, J. F., A. P Allen, and J. H. Brown. "Food Web Structure and Dynamics: Reconciling Alternative Ecological Currencies." This volume.

Hall, S. J., and D. G. Raffaelli. 1993. "Food Webs: Theory and Reality." *Adv. Ecol. Res.* 24:187–239.

Hartvigsen, G., and S. A. Levin. 1997. "Evolution and Spatial Structure Interact to Influence Plant-Herbivore Population and Community Dynamics." *Proc. Roy. Soc. Lond. B.* 264:1677–1685.

Hastings, A., and T. Powell. 1991. "Chaos in a Three-Species Food Chain." *Ecology* 72:896–903.

Hofbauer, J., and K. Sigmund. 1998. *Evolutionary Games and Population Dynamics.* Cambridge: Cambridge University Press.

Huxham, M., S. Beany, and D. Raffaelli. 1996. "Do Parasites Reduce the Chances of Triangulation in a Real Food Web?" *Oikos* 76:284–300.

Holland, J. 1975. *Adaptation in Natural and Artificial Systems.* Ann Arbor, MI: University of Michigan Press.

Holling, C. S. 1973. "Resilience and Stability of Ecological Systems." *Ann. Rev. Ecol. & System.* 4:1–23.

Holling, C. S., and L. H. Gunderson, eds. 2002. "Resilience and Adaptive Cycles. In *Panarchy: Understanding Transformations in Human and Natural Systems*, 25–62. Washington: Island Press.

Ives, A. R., and A. P. Dobson. 1987. "Antipredator Behavior and the Population-Dynamics of Simple Predator-Prey Systems." *Amer. Natur.* 130:431–447.

Jansen, W. 1987. "A Permanence Theorem for Replicator and Lotka-Volterra Systems." *J. Math. Biol.* 25:411–422.

Jansen, V. A. A., and G. D. Kokkoris. 2003. "Complexity and Stability Revisited." *Ecol. Lett.* 6:498–502.

Jen, E. 2003. "Stable or Robust? What's the Difference?" *Complexity* 8:12–18.

Jordano, P., J. Bascompte, and J. M. Olesen. 2003. "Invariant Properties in Coevolutionary Networks of Plant-Animal Interactions." *Ecol. Lett.* 6:69–81.

Kauffman, S. A. 1993. *The Origins of Order. Self-Organization and Selection in Evolution.* New York: Oxford University Press.

Kinzig, A. P., S. Pacala, and G. D. Tilman. 2002. *The Functional Consequences of Biodiversity: Empirical Progress and Theoretical Extensions.* Monographs in Population Biology. Princeton, NJ: Princeton University Press.

Kokkoris, G. D., A. Y. Troumbis, and J. H. Lawton. 1999. "Patterns of Species Interaction Strength in Assembled Theoretical Competition Communities." *Ecol. Lett.* 2:70–74.

Kokkoris, G. D., V. A. A. Jansen, M. Loreau, and A. Y. Troumbis. 2002. "Variability in Interaction Strength and Implications for Biodiversity." *J. Animal Ecol.* 71:362–371.

Kondoh, M. 2003. "Foraging Adaptation and the Relationship between Food-Web Complexity and Stability." *Science* 299:1388–1391.

Lafferty, K. D., R. F. Hechinger, J. C. Shaw, K. L. Whitney, and A. M. Kuris. 2005. "Food Webs and Parasites in a Salt Marsh Ecosystem." In *Disease Ecology: Community Structure and Pathogen Dynamics*, ed. S. Collinge and C. Ray. Oxford: Oxford University Press.

Law, R., and J. C. Blackford. 1992. "Self-Assembling Food Webs: A Global Viewpoint of Coexistence of Species in Lotka-Volterra Communities." *Ecology* 73:567–578.

Law, R., and R. D. Morton. 1993. "Alternative Permanent States of Ecological Communities." *Ecology* 74:1347–1361.

Lawton, J. H. 1999. "Are there General Laws in Ecology?" *Oikos* 84:177–192.

Lawton, J. H. 1989. "Food Webs." In *Ecological Concepts*, ed. J. M. Cherett, 43–78. Oxford: Blackwell Scientific.

Leaper, R., and M. Huxhman. 2002. "Size Constraints in Real Food Web: Predator, Parasite and Prey Body-Size Relationships." *Oikos* 99:443–456

Levin, S. A. 1998. "Ecosystems and the Biosphere as Complex Adaptive Systems." *Ecosystems* 1:431–436.

MacArthur, R. H. 1955. "Fluctuation of Animal Populations and a Measure of Community Stability." *Ecology* 36:533–536.

Martinez, N. D., R. J. Williams, and J. A. Dunne. "Diversity, Complexity, and Persistence in Large Model Ecosystems." This volume.

Martinez, N. D. "Network Evolution: Exploring the Change and Adaptation of Complex Ecological Systems over Deep Time." This volume

Martinez, N. D., B. A. Hawkins, H. A. Dawah, and B. P. Feifarek. 1999. "Effects of Sampling Effort on Characterization of Food-Web Structure." *Ecology* 80:1044–1055.

May, R. M. 1972. "Will a Large Complex System be Stable?" *Nature* 238:413–414.

May, R. M. 1973. *Stability and Complexity in Model Ecosystems.* Princeton, NJ: Princeton University Press.

McCann, K. S. 2000. "The Diversity-Stability Debate." *Nature* 405:228–233.

McCann, K., and A. Hastings. 1997. "Re-Evaluating the Omnivory-Stability Relationship in Food Webs." *Proc. Roy. Soc. Lond. B* 264:1249–1254.

McCann, K., A. Hastings, and G. Huxel. 1998. "Weak Trophic Interactions and the Balance of Nature." *Nature* 395:794–798.

McKane, A. J., and B. Drossel. "Models of Food Web Evolution." This volume.

Memmott, J., E. L. Berlow, A. Dobson, J. A. Dunne, R. V. Solé, and J. Weitz. "Biodiversity Loss and Ecological Network Structure." This volume.

Memmott, J., N. D. Martinez, and J. E. Cohen. 2000. "Predators, Parasitoids and Pathogens: Species Richness, Trophic Generality and Body Sizes in a Natural Food Web." *J. Animal Ecol.* 69:1–15.

Memmott, J., N. M. Waser, and M. V. Price. 2004. "Resilience of Pollination Networks to Species Extinctions." *Proc. Roy. Soc. Lond. Ser. B Biol. Sci.* 271:2605–2611.

Menge, B. A. 1995. "Indirect Effects in Marine Rocky Intertidal Interaction Webs: Patterns and Importance." *Ecol. Monogr.* 65:21–74.

Mitchell, M. 1996. *Introduction to Genetic Algorithms.* Cambridge, MA: MIT Press.

Moloney, C. L., and J. G. Field. 1991. "The Size-Based Dynamics of Plankton Food Webs. I. A Simulation Model of Carbon and Nitrogen Flows." *J. Plankton Res.* 13:1003–1038.

Montoya, J. M., M. A. Rodriguez, and B. A. Hawkins. 2003. "Food Web Complexity and Higher-Level Ecosystem Services." *Ecol. Lett.* 6:587–593.

Neubert, M. G., and H. Caswell. 1997. "Alternatives to Resilience for Measuring the Responses of Ecological Systems to Perturbations." *Ecology* 78(3):653–665.

Neutel, A., J. A. P. Heesterbeek, and P. C. de Ruiter. 2002. "Stability in Real Food Webs: Weak Links in Long Loops." *Science* 296:1120–1123.

Odum, E. 1953. *Fundamentals of Ecology.* Philadelphia, PA: Saunders.

Packer, C., R. D. Holt, P. J. Hudson, K. D. Lafferty, and A. P. Dobson. 2003. "Keeping the Herds Healthy and Alert: Implications of Predator Control for Infectious Disease." *Ecol. Lett.* 6:797–802

Pascual, M., J. A. Dunne, and S. Levin. "Challenges for the Future." This volume.

Peacor, S. D., R. L. Riolo, and M. Pascual. "Phenotypic Plasticity and Species Coexistence: Modeling Food Webs as Complex Adaptive Systems." This volume.

Peacor, S. D., and E. E. Werner. 2001. "The Contribution of Trait-Mediated Indirect Effects to the Net Effects of a Predator." *PNAS* 98:3904–3908.

Pimm, S. L. 1982. *Food Webs. Chapman and Hall*, 2d ed. Chicago: University of Chicago Press.

Pimm, S. L. 1984. "The Complexity and Stability of Ecosystems." *Nature* 307:321–326.

Pimm, S. L., and J. H. Lawton. 1977. "Number of Trophic Levels in Ecological Communities." *Nature* 268:329–331.

Pimm, S. L., and J. H. Lawton. 1978. "On Feeding on More than One Tophic Level." *Nature* 275:542–544.

Pimm, S. L., J. H. Lawton, and J. E. Cohen. 1991. "Food Web Patterns and Their Consequences." *Nature* 350:669–674.

Platt, T., and K. Denman. 1978. "The Structure of Pelagic Marine Communities." *Rapp. P. Reun. Cons. Int. Explor. Mer* 173:60–65.

Post, W. M., and S. Pimm. 1983. "Community Assembly and Food Web Stability." *Math. Biosci.* 64:162–192.

Ruiz-Moreno, D., M. Pascual, and R. Riolo. "Exploring Network Space with Genetic Algorithms: Modularity, Resilience, and Reactivity." This volume.

Solé, R. V., and J. M. Montoya. "Ecological Network Meltdown from Habitat Loss and Fragmentation." This volume.

Solé, R. V., D. Alonso, and A. McKane. 2002. "Self-Organized Instability in Complex Ecosystems." *Phil. Trans. Roy. Soc. Lond. B.* 357:667–681.

Sporns, O., and G. Tononi. 2002. "Classes of Network Connectivity and Dynamics." *Complexity* 7:28–38.

Strogatz, S. H. 2001. "Exploring Complex Networks." *Nature* 410:268–276.

Strogatz, S. 2003. *Sync: The Emerging Science of Spontaneous Order.* New York: Hyperion.

Warren, C., M. Pascual, K. D. Lafferty, and A. M. Kuris. 2005. "Parasitic Networks on Food Webs: Clustering, Nestedness and the Inverted Niche Model." Unpublished paper (in preparation).

Watts, D. J. 2003. *Six Degrees: The Science of a Connected Age.* New York: W.W. Norton and Company.

Watts, D. J. 1999. *Small Worlds: The Dynamics of Networks between Order and Randomness.* Princeton Studies in Complexity. Princeton, NJ: Princeton University Press.

Watts, D. J., and S. H. Strogatz. 1998. "Collective Dynamics of 'Small-World' Networks." *Nature* 393:440–442.

West, G. B., J. H., Brown, and B. J. Enquist. 1997. "A General Model for the Origin of Allometric Scaling Laws in Biology." *Science* 276:122–126

Wilke, C. O., and S. S. Chow. "Exploring the Evolution of Ecosystems with Digital Organisms." This volume.

Williams, N. 1997. "Fractal Geometry Gets the Measure of Life's Scales." *Science* (commentary) 276:34–35

Williams, R. J., and N. D. Martinez. 2000. "Simple Rules Yield Complex Food Webs." *Nature* 404:180–183

Wilmers, C. C., S. Sinha, and M. Brede. 2002. "Examining the Effects of Species Richness on Community Stability: An Assembly Model Approach." *Oikos* 99(2):363–367.

Worm, B., and J. E. Duffy. 2003. "Biodiversity, Productivity and Stability in Real Food Webs." *Trends in Ecol. & Evol.* 18:628–632.

Yodzis, P. 1981. "The Stability of Real Ecosystems." *Nature* 289:674–676.

Yodzis, P. 1982. "The Cmpartmentation of Real and Assembled Ecosystems." *Amer. Natur.* 120:551–570.

Yodzis, P., and S. Innes. 1992. "Body-Size and Consumer-Resource Dynamics." *Amer. Natur.* 139:1151–1173.

Structure of Complex Ecological Networks

"Este pensador observó que todos los libros, por diversos que sean, constan de elementos iguales: el espacio, el punto, la coma, las veintidós letras del alfabeto. También alegó un hecho que todos los los viajeros han confirmado: *No hay en la vasta Biblioteca, dos libros idénticos.*"

"This thinker observed that all books, however diverse, are made up of uniform elements: the period, the comma, the space, the twenty-two letters of the alphabet. He also adduced a circumstance confirmed by all travelers: *There are not, in the whole vast Library, two identical books.*"

— Jorge Luis Borges,
"La Biblioteca de Babel." *Ficciones*, 1956

The Network Structure of Food Webs

Jennifer A. Dunne

Descriptions of food-web relationships first appeared more than a century ago, and the quantitative analysis of the network structure of food webs dates back several decades. Recent improvements in food-web data collection and analysis methods, coupled with a resurgence of interdisciplinary research on the topology of many kinds of "real-world" networks, have resulted in renewed interest in food-web structure. This chapter reviews the history of the search for generalities in the structure of complex food webs, and discusses current and future research trends. Analysis of food-web structure has used empirical and modeling approaches, and has been inspired both by questions from ecology such as "What factors promote stability of complex ecosystems given internal dynamics and external perturbations?" and questions from network research such as "Do food webs display universal structure similar to other types of networks?" Recent research has suggested that once variable diversity and connectance are taken into account, there are universal coarse-grained characteristics of how trophic links and species

Ecological Networks: Linking Structure to Dynamics in Food Webs, edited by Mercedes Pascual and Jennifer A. Dunne, Oxford University Press.

defined according to trophic function are distributed within food webs. In addition, aspects of food-web network structure have been shown to strongly influence the robust functioning and dynamical persistence of ecosystems.

1 INTRODUCTION

This chapter describes research that seeks to characterize and model the structure of food webs, complex networks of feeding (*trophic*) interactions among diverse species in communities or ecosystems. In particular, I discuss studies that search for generalities in the network structure, also referred to as topology, of food webs. Over the last fifteen years there has been dramatic growth in the field-based documentation of food webs, as well as a general increase in data quality in terms of diversity and resolution. This is a welcome trend, and provides fuel for further statistical assessment, model testing, and theory development. This chapter reviews the history of and current trends in the data, analyses, and models of food-web network structure. I attempt to be relatively comprehensive within this narrow domain, but as with all reviews, some papers and topics are not addressed. In the interest of focus and length, studies focused on food-web dynamics are not covered here, although they have been used to explore dynamic constraints on structure and vice versa (Pimm and Lawton 1978; McCann and Hastings 1997). Reviews of dynamical food-web models and their uses and limitations (Lawton 1989; Drossel and McKane 2003; Dunne et al. 2005; McKane and Drossel Chapter 9; Martinez and Bascompte Chapter 12), as well as "network analyses" of flow networks, primarily applied to particular marine or estuarine systems (Wulff et al. 1989; Christensen and Pauly 1993), can be found elsewhere.

Research on food-web structure is but one example in a very broad cross-disciplinary research agenda on the structure of all types of networks, both biotic and abiotic (Strogatz 2001; Albert and Barabási 2002). Network research is often couched in the framework and language of statistical mechanics and graph theory. Indeed, research on food-web structure has foundations in graph theory (the random graph theory of Erdös and Rényi 1960 as cited by Cohen 1990; see also Cohen 1977b; Sugihara 1982; Kenny and Loehle 1991) as well as natural history (the food web for Bear Island described by Summerhayes and Elton 1923). The chapter by Cartozo et al. (Chapter 3) goes into detail about how food-web studies fit into the broader network topology framework, and it introduces basic graph theory definitions and properties of interest. However, this review is primarily situated within the history, language, and theory of ecology, and given this context I generally use terms such as *species* rather than *nodes* and feeding *links* rather than *edges*. The biological terms evoke a rich conceptual history in ecology and connect food-web research to related areas of study, such as evolutionary theory and conservation biology. However, more technical graph-theoretic terms do pop up, particularly in descriptions of more recent research. The review is divided into three sections called "Early Phase: Pioneering Research," "Middle Phase: Critique and Reassessment," and "Current Phase: New Models, New Directions." These phases correspond roughly to the late 1970s through the 1980s (Early Phase), the 1990s (Middle Phase), and the first half of the 2000s (Current Phase). In addition to many interesting questions, both basic and applied, that structural food-web research has been used to explore, its history provides a unique perspective on the interplay between theory, methods, and data, and on the role of critique. While the phases and topics emphasized within them are a subjective split of interrelated lines of research, they provide a useful way to organize a rich body of literature. The Early Phase and parts of the Middle Phase cover ground that is reviewed elsewhere (e.g., Lawton 1989; Cohen et al. 1990a; Pimm et al. 1991; Hall and Rafaelli 1993, 1997). However, a recent renewed interest in the network structure of food webs by ecologists as well as by physicists, mathematicians, biologists, and social scientists suggests that it would be wise to revisit early foundational work from a contemporary perspective. The final section, "Related Topics and Future Directions," discusses relevant topics neglected by the three main sections and how those topics relate to future research directions.

2 EARLY PHASE: PIONEERING RESEARCH

Descriptions of feeding relationships among species go back at least to the late 1800s (studies by Forbes from 1876 on, reprinted in 1977; Camerano 1880 as cited by Cohen et al. 1990a). By the 1910s researchers began to produce images not unlike food-web figures seen today in textbooks, such as a network of insect predators and parasites on cotton-feeding weevils (Pierce et al. 1912 as cited by Pimm et al. 1991) and the hypothetical animal-oriented descriptions of Shelford (1913). By the 1920s, the first relatively detailed empirical descriptions of terrestrial (Summerhayes and Elton 1923, 1928) and marine (Hardy 1924) food webs appeared. Elton (1927) coined the term *food chain*, and termed all the food chains in a community a *food cycle*, which we now call a *food web*. Many descriptions of food webs, both hypothetical and empirical, followed those early efforts.

However, it was not until the late 1970s that quantitative, comparative research on potential generalities in the network structure of food webs arose. Cohen published the first collection of food webs in 1978, comprised of 30 webs with binary links (i.e., links showing only the presence and direction of a feeding relationship, with no weighting for flow or strength) compiled from the literature. He considered 14 to be *community webs*, webs that attempt to be reasonably inclusive of species in a particular system, and 16 were *sink webs*, selective webs that focus on one or more predator species, their prey, their preys' prey, etc.

There are also *source webs*, selective webs that focus on one or more prey species, their predators, their predators' predators, etc. Both then and now, *species* in food webs, while often referring to actual biological species, in many cases represent other things: taxonomically related groups of species, all the way up to whole kingdoms (mites, arthropods, fungi); mixed groups of species (zooplankton); particular life-history stages (small-mouth bass young-of-year); parts of species (leaves, fruit); and non-living organic matter (detritus). Throughout this chapter, the term *species* will be used in the non-specific sense that includes all of these types of groupings within food webs, and the term *taxa* will sometimes be used in place of *species*.

In a strict sense, the trophic links in food webs are *directed* links, which means that a feeding relationship is directional (i.e., A eats B). However, *effects* of feeding move in both directions—A eating B has ecological and evolutionary implications for both species, at least when both "species" are living. Food-web interactions can be usefully represented as a matrix. The simplest way to build such a matrix is for rows to represent consumers/predators and columns to represent resources/prey. A "1" is assigned to the cell at the intersection of row i and column j if species i feeds on species j, and a "0" is assigned if species i does not feed on species j. Alternatively, rows can represent resources and columns consumers. Putting trophic relationships into a matrix format facilitates quantitative analysis of food-web structure. Larger datasets can be more efficiently represented using a two-column format, in which the first column lists the number of a consumer, and the second column lists the number of one of the resource species of that consumer.

2.1   COMPLEXITY-STABILITY AND FOOD-WEB STRUCTURE

Early collections of food webs provided, for better and for worse, empirical fodder for the complexity-stability debate. For several decades leading up to the 1970s, a dominant ecological paradigm was that complex communities are more stable than simple ones (Odum 1953; MacArthur 1955; Elton 1958; Hutchinson 1959). The argument in favor of complexity giving rise to stability in ecological communities was stated in a general way by MacArthur (1955), who hypothesized that "a large number of paths through each species is necessary to reduce the effects of overpopulation of one species." He concluded that "stability increases as the number of links increases" and that stability is easier to achieve in more diverse assemblages of species, thus linking community stability with both increased trophic links and increased numbers of species. This convention was challenged by May in a seminal paper (1972) and book (1973) using dynamical mathematical modeling methods. May conducted local stability analyses of randomly assembled community matrices and demonstrated that network stability decreases with complexity, following Gardner and Ashby (1970). May found that simple, abstract communities of interacting species will tend to transition sharply from stable to unstable behavior as the complexity of the system increases; in particular as the number of species (S), the connectance (C), *or* the average interaction strength (i) increase beyond critical values. May formalized this as a criterion that ecological communities near equilibrium will tend to be stable if $i(SC)^{1/2} < 1$. In May's framework, connectance refers to the probability that any two species will interact with each other; in a more technical sense, connectance refers to the percentage of non-zero elements in an interaction matrix (Gardner and Ashby 1970). The measure S, the number of species in a food web, or more generally, the number of nodes in a network, will appear repeatedly throughout this review and will typically be referred to as species diversity or richness. "Diversity" and "richness" can also be quantified using more complex formulations not covered here.

Several papers since May (1972, 1973) have pointed out flaws and limitations in his analysis (e.g., Lawlor 1978; Cohen and Newman 1985a; Taylor 1988; Law and Blackford 1992; Haydon 1994). For example, patterns of species interactions are not random, and varying such patterns can have a significant impact on dynamics. However, May's criterion and the general question of how diversity is maintained in natural ecosystems provided a framework on which to hang some readily accessible empirical data, namely the numbers of links and species in food webs. How does this work? Given a particular interest in species diversity (S), and assuming that average interaction strength (i) is constant, May's criterion suggests that communities can be stable given increasing diversity, as long as connectance decreases. This can be empirically demonstrated using food-web data in three similar ways, by showing that (1) C hyperbolically declines as S increases, so that the product SC remains constant, (2) the ratio of links to species (L/S), also referred to as link or linkage density, remains constant as S increases, or (3) L plotted as a function of S on a log-log graph, producing a power-law relation of the form $L = aS^b$, displays an exponent of $b = 1$ (the slope of the regression) indicating a linear relationship between L and S.

In an empirical framework, connectance is measured as the proportion of potential links among species that are actually realized. The simplest way of expressing this is $C = L/S^2$ (*directed connectance*; Martinez 1991), where the numerator gives the number of observed links and the denominator includes all potential directed trophic links among S species, equal to S^2. Food-web connectance has also been measured in more complex ways that exclude particular kinds of trophic links. For example, another measure is $C = L/[S(S-1)/2]$ which excludes all cycles, also called loops (Rejmánek and Starý 1979). Within food webs, cycles of length 1 (A eats A) are referred to as cannibalism, cycles of length 2 (A eats B eats A) are referred to as mutual predation, and even longer cycles are possible (e.g., A eats B eats C eats A). The $S - 1$ part of the denominator excludes the main diagonal of an S by S matrix, thus eliminating cannibalism links, and dividing $S(S-1)$ by 2 excludes all other cycles. Thus, $S(S-1)/2$ constrains the zone of potential directed links in an S by S matrix to a triangle on one side of the main diagonal. It is the equivalent of counting the total number of possible undirected interactions (i.e., where "A interacts with B" could mean that A eats B, B eats A, or both eat each other) in an S by S matrix, excluding intraspecific interactions, and thus is sometimes referred to as *interactive connectance*. However, if cycles do occur in food webs, $C = L/[S(S-1)/2]$ exaggerates connectance, since cycling links are counted in the numerator but are excluded from the denominator. This can be avoided by calculating interactive connectance based on undirected links, such that $C = L/[S(S-1)]$ (Warren 1990; Martinez 1991). How do these expressions relate to May's criterion and the potential for demonstrating stability with increasing diversity? Given the form $C = L/S^2$, the hyperbolic decline of C with increasing S (so that SC is constant) is mathematically equivalent to constant L/S, since $SC = L/S$ is the same as $C = L/S^2$. Given the forms $C = L/[S(S-1)/2]$ or $C = L/[S(S-1)]$, the same equivalence between the hyperbolic decline of C and constant L/S occurs under the condition that S is large, that is, when $(S-1)/S$ approximates 1 (Macdonald 1979). This means that when S is small ($S < 20$), different forms of C can alter which hypothesis about variability of C is accepted or rejected (Martinez 1995).

Quickly following the publication of the first catalog of food webs (Cohen 1978), examples were published documenting all three ways of empirically corroborating the potential for stability with increasing diversity, assuming constant interaction strength (a big assumption that will be discussed briefly in Related Topics and Future Directions). MacDonald (1979) analyzed Cohen's data and found that mean L/S of the 30 webs was 1.88 ($SD = 0.27$). Independently, Rejmánek and Starý (1979) compiled and analyzed 31 plant-aphid-parasitoid source webs and reported a hyperbolic relationship between S and C (using $C = L/[S(S-1)/2]$) with a central tendency of $C = 3/S$ (all data points fell between $C = 2/S$ and $C = 6/S$) corresponding to $L/S = 1.5$ (MacDonald 1979). Briand (1983) analyzed 40 community webs including 13 from Cohen (1978), and using log-log regression analysis he found that trophic links increase as a nearly linear function of S, with $b = 1.10$ ($L = 1.3S^{1.10}$). In these and other early studies (e.g., Pimm 1982; Auerbach 1984), the results seemed to indicate that connectance decreases with species richness and L/S is approximately constant with a value between 1 and 2. The ecological interpretation is that species tend to eat the same small number of prey (1 to 2 species on average) regardless of the diversity of the food web (Pimm 1982). Given May's criterion, these results provided empirical support for the possibility of stable, diverse ecosystems.

Ensuing studies tested these findings with new data sets and introduced trophic aggregation. Expanding Briand's 40 food webs, Cohen and Briand (1984) reported for 62 webs that L/S is "roughly independent of variation in S" with a value of 1.86 ($SD = 0.07$). Unlike previous studies, they used trophic species aggregation (Briand and Cohen 1984; see also Sugihara 1982; Yodzis 1982), in which species that share the same set of predators and prey in a particular food web are lumped into a single *trophic species*. This was meant to reduce methodological artifacts due to researchers' tendencies to resolve higher trophic level taxa more finely than lower trophic level taxa (Pimm 1982), which can add noise to trends and bias results (Briand and Cohen 1984). Sufficiently convinced by the generality of the data that supported both scale invariance of L/S and its value of ~ 2, Cohen and Newman (1985b) began to refer to the relationship as the *link-species scaling law*, and further corroborated it with an expanded set of 113 webs (Cohen et al. 1986) that will be referred to throughout this chapter as the *113 web catalog*.

Using a separate set of 60 relatively well resolved insect-dominated food webs, drawn from a catalog of 95 such webs (Schoenly et al. 1991), Sugihara et al. (1989) reported SC (equivalent to L/S) as "roughly independent of species number" for the subset of 41 webs with 10 or more species, ranging up to 87 species. While they did not use trophic species as the unit of analysis, they did follow methods introduced by Martinez (1988) and conducted an aggregation study of the 41 web subset, lumping species by trophic similarity until half of the original species remained. While the data showed that SC "tended to fall slightly with increasing aggregation," they nevertheless concluded that SC was robust to aggregation. The aggregation study and its conclusions were meant to assuage concerns (Pimm 1982; Paine et al. 1988) that variable resolution in available data might create false patterns. However, the aggregation criterion used by Sugihara et al. (1989) resulted in about three-quarters of the insect webs being aggregated only to the trophic species level, so like Cohen and Briand they were primarily eliminating topological redundancy (Martinez 1993b). The 113 web catalog, the 60 insect webs, and other food webs were ultimately compiled in the ECOWeB database in a machine-readable format to facilitate analyses by other researchers (Cohen 1989).

2.2 EMPIRICAL REGULARITIES AND SCALE INVARIANCE

As the previous section suggests, a great deal of focus on constant linkage density (L/S) resulted from its connection to fundamental theory regarding ecosystem stability and diversity/complexity, a key ecological issue that continues to endure (McCann 2001). However, the first collection of food webs was compiled not to test the relationship between L/S and S, but to look for other empirical regularities in the network structure of trophic interactions (Cohen 1978). Based on 14 community webs, Cohen (1977a) reported a ratio of prey to predators of $\sim 3/4$, constant across food webs with variable S. This ratio had been explored previously in other empirical datasets (e.g., Evans and Murdoch 1968; Arnold 1972; Cameron 1972). Cohen (1977b) also found that most of the 30 sink and community food webs are *interval*. This means that all of the species in a food web can be placed in a fixed order on a line such that each predator's set of prey forms a single contiguous segment of that line. Intervality suggests that trophic niche space can be represented by a single dimension. Why this might be the case, or what the single dimension might represent, continues to be unclear (Williams and Martinez 2000). Other types of graph-theoretic properties quantifying how diets overlap in food webs such as *triangulation* have been explored (Sugihara 1982; Sugihara et al. 1989; Cattin et al. 2004).

A great deal of research following Cohen (1977a,b) focused on patterns of food-web network structure, and had more in common with Cohen's initial search for empirical regularities than the search for constant L/S in the service of complexity-stability theory. From this viewpoint, L/S is just one of many properties of food-web structure, based on analysis of binary links, which may have a general value or central tendency across food webs with varying diversity. These types of food-web patterns, in which a property is found either to be constant, or under a weaker standard to not change systematically as the number of species across food webs changes, came to be referred to as *scale-invariant patterns* (Briand and Cohen 1984) or *scaling laws* (Cohen and Newman 1985b). Such scale-invariant patterns can be thought of as extremely general regularities that are theoretically valid from the smallest food webs to food webs that comprise the entire planet (Martinez and Lawton 1995).

Additional scale-invariant patterns found using early food-web data included: previously mentioned predator-prey ratios; constant proportions of top species (T, species with no predators), intermediate species (I, species with both predators and prey), and basal species (B, species with no prey), collectively called *species scaling laws* (Briand and Cohen 1984); and constant proportions of T-I, I-B, T-B, and I-I links between T, I, and B species, collectively called *link scaling laws* (Cohen and Briand 1984). Other general properties of food webs were thought to include: food chains are short (Elton 1927; Hutchinson 1959; Pimm 1982; Cohen et al. 1990a); cycling/looping is rare (Cohen and Newman 1985b); compartments, or subwebs with many internal links that have few links to other subwebs, are rare (Pimm and Lawton 1980); omnivory, or feeding at more than one trophic level, is uncommon (Pimm and Lawton 1978); and webs tend to be interval, with instances of intervality decreasing as S increases (Cohen 1977b; Yodzis 1984; Cohen and Palka 1990). Most of these patterns were reported for the 113 web catalog (Cohen et al. 1986, 1990). Select patterns, such as short food chains, constant predator-prey ratio, and scale-invariant fractions of T, I, and B species, were also documented in a subset of 41 of 60 insect webs with 10 or more species (Sugihara et al. 1989).

2.3 THE CASCADE MODEL

The many properties being proposed and explored threatened to become a kind of stamp collection of food-web patterns, with no particular rhyme, reason, or organizing principals. However, in a series of six papers published from 1985 to 1990 with the common title "A Stochastic Theory of Community Food Webs," Cohen and colleagues sought to unify food-web patterns through a simple model called the *cascade model*. By ignoring dynamics and using a stochastic, binary link approach similar to that of random graph theory (Erdös and Rényi 1960), the cascade model sought to explain "the phenomenology of observed food web structure, using a minimum of hypotheses" (Cohen et al. 1990a). A number of other simple models for generating food-web structure were explored prior to the cascade model (Cohen 1978; Pimm 1982; Sugihara 1982; Pimm 1984; Yodzis 1984) and several variations were explored concurrently with the cascade model (Cohen and Newman 1985b; Cohen 1990). Most performed poorly in predicting empirical trends, and only one appeared to perform as well as the cascade model (Cohen 1990).

The cascade model is based on two parameters, species richness S and link density L/S. The model distributes species and feeding links stochastically, subject to two simple constraints: species are randomly placed in a one-dimensional feeding hierarchy, and species can only feed on species that are lower in the hierarchy than themselves (Box 1). This ensures a "triangular" matrix that prohibits cycles or loops, including cannibalism. The cascade model is simple enough that it is analytically tractable, although as a stochastic model it can also be explored through computational approaches such as numerical simulation. To explore whether the cascade model reproduced patterns of food-web structure in 62 community webs, Cohen and Newman (1985b) assumed constant L/S and they tuned the value of L/S to an average across the empirical data. They found that the cascade model successfully reproduced the qualitative patterns of both species and link scaling "laws" in all but the smallest webs. While it produced quantitatively similar values to those observed for species and link proportions, the cascade model was less successful in explaining variation in the data (Cohen and Newman 1985b).

Cohen et al. (1985) also tested whether the cascade model could reproduce the values of properties for particular food webs by tuning S and L/S to the values for each of the 62 webs. They found that the cascade model described the proportions of intermediate species (I) and B-I, I-I, and I-T links well. In the expanded 113 web catalog, Cohen et al. (1986) found that the cascade model made good predictions of numbers of food chains of each length (the *frequency distribution*) in most webs, while describing mean chain lengths adequately and chain-length variance less well. The cascade model also gave good qualitative support, and reasonable quantitative support, to the frequency of interval webs (Cohen and Palka 1990). In sum, the cascade model, a simple, stochastic, analytically tractable model, appeared remarkably successful at generating network topology and trends similar to those observed in empirical data at both "coarse-grained" and more detailed levels. It was suggested that the feeding hierarchy assumption of the cascade model might reflect natural processes such as body size constraints on feeding (Warren and Lawton 1987). Perhaps most importantly, the cascade model provided the first explicit and quantitative hypothesis that the network structure observed in early food-web data was not only non-random, but might be governed by simple rules.

3 MIDDLE PHASE: CRITIQUE AND REASSESSMENT

A rosy picture was thus painted by early practitioners. Empirical structural data appeared to corroborate aspects of dynamical complexity-stability theory; network structure appeared to be well-described by multiple empirical scaling laws; and a simple, stochastic graph-theoretic model based on the link-species scaling law and a seemingly ecologically reasonable principle of hierarchical feeding predicted the phenomenology of food-web patterns. However, the hounds were baying at the door. From the beginning, most of the purveyors of structural food-web research pointed out some of the limitations in the data they analyzed. The most obvious issue was that most of the food webs analyzed had very low diversity compared to the biodiversity known to be present in ecosystems. The webs of the 113 web catalog (Cohen et al. 1986) have 5 to 48 original species (mean = 17) and 3 to 48 trophic species (mean = 17), while the 60 insect webs in Sugihara et al. (1989) have 2 to 87 original species (mean = 22) and 2 to 54 trophic species (mean = 12). Another obvious issue was the highly uneven resolution and representation of "species" in most early food webs. Many types of organisms are aggregated, underrepresented, missing altogether, or misrepresented as basal species because no prey items were recorded for them. The webs of the 113 web catalog were culled from the literature, where they had been put together by a wide array of researchers, using a variety of methods, and for many different purposes, which did not include quantifying or testing structural food-web patterns. The 113 web catalog data were also explicitly purged of cannibalistic links. Since the cascade model excluded cannibalism, this increased the fit between the model and the data.

These and other methodological issues were taken up and amplified in several serious, and to many, devastating critiques of the adequacy of the data, casting doubt on the entire research program (Paine 1988; Polis 1991; Hall and Raffaelli 1993, 1997; Winemiller and Polis 1996). The most prominent early critique was by Paine (1988), who suggested that "future connectance-based development, even from sanitized webs, will not be enormously profitable," due to the possibility of significant spatial and temporal variation in diets as well as idiosyncrasies in how researchers ascribe trophic links. He suggested that a tendency of researchers to describe trophic links more completely in small versus large webs, given the greater effort required to describe interactions in speciose systems in detail, could alone account for an apparent hyperbolic decline of connectance (C) with diversity (S).

Meanwhile, potential problems with the "conventional" view of C and L/S were emerging. Recall that the hyperbolic decline of C with S, constant L/S, and log L versus log S showing a power-law exponent of 1 all express the same thing: a linear increase of links with species in food webs. Thus, given May's stability criterion, constant L/S associated with increasingly low C in increasingly diverse communities was thought to be a condition of stability. Empirical evidence began to point towards trends of higher connectance than expected in apparently stable or persistent natural communities across gradients of diversity. Cohen and colleagues, the strongest proponents of a linear relationship between L and S, noted that they could not exclude a nonlinear relationship like that hinted at by Briand (1983). When regressing $\log L$ versus $\log S$ for the 62 and 113 web catalogs, they found an exponent of $b = 1.36$. This makes $L^{3/4}$ rather than L^1 proportional to S, meaning links increase faster than species (Cohen and Briand 1984; Cohen et al. 1986; see also Schoener 1989). Based on relatively detailed trophic information for 24 species in an English stream, Hildrew et al. (1985) found higher C than that reported for all but 2 of 40 webs analyzed by Briand (1983). Ten food webs based on detailed, highly resolved trophic interactions among 36 species in an English pond also displayed high C (Warren 1989). Those were the first published data explicitly collected to test food-web patterns. A summary pond web including all species displayed higher C than less diverse subwebs based on different habitats and sampling times. Detailed tropical fish sink webs (i.e., sampling focused on all fish present, their gut contents, prey lists for their prey, etc.) for swamps and streams in Costa Rica and Venezuela included 58 to 104 species and showed a trend of increasing C with S (Winemiller 1989, 1990).

There were also hints that other scale-invariant scaling "laws" had been overstated. A look at figures purporting to show scale-invariance of food-web properties provides little evidence supporting the strong version of scale invariance. In the classic scale invariance studies (Cohen and Briand 1984; Briand and Cohen 1984), a typical figure, such as percent of top species plotted as a function of S, displays a cloud of data points on which a line whose slope is constrained to be zero is superimposed. The height of this horizontal line is determined by the mean value of the property in question, calculated over all of the food webs. No regression is calculated or plotted; instead, the approximate visual "fit" of the line to the central tendency of the highly variable data is taken as evidence of scale invariance. In effect, a weak, non-statistical finding of "no relationship with S" was interpreted as evidence for a strong claim of the presence of scale-invariant scaling laws (Hall and Raffaelli 1993). In addition, exceptions to other food-web "generalities" started to appear. For example, detailed field-based food-web data suggested that food chains could be longer than previously claimed, and that omnivory and cannibalism might not be rare in some systems (Hildrew et al. 1985; Sprules and Bowerman 1988; Warren 1989).

By the early 1990s, most researchers readily acknowledged problems with the data and the potential impacts on food-web "laws," particularly constant L/S (Pimm et al. 1991). The question of where those problems might lead was summed up cogently in an excellent early review by Lawton (1989): "Confronted with limited data of highly variable quality, hardly any of which is really good, food web studies face either hand-wringing paralysis, or cautious efforts to see what can be discovered in the existing information. If nothing else, the latter course of action should serve as a spur to gather more and better data, particularly if published webs reveal evidence of interesting regularities and patterns in nature. However, we have to accept that some of the patterns may eventually prove to be artifacts of poor information." Indeed, a new wave of empirical food-web structure research was inspired by the inadequacy of the early data, the intriguing possibility of either corroborating or overthrowing previous theory or patterns, and the potential for describing new generalities. This research was fueled by improved data and more sophisticated methods, particularly concerning resolution of taxa, sampling effort, and sampling consistency.

3.1   IMPROVED DATA

An entirely new level of empirical detail characterizing diverse food webs was presented in 1991 in two seminal papers (Polis 1991; Martinez 1991). Polis (1991) published a dizzying array of trophic information, compiled over nearly two decades, for the Coachella Valley desert in California, whose biota include at least 174 species of vascular plants, 138 vertebrate species, 55 arachnid species, thousands of insect species including parasitoids, and unknown numbers of microorganisms, acari, and nematodes. Rather than trying to create a complete food web including all species, he compiled a number of detailed subwebs (a soil web, a gall web, a parasitoid web, a scorpion-focused web, a carnivore web, and a predaceous arthropod web) to demonstrate the enormous trophic diversity and complexity found in a type of ecosystem typically considered to be relatively simple and species-poor. Each subweb is more diverse than many of the community webs in the 113 web catalog and is also more complex in terms of number and density of feeding interactions. On the basis of the subwebs and a simplified, aggregated 30 taxa web of the whole community, Polis (1991) concluded, "most cataloged webs are oversimplified caricatures of actual communities" and are "grossly incomplete representations of communities in terms of both diversity and trophic connections." Coachella Valley web properties include frequent omnivory, cannibalism, and looping; a high degree of interaction among species (i.e., L/S close to 10); and a nearly complete lack of top species, since few species completely lack predators or parasites. These and other properties contradicted accepted food-web patterns, and Polis (1991) suggested that "theorists are trying to explain phenomena that do not exist."

Martinez (1991) compiled a detailed community food web for Little Rock Lake, Wisconsin, in one of the earliest studies to explicitly test food-web theory and patterns (see also Martinez 1988; Warren 1988, 1989). By piecing together diversity and trophic information from multiple investigators actively studying the biota of Little Rock Lake, he was able to produce a relatively complete and highly resolved food web of 182 taxa, most identified to the genus, species, or ontogenetic life-stage level, including fishes, copepods, cladocera, rotifers, diptera and other insects, mollusks, worms, porifera, algae, and cyanobacteria. The resulting 182 original-species web (i.e., the web that includes whatever taxa and links the original investigator reported) and the 93 trophic-species web generated from it (i.e., the web that results from trophic species aggregation of an original-species web) were by far the most diverse and complete depictions of the community food web of a complex ecosystem yet published. Previously, the largest community food webs described were an 87 original-species web of insects associated with a felled oak log (Sugihara et al. 1989) and a 48 trophic-species web of the Sonora Desert (Briand and Cohen 1987). In the 93 trophic-species Little Rock Lake web, $L/S = 11$, looping is common, food chains are long, and there are very few top species (see also Hildrew et al. 1985; Sprules and Bowerman 1988; Warren 1989; Polis 1991). This structure is quite different from patterns observed in prior web catalogs and patterns predicted by the cascade model (Cohen et al. 1990a).

While other detailed food webs were published at about the same time as the Little Rock Lake food web (Martinez 1991), they either were compiled for less diverse systems (Warren 1989); focused on subwebs based on particular sub-habitats, species, or substrates (Winemiller 1989, 1990; Polis 1991; Schoenly et al. 1991; Havens 1992), or had less even representation of taxa resulting from strict criteria for designating links (Hall and Rafaelli 1991). Because the Little Rock Lake web was based on expert knowledge developed over many years of known or probable feeding relationships, it has been suggested that this web overestimates feeding links compared to food webs based on a discrete set of observations, for example, one- or few-times sampling of species and their gut contents coupled with feeding trials (Hall and Raffaelli 1997). Indeed, "cumulative" webs like the Little Rock Lake web have been shown to alter S, L, and other web properties compared to "snapshot" webs (Schoenly and Cohen 1991; and see Sampling Consistency section). For the purposes of discerning coarse-grained patterns of network structure in food webs, such cumulative webs may be more useful than snapshot webs. Approaches that narrowly constrain spatial and temporal boundaries of sampling may miss structurally and dynamically important species and links that are uncommon or rare (Martinez and Dunne 1998); for example, a little used food resource that becomes crucial during periods of scarcity of other resources. Regardless of the inconsistencies, flaws, and limitations still to be found among the second wave of data, all of the datasets mentioned, and others not mentioned, represent significant improvements over the 113 web catalog in terms of field-based observation, higher and more even resolution, and/or greater diversity. Most of these improved datasets were specifically collected to test different methodological aspects of analysis of food-web structure, as detailed below.

3.2   SPECIES AGGREGATION

In addition to setting a new standard for compiling community food-web data, Martinez (1988, 1991) was the first researcher to look systematically at the effects of variable species resolution and aggregation on the network structure of food webs. While trophic species aggregation (Cohen and Briand 1984) is based on 100% trophic equivalence of species, the threshold for aggregation can be relaxed to reduce the resolution of food-web data by progressively lumping taxa based on less and less trophic similarity. There are different indices that can be used to quantify similarity between objects, and the Jaccardian similarity index (Jaccard 1900) is probably the best known and most widely used in food-web research (Martinez 1988, 1991, 1993; Sugihara et al. 1989; Yodzis and Winemiller 1999). Within a food-web context, two basic choices are necessary for deciding how to aggregate taxa: whether to use an additive versus multiplicative Jaccardian index to define similarity between each pair of species, and how to define the similarity of two aggregates based on pairwise similarities of species in the aggregates (Yodzis and Winemiller 1999). Martinez (1988, 1991) used an additive Jaccard index to determine similarity of pairs of species in Little Rock Lake, and then aggregated taxa based on an *average-linkage method* that calculates similarity between two aggregates as the average of similarity indices across all possible pairs between the aggregates. Interestingly, a later detailed study of trophospecies aggregation methods concluded that out of 12 combinations of the two Jaccard similarity indices and six cluster linkage methods, the additive average-linkage method performed better than the other methods (Yodzis and Winemiller 1999).

Martinez (1991) explicitly tested the hypothesis that patterns observed in early food-web catalogs are an artifact of the low resolution and high aggregation of the data. Using the detailed Little Rock Lake food web, the aggregation methods discussed, and three criteria for designating links between aggregates (see Sampling Effort section), he created three sequences of increasingly aggregated versions of the original 182 species web. The sequences end when species in the aggregates share only 10% of their predators and prey, resulting in webs with nine highly aggregated taxa. In effect, trophic aggregation reduces the Little Rock Lake food web to levels of resolution and diversity comparable to those in the 113 web catalog. Martinez then analyzed how different food-web properties changed, or do not change, with increasing aggregation and thus decreasing S. For example, links per species, quite high in the 93 trophic-species web ($L/S = 11$), drops steadily as aggregation increases, so that the 9 to 42 species versions of the Little Rock Lake web display similar L/S (~ 1 to 4) and diversity to the webs in the 113 web catalog (Cohen et al. 1986). Chain-length statistics, trophic-level statistics, and several proportions of links and species are also very sensitive to aggregation (table 1), and display comparable values to earlier data once diversity is reduced to similar levels. Based on these results, Martinez (1991) argued that "most published food-web patterns appear to be artifacts of poorly resolved data." The main property relatively robust to increasing aggregation and decreasing diversity is directed connectance, $C = L/S^2$, which displays values that hover near 0.10 until S drops below ~ 20. A trophic resolution study performed on 11 of 60 insect webs (Sugihara et al. 1989; Schoenly et al. 1991) with 20 or more trophic species produced similar results: compared to poorly resolved versions of webs, higher resolved webs have higher L/S, I, I-I and mean chain length, and lower T, B, and T-B (Martinez 1993b). Directed connectance, predator-prey ratio, and I-I link ratio are less sensitive to aggregation than other properties.

Hall and Raffaelli (1991) also conducted an aggregation study, in this case lumping species in a 92-species food web of the Ythan Estuary based on taxonomic rather than trophic similarity. They created four increasingly aggregated versions of the original web through qualitative decisions about how to lump taxonomically related taxa at each level. Similar to Martinez (1991) they found that percent basal species increases and mean chain length decreases with aggregation (lower S), but unlike Martinez (1991), most properties appear relatively insensitive to aggregation. Their web, while more speciose than webs in the 113 web catalog, was biased in similar ways, which may have impacted its sensitivity to lumping. There is a large number of top species (26 = 28%), and a small number of basal species (3 = 4%). Also, shorebirds are highly resolved, have no predators, and comprise more than a quarter of all taxa, while lower trophic level taxa are increasingly aggregated. Sugihara et al. (1997) conducted both taxonomic and trophic aggregation analyses on a set of 38 of 60 insect webs (Schoenly et al. 1991) and concluded that most properties except for L/S and B are insensitive to changing S due to aggregation. The sensitivity of L/S in the insect webs was counter to Sugihara et al. (1989) but in line with Martinez (1993b). The lack of sensitivity in other properties is likely because the most aggregated versions of the insect webs usually retained more than 50% of the original S. Sugihara et al. (1997) acknowledged, "after a few food webs had been aggregated to 70–89% of their original size, even the sturdiest food web properties became sensitive to these coarser degrees of data resolution." This corroborated results reported by Martinez (1991, 1993b). A potentially confounding aspect of both taxonomic aggregation analyses (Hall and Raffaelli 1991; Sugihara et al. 1997) is their use of a maximum linkage criterion for designating links between aggregates, which is a weak standard that tends to dampen changes in the value of structural properties with S (see Sampling Effort section).

3.3   SAMPLING EFFORT

A question related to the effects of species aggregation on food-web structure is, How do different observation thresholds for links affect observed patterns? Winemiller (1989, 1990) was the first to explicitly quantify these sampling effort effects. In constructing a set of detailed tropical fish sink webs, he recorded not only the presence of feeding links but also the volumetric fraction of each prey species in a predator's diet, usually based on gut content analysis. This volumetric fraction was used as an estimate of the relative strength of the predator-prey interaction, which then allowed the construction of webs that include all links versus webs that exclude "weak" trophic links. Winemiller (1989), who reported a strong positive relationship between C and S when all links are included, found that even when links representing < 1% of gut contents are excluded, there is still a slight positive relationship. Using a range of link thresholds from 0 to 4.5%,

Winemiller (1990) found that food-web properties are very sensitive to changing thresholds of link inclusion, with rapid changes between 0 and 1% thresholds that level off as thresholds approach 4.5%. Various food-web properties based on early catalogs look more like values observed for the detailed fish food webs at 4 or 4.5% link exclusion thresholds, suggesting that "food-web diagrams taken from the literature tend to depict only the strongest feeding interactions" (Winemiller 1990). Connectance increases as more links are included. Goldwasser and Roughgarden (1997) also found that C increases as they include rarer diet items in a food web focused on Anolis lizards on the island of St. Martin. In general, L/S decreases in its sensitivity to S with increasing link thresholds, and the mean value of L/S also decreases (Winemiller 1990; Tavares-Cromar and Williams 1996; Bersier et al. 1999).

Martinez looked at a slightly different aspect of link threshold effects on food-web structure by examining different criteria for assigning a feeding link between aggregates of species in resolution studies of the Little Rock Lake food web (Martinez 1991) and 11 insect webs (Schoenly et al. 1991) with 20 or more trophic species (Martinez 1993b). The strictest criterion is *minimum linkage*, which requires that every species in an aggregate be linked to every member of another aggregate for them to be linked. The weakest criterion is *maximum linkage*, which requires that only one member of each aggregate share a link for them to be linked. An *average linkage* criterion rests between the two; for example, two aggregates are linked if half their species share links. For all of the webs examined, many food-web properties are quite sensitive to linkage criteria, especially to minimum and average criteria (Martinez 1991, 1993b). Maximum linkage generally impacts the values of food-web properties the least across a range of aggregation. Some structural properties are more sensitive to linkage criterion choice than to aggregation. This means that webs with similar S display a wider range of values across different linkage criteria than cases in which S is allowed to vary for a particular linkage criterion (table 1).

Sampling effort potentially impacts the inclusion of both links *and* species in food webs. For example, source webs (one or more basal species, their consumers, their consumers' consumers, etc.) are easier to compile than community webs, but, by definition, undersample species and links of the community within which they are embedded. About half of the 60 insect webs analyzed by Sugihara et al. (1989) and reanalyzed elsewhere have only a single basal "species" (often carrion or dung) and are, in effect, source webs. Detailed community food webs have been used to show that many properties are indeed sensitive to the inclusion of increasing numbers of source species (Hawkins et al. 1997); for example L/S tends to increase. The study that perhaps best captures the impacts of sampling effort as it might actually manifest in field observations was conducted by Martinez et al. (1999). They analyzed a detailed grass source web (10 grass species plus 77 endophytic insect species, including herbivores and parasitoids, living inside the grass stems) that included data on the frequency with which each consumer and each feeding relationship were observed. They simulated increases

TABLE 1 Summary of sensitivity of selected food-web properties in the Little Rock Lake food web (Martinez 1991, table 2) to species aggregation and linkage criteria. For aggregation, *High* refers to a large and clear systematic change across variable S, *Medium* refers to a substantial change, and *Low* refers to a relatively small change. For linkage criteria, *High* means that the property is more sensitive to change in linkage criteria than to aggregation, *Medium* means that the sensitivity of the property is similar to both linkage criteria and aggregation, and *Low* means that the property is less sensitive to linkage criteria than to aggregation. The last seven properties are proportions relative to the total number of species in the web (Top, Intermediate, Basal) or the total number of links in the web (T-I, T-B, I-I, I-B).

Property	Sensitivity to Aggregation	Sensitivity to Linkage Criteria
Interactive connectance	Medium	High
Directed connectance	Low	Medium
L/S	High	Low
SC	High	Low
Average chain length	Medium	Low
Maximum chain length	High	Medium
Predator/prey ratio	Low	Medium
Top species	High	High
Intermediate species	Low	Medium
Basal species	High	Medium
T-I links	High	High
T-B links	High	High
I-I links	Low	High
I-B links	Medium	Medium

in sampling effort in terms of inclusion of both links and species in different versions of the food web, and looked at the impacts on observed connectance. In this case, directed connectance initially decreases with lower thresholds for including species and links, but quickly reaches an asymptote beyond which C changes very little with increasing sampling effort. This asymptote only appears among trophic-species versions of the food webs, but closely approximates C of the original-species web.

3.4 SAMPLING CONSISTENCY

One of the critiques of the 113 web catalog was that sampling methodology varied widely. The first researcher to address this issue explicitly in the context of testing structural patterns was Havens (1992). He constructed food webs representing the pelagic communities of 50 small lakes and ponds in the Adirondacks,

New York, using consistent sampling methods and identical linkage criteria. The number of taxa ranges from 10 to 74, and they are resolved to the genus, species, or ontogenetic life-history stage level. Havens used a method that can be referred to as "subsampling from a metaweb." Here, *metaweb* refers to a master web that includes all of the taxa found among multiple similar habitats and all possible feeding links if all of the species co-occurred in a single habitat. In this case, the metaweb includes 220 taxa found across all of the lakes and feeding links among them (see also Sprules and Bowerman 1988). The 50 individual webs were created using species lists for each lake and then deriving feeding interactions from the metaweb. The main drawback of this approach is that it assumes that predator selectivity does not vary among habitats (Havens 1992); that is, if species A eats species B in lake X, A will also eat B if they are both present in lake Y. While the strength or magnitude of interactions between A and B will almost certainly vary from lake to lake, it is less likely that the presence of particular feeding interactions will vary (Havens 1992), especially when the food webs are of the same habitat type and within the same climatic and geographic region.

Havens (1992) found that L/S increases fourfold over the range of diversity of the webs, providing more evidence that constant L/S is related to deficiencies in the earlier data and "does not reflect a real ecological trend." He also reported that fractions of species and links show no significant trends with S, and concluded that those properties are scale invariant, consistent with Cohen and Briand (1984). However, compared to empirical values (Briand and Cohen 1984) and values predicted by the cascade model (Cohen and Newman 1985b), the lake food webs display central tendencies much lower for T and T-I (similar to Warren 1989; Martinez 1991) and higher for B and I-B.

Havens's approach is only one way to create multiple food webs in a consistent manner. Another common method is simply to chose a type of habitat, and to create food webs for particular instances of that habitat (e.g., multiple streams within a region) using consistent field sampling and trophic link attribution methods at each site (Winemiller 1990; Townsend et al. 1998; Schmid-Araya et al. 2002). This approach has also been used to create multiple time-specific webs at particular sites in order to look at temporal resolution issues (Baird and Ulanowicz 1989; Schoenly and Cohen 1991; Tavares-Cromar and Williams 1996). Composite webs that integrate over time will necessarily have greater S and L than the time-specific webs on which they are based. They have also been used to demonstrate the sensitivity of other structural food-web properties to temporal variation and integration (Schoenly and Cohen 1991; Deb 1995; Tavares-Cromar and Williams 1996; Thompson and Townsend 1999).

3.5   ABANDON SHIP?

Improved data, as well as studies that looked at species aggregation, sampling effort, and sampling consistency indicated that many conventionally accepted patterns of food-web structure, especially constant L/S, were almost certainly artifacts of weak data or methodology. For many researchers, the array of issues being brought to light was enough for them to wash their hands of the entire research program. This attitude was reflected in the change in focus between the first and second major international symposia on food-web research. The first symposium, sponsored by the U.S. Department of Energy and organized by Oak Ridge National Laboratory, occurred in October 1982 and was followed a year later by a proceedings (De Angelis et al. 1983). This landmark meeting was dominated by talks relating to theory, dynamical and structural models, and topological patterns. The second major symposium on food webs, convened in September 1993 at Colorado State University's Pingree Park Conference Center, with a proceedings following a few years later (Polis and Winemiller 1996), was dominated by talks on empirical and experimental research. While several of the talks and resulting book chapters discussed food-web structure in the context of particular systems, there was almost no treatment of general properties or patterns, apart from a review of criticisms of such research in the introduction (Winemiller and Polis 1996). As Paine notes in the preface, "web metrics...are pleasantly inconspicuous." Even dynamical modeling was sidelined—of 37 main chapters, only 4 focused primarily on dynamical models (Abrams 1996; Arditi and Michalski 1996; Hastings 1996; Yodzis 1996). The overwhelming attention on experimentation and describing the impact of variability (e.g., spatial, temporal, environmental, habitat) on the dynamics of particular food webs or sets of trophic relationships came as a relief to many dissatisfied with the earlier work, which was often viewed as a gross oversimplification of the natural world. The pendulum in food-web research had swung sharply in favor of the WIWAC school—the "world is infinitely wonderful and complex" (Lawton 1995), and away from the search for generalities.

Nevertheless, over the course of the 1990s, ecologists continued to put together improved, detailed, field-based data on food-web network structure. An article co-authored by 24 top food-web researchers suggested a variety of ways for "improving food webs" with a focus on better data collection (Cohen et al. 1993). However, there is no universal correct way to compile a food web: all food webs will reflect the focus of researchers on particular methods, taxa, habitats, questions, and spatial and temporal boundaries and scales. The promise that structural food-web research holds is that by stepping far enough back from noisy details of particular systems, coarse-grained attributes and generalities will emerge, despite inherent variability and noise associated both with particular systems and with particular research approaches and agendas. In effect, this is a statistical mechanics approach to ecology. In this spirit, a few researchers continued to explore food-web patterns to call into question or corroborate old "laws" and theory, and also to offer alternative hypotheses about the network structure of food webs based on improved data and/or statistical analyses.

3.6   SCALE DEPENDENCE

One alternative hypothesis to scale-invariant patterns is "scale dependence." In his Ecological Society of America MacArthur Award Lecture article, Schoener (1989) suggested that "S-independence" of food-web properties including L/S, prey-predator ratios, and fractions of basal and top species was unlikely in both principle and practice. In other words, rather than being scale-invariant, many food-web patterns are likely to be *scale-dependent*. Schoener explored a simple conceptual model based on an extension of Pimm's (1982) notion that species should be limited in the number of prey species they can consume. Schoener suggested that not only may the number of prey species (*generality*) be constrained, but the number of predator species against which a species can defend (*vulnerability*) may also be constrained. He suggested that as food-web S increases, the vulnerability of a species also increases. The consequences of these basic assumptions include scale-dependence of many food-web properties. Schoener (1989) tested those predictions with a statistical analysis, which Cohen and colleagues had avoided, of 98 webs drawn from the 113 web catalog and from source papers for particular webs in the catalog. He found the data generally agreed with the S-dependent predictions, not with scale-invariance. The data also supported the basic assumptions of his conceptual model: generality does not increase with S, while vulnerability does.

Several studies previously mentioned utilized improved data as well as novel methodological analyses and found that many purported scale-invariant properties of food webs appeared to be attributes of poorly resolved webs, with quite different values observed in highly resolved or sampled webs (Warren 1989, 1990; Winemiller 1990, 1991; Martinez 1991, 1993b; Hall and Rafaelli 1991; Polis 1991; Hawkins et al. 1997). This provided some empirical support for "scale-dependence" of most food-web properties. Additional, more explicit support came in the form of reanalysis of Havens' 50 lake webs (Martinez 1993a) and subsets of Schoenly et al.'s (1991) insect webs (Martinez 1994). The reanalyses, which used more appropriate statistical tests and/or trophic species aggregation, reported significant scale-dependent trends of species and link proportions across food webs, similar to what Schoener (1989) found in his reanalysis of the 113 web catalog. The scale-dependent trends reported by Martinez (1993a, 1994) were actually present in data originally used to support scale invariance, but either the trends were weak and lacked significance, or significance tests were not conducted properly, if at all (Havens 1992; Sugihara et al. 1989). More sophisticated statistical methods based on generalized linear models also suggested that fractions of species are sensitive to S, based on a subset of 61 of the insect webs (Murtaugh and Kollath 1997).

An analysis of observed patterns of species fractions (T, I, B) at local scales and expected patterns of species fractions at regional and global scales based on biodiversity estimates suggests a limit to food-web scale dependence (Martinez and Lawton 1995). Empirical webs with less than a hundred species appear to display strong scale dependence. However, hypothetical webs with more than a thousand species appear to display scale invariance due to unchanging fractions of species: no top species (due to cannibalism, parasitism, mutual predation, and longer loops), a small percentage of basal species, and a large majority of intermediate species. In essence, given that researchers are unlikely to compile detailed food webs with more than a thousand species, the food webs they can describe are likely to exhibit scale dependence. Using a similar concept, Bersier and Sugihara (1997) attempted to rehabilitate scale invariance for webs with S of $\sim$ 10 to 100, arguing that species and link fractions in 60 insect food webs (Sugihara et al. 1989) display *scaling regions*: one region of scale dependence and one region of scale invariance. However, in this case, webs containing 12 or fewer original species or 7 or fewer trophic species are suggested to exhibit scale dependence, while webs with more species are suggested to exhibit scale invariance. These cut-off points are one to two orders of magnitude lower than those suggested by Martinez and Lawton (1995). At this point, the massaging and re-massaging of the Schoenly et al. (1991) insect web catalog is more confusing than convincing. Regardless of what is thought of the various statistical approaches, the fact that the insect webs are not comprehensive community webs, but are mostly source/substrate webs with mean original S of 22 and mean trophic S of 12, should lessen their importance in the overall assessment. The balance of evidence suggests that early patterns of scale invariance are artifacts of poorly resolved data, and that scale dependence of most properties is likely to be observed across higher quality datasets, at least within the range of diversity (i.e., $S < 1000$) that ecologists are likely to be able to sample in detail.

3.7 CONSTANT CONNECTANCE

Just as improved data and methodological analyses highlighted problems with scale invariance, they also showed problems with constant L/S and hyperbolic decline of C with S, as discussed previously. L/S was shown to be much greater than 2 for a variety of improved food-web datasets (e.g., Polis 1991; Martinez 1991; Hall and Raffaelli 1991). A set of fish sink webs with $\sim$ 20 − 120 S displayed strongly increasing C with S, although when links representing 1% or less of diet were excluded, C increased only slightly with S (Winemiller 1989). Several aggregation and sampling effort analyses suggested that C is relatively robust to changes in S (Warren 1989; Martinez 1991, 1993b; Martinez et al. 1997) while L/S is not, except possibly at low levels of link sampling (Bersier et al. 1999). Consonant with the former studies, an alternative hypothesis of "constant connectance" emerged out of new data and analyses (Martinez 1992). The mathematical difference between constant C and constant L/S can be simply stated using a log-log graph of links as a function of species (fig. 1). As discussed previously, if a regression can be reasonably fit to the data, it produces a power law of the form $L = aS^b$. In the case of the link-species scaling law, $b = 1$, which means that $L = aS, L/S = a$, and thus L/S (directed connectancce) is constant. In the case of constant connectance, $b = 2$, which means that $L = aS^2, L/S^2 = a$, and thus L/S^2 (directed connectance) is constant. Instead of L/S being constant, L/S increases as a fixed proportion of S. One ecological interpretation of constant connectance is that consumers are likely to exploit an approximately constant fraction of available prey species, so as diversity increases, the number of links per species increases (Warren 1990).

Do data support constant connectance? For a set of 15 webs derived from an English pond (11 habitat webs, 4 arbitrary subwebs, and a summary composite web), Warren (1990) reported $b = 2$, and more specifically $L = 0.24S^2$. He suggested that within a community, "increasing S should produce a curvilinear relationship whereby L is roughly proportional to S^2." He hypothesized that the value of that proportionality will be greater for subwebs dominated by generalist feeders and lower for those dominated by specialists. Warren (1990) suggested that looking at the $L - S$ relationship across communities would result in a great degree of scatter and only a slight upward trend due to sampling effects, a hypothesis that appeared consistent with previous food-web data. However, another study hypothesized that the proportionality of L with S^2 does hold across communities (Martinez 1992). Trophic species versions of 175 webs display $b = 1.54$ ($R^2 = 0.93$), smaller than $b = 2$ expected for constant C (Martinez 1992). The 175 webs, which include the 113 web catalog, the 60 insect webs, the Little Rock Lake food web, and a food web of the island of St. Martin (Goldwasser and Roughgarden 1993), are dominated by the earlier, more poorly resolved data. To address this, Martinez (1992) conducted an analysis of a subset of the 12 most "credible" datasets in terms of resolution and completeness, representing the whole range of S (2 to 93), and found $b = 1.73$ ($R^2 = 0.98$). Further distillation of the data by eliminating food webs with less than 10 species resulted in 5 webs with $b = 2.04$ ($R^2 = 0.95$).

Based on these and other analyses, Martinez (1992) argued that the true value of b in highly resolved webs across a wide range of S is likely closer to 2 than to 1, suggesting "roughly constant connectance within relatively homogeneous environments." If connectance is roughly constant ($b \approx 2$), mean C is quantified by the value of a, which is 0.014 ($SD = 0.06$) for all 175 webs and 0.11 ($SD = 0.03$) for the five web subset, suggesting that 0.11 is the best estimate of mean C. This indicates that approximately 10% of all possible feeding links, including cannibalism and other loops, are actually realized in food webs compiled "within" habitats where species are likely to be relatively well mixed. Food webs compiled across obvious environmental boundaries (e.g., a lake and its surroundings) are likely to have lower C, since there will be species that never encounter one another and thus have no chance of a feeding relationship. Also, connectance will vary to the degree that specialists, generalists, or omnivores are prevalent (Warren 1990).

The hypothesis of constant connectance, or $b = 2$, has been called into question by a few later studies, although in some cases a second look suggests otherwise. Havens (1992) reported $b = 1.4$ for 50 pelagic lake food webs. Reanalysis without forcing the regression through 0 showed that the data display $b = 1.9$ ($R^2 = 0.92$) (Martinez 1993a) and mean C of 0.10 ($SD = 0.02$). These results corroborate prior predictions of the likely values of both a and b (Martinez 1992). Using a set of 22 stream food webs with 22 to 212 species, Schmid-Araya et al. (2002) reported $b = 1.30$, lower than the value of $b = 1.36$ for the 113 web catalog, as well as relatively low C that ranges from 0.03 to 0.12. The webs focus on algae, micro- and in some cases meiofauna, and macroinvertebrates, with vertebrates excluded. It is unclear how many and what types of vertebrates might be a part of any of the food webs. The exclusion of higher trophic-level taxa, such as generalist opportunistic feeders on invertebrates, would tend to decrease C and could also lower the value of b. However, it may be that low mean C and different scaling of C are characteristic of stream webs.

Figure 1 shows the relationship between links and species for 19 relatively diverse food webs from a variety of habitats (e.g., pond, lake, stream, desert, grassland, rainforest, coral reef, marine shelf), with trophic species of 25 to 172 (refer to Dunne et al. 2002b, 2004 for information on individual webs). For this set of data, $b = 1.5$, but there is a large amount of scatter ($R^2 = 0.68$) (see also Montoya and Solé 2003). Directed connectance ranges from 0.03 to 0.3, with a mean of 0.13. The regression line for the data does not deviate strongly from the predicted line for constant connectance, but does obviously deviate from the line predicted by the link-species scaling law (fig. 1). In any case, these data are clearly too variable to convincingly demonstrate a particular $L - S$ relationship, other than not supporting constant L/S, perhaps because they represent such widely different ecosystems and methods (Warren 1990). Other studies have questioned the fit of constant connectance to empirical data, for example by conducting more appropriate statistical analysis of existing data (Murtaugh et al. 1998) or by doing detailed studies of consumer guild diets (Winemiller et al. 2001). The latter approach may not be a good way to test for constant connectance since it focuses on the trophic breadth of particular kinds of organisms, and C is a global property of a food web calculated across many types of species. Whether values of b are closer to 1.5 or 2, directed connectance for resolved community webs appears constrained to ~ 0.03 to 0.3 out of a possible range of 0 to 1. The central tendency across communities appears to be ~ 0.10 to 0.15, much lower than a null expectation of 0.5 (Kenny and Loehle 1991).

4 CURRENT PHASE: NEW MODELS, NEW DIRECTIONS

The cascade model was largely neglected during the 1990s as researchers either focused on methodological issues or veered entirely away from structural food-web analyses, especially research on the relationship of dynamical stability to diversity and connectance. Given that the cascade model was developed with data that turned out to be poorly resolved, it was unlikely to survive testing with improved data that differed dramatically from earlier data. Indeed, its fun-

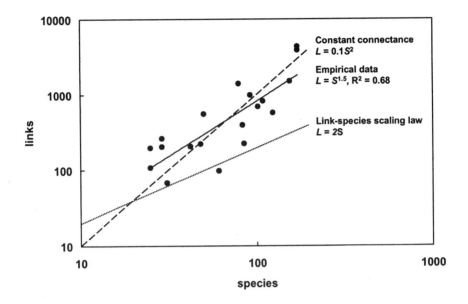

FIGURE 1   The relationship of links to species for 19 trophic-species food webs from a variety of habitats (black circles). The solid line shows the log-log regression for the empirical data, the dashed line shows the prediction for constant connectance, and the dotted line shows the prediction for the link-species scaling law.

damental assumption of no looping, particularly cannibalism, was shot down repeatedly in improved datasets. It was also unclear whether the cascade model even described early data well (Schoener 1989). A graph theoretic analysis suggested that the relationship between connectance and species in early empirical webs that appeared to support the cascade model could not be distinguished from the relationship generated by a random model that included sampling effects, although other aspects of potentially non-random network structure were not considered (Kenny and Loehle 1991). Another analysis looked at the goodness of fit of the cascade model to the early ECOWeB data and found that, while it did seem to capture the central tendency of the data, it failed to characterize variability. Namely, the data were over-dispersed in relation to model predictions, leading to rejection of the model and a modified form of the model at "essentially any significance level" (Solow 1996). Furthermore, the random distribution of links by the cascade model, albeit within a constrained portion of an interaction matrix, resulted in less species lumpiness than found in the Little Rock Lake and larger ECOWeB food webs. In other words, there is more overlap of predators and prey for species in diverse empirical webs than in simulated webs based on a random distribution of links (Solow and Beet 1998).

4.1   NICHE MODEL

A new food-web structure model, in the tradition of the phenomenological, graph-theoretic, stochastic cascade model, was proposed by Williams and Martinez (2000) (Box 1). This *niche model* addresses several limitations of the cascade model, particularly the assumption of link-species scaling, the exclusion of looping, and the lack of trophic overlap among species. However, the niche model retains much of the simplicity and tractability embodied by the earlier model. As in the cascade model, the niche model has two tunable parameters that determine the number of species and links. In the case of links, instead of assuming constant L/S (Cohen and Newman 1985b), Williams and Martinez used directed connectance C (L/S^2) as the link parameter, allowing it to vary (Cohen et al. 1985). Thus, Williams and Martinez (2000) made no assumption about link-species scaling, and did not replace it with any other hypothesis about the relationship between L and S. However, their choice of C reflected the notion that C is independent of S (Martinez 1992, 1993a) just as Cohen and Newman's (1985b) choice of L/S reflected the notion that L/S is independent of S (Cohen and Briand 1984). Beyond the two input parameters S and C, the niche model, like the cascade model, orders species along a single dimension. However, instead of the cascade model's simple rank-ordering of species, the niche model assigned each species a uniformly random *niche value* along a line, and that niche value corresponds to the position of each species on the line. To distribute links, each species is assigned a feeding range that represents an interval of the line whose midpoint is a uniformly random number less than the niche value of the species possessing the range. All species that fall in this range are eaten by the consumer species (Box 1). Feeding range sizes are drawn randomly from a beta distribution to produce a C close to the target C (Williams and Martinez 2000).

The distribution of feeding links in the niche model leads to several outcomes: (1) cannibalism and feeding on species with higher niche values can occur, (2) there is higher trophic overlap than in purely stochastic link distribution schemes, since species with similar niche values are more likely to share consumers, and (3) food webs are rendered interval due to contiguous feeding by each consumer on resources within a single range or segment of the line. The first two outcomes, along with using variable C as a parameter, address the three main limitations of the cascade model. The third outcome represents an acknowledged limitation of the niche model (Williams and Martinez 2000), since diverse empirical food webs are known to not be interval (Cohen and Palka 1990). Williams and Martinez (2000) argued that intervality is a delicate property that is easily broken, often by the loss of just one link in a web. They hypothesized that quantitative measures of intervality will show that the degree of intervality, rather than just the presence of intervality, is quite high in empirical food webs.

In addition to these model innovations, Williams and Martinez (2000) also introduced the use of numerical simulations to compare statistically the ability of the niche model and alternate network models to fit empirical food-web data. They standardized the form of multiple models by creating versions of the cascade and a random model that use variable S and C as input parameters. This ensures that the three models vary only in how they distribute links among species (Box 1). Because of stochastic variation in aspects of how species and links are distributed in any particular model web, analysis begins with the generation of hundreds to thousands of model webs with the same S and similar C as an empirical food web of interest. Model webs that fall within 3% of the target C are retained. Species that are either disconnected (i.e., species that lack links) or trophically identical to other species, or webs that contain such species, are eliminated and replaced. The exclusion of trophically identical species means that model webs are most fruitfully compared to trophic-species versions of empirical food webs, rather than original-species webs, unless most or all of their species are already trophically distinct. Additionally, in the cascade and niche models, no prey are assigned to the species with the lowest niche value, ensuring there is at least one basal species per web. Once a set of model webs is generated, model means and standard deviations are calculated for each food-web property of interest, which can then be compared to empirical values. Raw error, the difference between the value of an empirical property and a model mean for that property, is normalized by dividing it by the standard deviation of the property's simulated distribution. This approach allows assessment not only of whether a model over- or underestimates empirical properties as indicated by the raw error, but also to what degree a model's mean deviates from the empirical value. Normalized errors within ±2 are considered to indicate a good fit between the model prediction and the empirical value (Williams and Martinez 2000).

All three models were evaluated by how well they fit up to 12 structural properties of seven empirical community food webs drawn from the expanding set of improved food-web datasets, with 25 to 92 trophic species and 0.061 to 0.32 connectance (Williams and Martinez 2000). The properties include old standbys such as T, I, and B as well as other properties of interest in food-web research (proportions of omnivores, cannibals, and species in loops; chain length properties; trophic similarity; variation of generality and vulnerability). Link proportions were excluded due to their strong correlation with T, I, and B. In summary, the random model performed poorly, with an average normalized error (ANE) of 27.1 ($SD = 202$), the cascade model performed much better, with an ANE of -3.0 ($SD = 14.1$), and the niche model performed an order of magnitude better that that, with an ANE of 0.22 ($SD = 1.8$). Only the niche model falls within ±2 ANE and is considered to show a good fit to the data. Not surprisingly, there is variability in how all three models fit different food webs and properties. For example, the niche model generally overestimates food-chain length. Specific mismatches are generally attributable either to limitations of the models or biases in the data. A separate test of the niche and cascade models with three marine food webs, a type of habitat not included in the original analysis, obtained similar results (Dunne et al. 2004). That test added four properties to the analysis: proportion of herbivores, mean trophic level, clustering coefficient, and characteristic path length. The latter two properties are borrowed from "small-world" network research and will be discussed more below (see Small-World Properties and Degree Distribution section). The niche model's performance on the new properties is similar to previous properties. For example, the niche model tends to systematically underestimate the proportion of herbivores, which partly explains its overestimation of food-chain length.

Like the cascade model, but taking into account improved food-web data, these niche model studies demonstrate that the structure of food webs is far from random, and that simple link distribution rules can yield apparently complex network structure, comparable to that observed in empirical data. In addition, like the cascade model, the niche model is simple enough that it is analytically solvable, leading to theoretical predictions similar to trends in numerical simulations and empirical data (Camacho et al. 2002a,b). The hypothesis of scale dependence suggests that there are not simple generalities or constant values of network properties that hold across food webs with varying S. However, the concordance of the niche model with empirical data suggests that once variable S and C are taken into account, there appear to be universal coarse-grained characteristics of how trophic links and species (defined according to trophic function) are distributed within food webs across a wide array of habitats. Of course, the current success of the niche model needs to be taken in context, since, like the cascade model it orients itself to data that are flawed and limited. New data and better understanding about key processes, properties, and interactions in food webs, or more broadly, ecological networks, may lead to more efficacious approaches or to the rejection of any general model. Martinez et al. (Chapter 6) and others have already proposed some variations on the niche model as discussed below.

4.2 NESTED-HIERARCHY MODEL

Another simple topological model similar to the niche and cascade models, the *nested-hierarchy model*, addresses the intervality limitation of the niche model (Box 1, Cattin et al. 2004). Following methods introduced by Williams and Martinez (2000), Cattin et al. use S and C as input parameters, distribute species niche values randomly along a line, and stochastically assign the number of prey items for each species using a beta distribution. They then distribute links, starting with the species with the lowest niche value, in a way that (1) avoids creating interval webs, (2) generates webs with trophic overlap (Solow and Beet 1998; but see Stouffer et al. 2005)), and (3) allows a low probability of looping. First, a link is randomly assigned from a consumer species i to another species j with a lower niche value. If that resource species j is also fed upon by other species, the consumer species i's next feeding link is randomly selected from the pool of resource species of a set of consumer species defined as follows: they share at least one prey species, and at least one of them feeds on resource species j. If more feeding links are required, links are randomly assigned to species without predators and with lower niche values. If yet more feeding links are needed, links are randomly assigned to species with equal or higher niche values.

Cattin et al. (2004) compared the fit of this model and the niche model to the 7 webs and 12 properties analyzed by Williams and Martinez (2000) plus two additional properties that reflect aspects of intervality. They compared model means and standard deviations with empirical values, but did not calculate normalized error. They concluded that the two models perform comparably on the original 12 properties, but the nested-hierarchy model does better for the two intervality properties. This is not surprising since the niche model necessarily returns 0 values for the intervality properties. A re-analysis of Cattin et al.'s results illuminates some overstated or incorrect claims in their abstract, namely that the nested-hierarchy model "better reflects the complexity and multidimensionality of most natural systems," and that the niche model fails "to describe adequately recent and high-quality data." Martinez and Cushing (Box A) suggest that Cattin et al. base their claims on selective favoring of intervality, and gloss over other details of fit. Indeed, these types of analyses will always be influenced by how much importance is ascribed to particular properties (Cattin et al. Box B). However, any slight difference in the performance of the niche and nested-hierarchy models, which could be assessed more rigorously using normalized error, is not comparable to the order of magnitude improvement of the niche model over the cascade model, the quantitative assessment of which did not even include the cascade model's failure to generate cannibalism or other looping (Williams and Martinez 2000). Indeed, the fit of the nested-hierarchy model to intervality properties may come at the expense of reduced fit to other properties (Martinez and Cushing Box A). Another way the niche model could be modified to break intervality is by generating feeding ranges that are slightly larger, and by making the probability that the consumer species eat species within their feeding ranges slightly less than 100% (R. J. Williams, personal communication). However, all previous attempts to modify the niche model based on ecological understanding to improve its overall fit to data failed (unpublished data, Williams and Martinez).

4.3 GENERALIZED ANALYTICAL MODEL

An analytical study building on Camacho et al. (2002a,b) provides a compelling reason why the nested-hierarchy model does not generally improve on the niche model. Although the nested-hierarchy model "appears to be quite different in its description, it nevertheless generates webs characterized by the same *universal* distributions of numbers of prey, predators, and links" (Stouffer et al. 2005, original emphasis). Stouffer et al. found that only two conditions must be met for network models to reproduce several central properties of currently available improved food-web data: (1) species niche values form a totally ordered set, and (2) each species has a specific probability, drawn from an approximately exponential distribution, of preying on species with lower niche values. The first

Box 1: Models of Food-Web Structure

Four simple, stochastic models that have been proposed to generate and predict the network structure of empirical food webs are described. The models share two empirically quantifiable input parameters: (1) S, the number of species or taxa in a web, and (2) C, connectance, a metric determined by the number of links and species in a food web. There are S^2 possible and L actual links in a particular food web, and directed connectance C is defined as L/S^2, or the proportion of possible links that are actually realized. The models differ in the rules they use to distribute links among species, as follows:

- **Random Model (inspired by Erdös and Rényi 1960; see also Cohen 1977b)**

  Any link among S species occurs with the same probability P equal to C. This creates food webs as free as possible from biological structuring.

- **Cascade Model (modified from Cohen and Newman 1985b)**

  Each species is assigned a random value drawn uniformly from the interval $[0,1]$. Each species has the probability $P = 2CS/(S-1)$ of consuming species with values less than its own. This creates a feeding hierarchy and disallows cannibalism and feeding on species higher in the hierarchy, as illustrated in the following diagram.

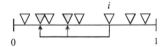

Species i feeds on two species, shown in grey, with lower values.

This formulation is a modified version (Williams and Martinez 2000) of the original cascade model (Cohen and Newman 1985b), which tuned L/S (link density) to the data by using an average across empirical webs to look at scaling patterns generated by the model. This assumed constant L/S and ensured that C declined hyperbolically with S, as suggested by early theory and data. Recent studies use variable C as the parameter that constrains the number of links in model webs, employing a similar approach to Cohen et al. (1985) who tuned L/S to values for particular webs.

- **Niche Model (Williams and Martinez 2000)**

  As in the modified cascade model, each species is assigned a random value drawn uniformly from the interval $[0,1]$, referred to as the species' niche value, n_i. Each species consumes all species within a range of niche values r_i. The size of r_i is randomly assigned using a beta function, producing a C close or identical to the target C. The center of the range c_i is drawn uniformly from the interval $[r_i/2, n_i]$ or $[r_i/2, 1 - r_i/2]$ if $n_i > 1 - r_i/2$. This keeps all of the feeding range within $[1,0]$ and places the center of a species' range lower than its niche value.

continued on next page

Box 1 continued

As shown in the diagram below, the c_i rule relaxes the strict feeding hierarchy of the cascade model by allowing up to half of the feeding range to include species with niche values $\geq n_i$, thus permitting cannibalism and feeding on species with higher niche values.

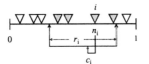

Species i feeds on all four species, shown in grey, within its feeding range r_i. This includes a cannibalistic link to itself and a link to a species with a higher niche value.

- **Nested-Hierarchy Model (Cattin et al. 2004)**

  Following the cascade and niche models, each species is assigned a random niche value $[0, 1]$. Like the niche model, the number of prey items for each species is drawn randomly from a beta distribution that constrains C to be close to the target. Feeding links are assigned in a multistep process. First, a link is randomly assigned from species i to a species j with a lower niche value. If that prey species j is also fed upon by other species, the next feeding link for species i is selected randomly from the pool of resource species fed on by a set of consumer species defined as follows: they share at least one prey species, and at least one of them feeds on species j. If more feeding links are required, links are randomly assigned to species without predators and with niche values $< n_i$. If more feeding links are required, links are randomly assigned to species with niche values $\geq n_i$. Thus, the model relaxes the contiguous feeding, and thus intervality, of the niche model.

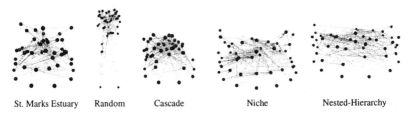

St. Marks Estuary   Random   Cascade   Niche   Nested-Hierarchy

The food web for St. Marks Estuary (Christian and Luczkovich 1999), and an example of four types of model webs with the same S and C as the empirical web. Images produced with FoodWeb3D, written by R. J. Williams and available at www.foodwebs.org.

condition was met by the cascade model (Cohen and Newman 1985b) and is equivalent to their rule that species are ordered along a single dimension. The niche model adopted that rule by assigning each species a unique niche value drawn randomly from the interval $[0, 1]$. The second condition was newly embodied in the niche model through its use of a beta distribution as the means by which the size of feeding ranges is distributed (Williams and Martinez 2000). Stouffer et al. (2005) note that the beta distribution is a type of exponential distribution. They show that the specific form of the exponential distribution does not alter predictions about the distributions of numbers of prey and predators, nor does it alter a number of network structure properties derivable from those two distributions. Their focus on numbers of prey and predators as two fundamental, non-identical distributions relates back to Schoener's (1989) work discussing the importance of understanding differences in vulnerability (number of predators) and generality (number of prey) for developing simple models of food-web structure, differences reflected in improved empirical data and the niche model (Williams and Martinez 2000).

Ordered niche values and the beta distribution (Williams and Martinez 2000) were adopted by the nested-hierarchy model (Cattin et al. 2004). The nested-hierarchy model has different link distribution rules that are meant to mimic phylogenetic constraints. However, Stouffer et al. (2005) point out that Cattin et al.'s distribution rules ensure that a species is assigned prey essentially randomly from the set of species with lower niche values, as constrained by the beta distribution. As a result, the nested hierarchy model returns the same distributions of numbers of predators, prey, and links among species as does the niche model. Stouffer et al. (2005) test their hypothesis about the centrality of the two conditions by modifying the cascade model to meet the second condition. This modified cascade model also produces the same general analytical expressions, or universal functional forms, for distributions as the niche and nested-hierarchy models. Based on their analysis, Stouffer et al. consider the nested-hierarchy and modified cascade models to be "randomized" versions of the niche model, which was the first model to embody the two fundamental conditions they identify as central to model representations of empirical food-web network structure.

Several common food-web properties previously used to assess particular models (Williams and Martinez 2000; Cattin et al. 2004) can be derived from the analytical expressions for the distributions of numbers of prey and predators, including fractions of top, basal, and cannibalistic species, and standard deviations of vulnerability and generality (Stouffer et al. 2005). Other derivable properties of potential interest include the *correlation coefficient* between each species' number of prey and predators, and *assortativity*, the probability with which nodes with high degree (many links) link to other high-degree nodes (Newman 2002). Food webs tend to be negatively assortative, or disassortative (Newman 2002; Stouffer et al. 2005). With reference to this set of properties, and drawing on 15 of 19 improved food webs described in Dunne et al. (2002ab, 2004), Stouffer et al. (2005) find that empirical values for 11 of the 15 food webs are well described by the model's analytical expressions and numerical simulations. Stouffer et al. (2005) also added the potential for cannibalism (i.e., feeding on species with equal niche values) to the modified cascade model to make it even more comparable to the other two models. However, allowing for a low probability of cannibalism, or a low probability of feeding on species with higher niche values as in the niche and nested-hierarchy models, does not alter analytical predictions of predator and prey distributions and thus is not included in condition two (Stouffer, personal communication).

What Stouffer et al. (2005) do not address are other aspects of the models and data, particularly with regard to looping, intervality, omnivory, herbivory, food-chain statistics, trophic level, and other properties of ecological interest (Williams and Martinez 2000, 2004b; Dunne et al. 2004a,b; Cattin et al. 2004). While they appear to have identified fundamental aspects of species and link distributions that underlie and emerge from currently successful models of food-web network structure, not all properties of ecological interest are derivable from distributions of numbers of predators and prey. Differences and similarities among particular models with respect to such properties are not addressed by the general analytical model (Stouffer et al. 2005). For example, the nested-hierarchy and modified cascade models allow for non-interval webs, and the niche and nested-hierarchy models allow for feeding on species with higher niche values. These particular aspects of the models may not significantly impact the overall distribution of numbers of predators and prey, but will affect how well the models capture other quantifiable and ecologically interesting variability in food-web network structure. Also, while Stouffer et al. (2005) have demonstrated core conditions and universal functional forms for some aspects of food-web structure that emerge out of the niche model, how well those conditions and functional forms continue to fit the data depends on evolving data availability and standards. Access to more comprehensive data that is more highly and evenly resolved, or a switch in focus to other more ecologically compelling ways to slice and dice data, may necessitate the development of some other approach to modeling ecological network structure. Whether and how the simple rules that appear to generate food-web-like topology actually connect back to ecological, evolutionary, thermodynamic or other principles remains a wide-open area of inquiry. This, more than anything, is likely to impact how ecological network data is modeled in the future.

4.4   SMALL-WORLD PROPERTIES AND DEGREE DISTRIBUTION

The resurgence of interest in the late 1990s across disciplines in describing general properties of the network structure of everything from social groups to WWW page links to power grids to transportation systems to metabolic pathways to scientific citations, brought to light a few topological properties that had not previously been explicitly evaluated for food webs. It also brought a new question to the table: do food webs have similar topology to other biotic, social, and abiotic networks? A paper published by Watts and Strogatz (1998) led the resurgence by bringing the notion of "small-world" network structure to the foreground. They suggested that most real-world networks look neither like randomly connected graphs (Erdös and Rényi 1960) nor regularly connected lattices in which every node has the same number and pattern of links. Instead, real-world networks appear to combine aspects of both—they have high clustering, like regular lattices, but short paths between nodes, like random graphs. These features are typically expressed by two properties: clustering coefficient, or the average fraction of pairs of nodes connected to the same node that are also connected to each other, and characteristic path length, or the average shortest distance between pairs of nodes. Work initiated by Barabási and colleagues (as reviewed in Albert and Barabási 2002) suggested that most real-world networks also display power-law degree distributions, which refers to the distribution of the number of links per node. Regular lattices display a constant distribution of links among nodes, while random graphs display a Poisson distribution of links among nodes. Most empirical networks appear to display a highly uneven power-law or scale-free distribution of links among nodes, with most nodes having few links and a few nodes having a very large number of links.

Several papers published in 2002 considered the question of whether empirical food webs display small-world, scale-free network structure similar to many other real-world networks (also reviewed in Cartozo et al. Chapter 3). Using original species versions of three relatively diverse and well-resolved webs, Montoya and Solé (2002) suggested that food webs do tend to display small-world, scale-free structure, although the Little Rock Lake food web did not fit those patterns well. Looking across trophic-species versions of the seven community webs analyzed by Williams and Martinez (2000), Camacho et al. (2002b) contradicted Montoya and Solé (2002) by concluding that clustering coefficients of food webs appear similar to random expectations, less than the clustering observed in small-world networks. Using cumulative rather than density distributions due to the noisiness of the data (Amaral et al. 2000), Camacho et al. (2002b) also concluded that food webs do not display scale-free distributions of links, irrespective of whether total links, number of predators, or number of prey are considered. However, when the distributions are normalized for linkage density by dividing the number of links by $2L/S$, empirical food-web link distribution data appears to display universal functional forms (Camacho et al. 2002b). For example, cumulative degree distribution, assessed using data pooled across six of seven food webs, shows a systematic exponential decay in its tail. This type of distribution, less skewed than a power-law distribution, has been observed in a few other "real-world" networks (Amaral et al. 2000). Clustering coefficient and path length also appear to follow universal functional forms that scale with linkage density (Camacho et al. 2002b). Numerical and analytical predictions of the niche model (Williams and Martinez 2000) fit empirical data quite closely for clustering coefficient, characteristic path length, and distributions of numbers of predators, prey, and links (Camacho et al. 2002b; Williams et al. 2002; Dunne et al. 2004; Stouffer et al. 2005).

Dunne et al. (2002a) attempted to resolve differences between Montoya and Solé (2002) and Camacho et al. (2002b) by examining a larger array of 16 trophic-species food webs, including those analyzed in the other two studies. Corroborating aspects of Camacho et al. (2002b), Dunne et al. found that most food webs display low clustering coefficients and non-power-law degree distributions, in particular less skewed exponential and uniform distributions. However, they also found that webs with very low connectance (e.g., the Scotch broom source web with $C = 0.03$) were more likely to display both higher-than-random clustering and power-law degree distributions, consistent with the small-world, scale-free structure of many other types of networks, and as reported for food webs by Montoya and Solé (2002). Using linkage density normalization to overlie cumulative degree distributions of the 16 webs, Dunne et al. (2002a) concluded there was too much variation in the data to support the notion of a universal functional form (Camacho et al. 2002b; Stouffer et al. 2005), and pointed out that such variation can be masked by pooling the data and excluding datasets that don't fit the pattern well. However, the data are obviously constrained within a region that is not power-law in its form. Both the tendency for much, but not all, improved food-web data to converge on universal functional forms that scale with linkage density (Stouffer et al. 2005), as well as potentially systematic deviations from or variability around those central tendencies (Dunne et al. 2002a), are important research issues that can reveal interesting insights at different levels of analysis. For example, while the niche model produces network structures that have exponential degree distributions (Camacho et al. 2002b), individual empirical webs can show other distributions, particularly uniform distributions (Dunne et al. 2002a). A simple model that starts with a randomly linked "regional" pool of species, and then creates "local" food webs via random immigration from the regional pool coupled with random extinctions from the local web, produces significant percentages of webs with exponential as well as uniform degree distributions (Arii and Parrott 2004). This study highlights simple assembly mechanisms that can produce variable degree distributions, mechanisms that may be relevant for empirical webs.

All studies looking at small-world structure in food webs (Montoya and Solé 2002; Camacho et al. 2002b; Dunne et al. 2002a, 2004; Williams et al. 2002) have reported short path lengths similar to random expectations (i.e., "two degrees of separation"; Williams et al. 2002), consistent with one aspect of small-world structure. However, apart from path length, most currently available food-web data clearly deviate from the small-world, scale-free topology observed for other biotic and abiotic networks (Camacho et al. 2002b; Dunne et al. 2002a). Compared to other networks, food webs have low diversity and high connectance, which appear to be drivers of clustering coefficient and degree distribution patterns. The ratio of observed clustering coefficient to the random expectation, which is 1 for many food webs, scales approximately linearly with network size among a wide range of biological, social, and technological networks (Dunne et al. 2002a). Degree distribution is related to connectance, with networks that show power-law distributions being much more sparsely connected (i.e., lower C) than most food webs. While many empirical food webs display exponential degree distributions, higher connectance webs often display less skewed uniform distributions (Dunne et al. 2002a).

4.5 UNIVERSAL PATTERNS?

In addition to the question of whether food webs display small-world, scale-free topology, researchers have considered other ways of identifying general topological patterns in binary-link food webs, often borrowing methods from other types of network research. Research on scale-dependence of empirical food-web properties, the niche model and its offshoots, and analyses of small-world structure and degree distribution all suggest that food webs generally do not have scale-invariant patterns conventionally understood to be "universal." Instead, food webs tend to display scale-dependent patterns that can be characterized, depending on how much variability is ignored, as universal functional forms (Camacho et al. 2002b). These scale-dependent patterns or functional forms emerge once data from different webs are normalized for link density, $2L/S$, the relationship between the number of links (L) and species (S) in a web (Camacho et al. 2002b; Dunne et al. 2002a; Stouffer et al. 2005).

However, other studies suggest that there are universal topological patterns in empirical food webs that hold regardless of S, L, or their relationship. Milo et al. (2002) developed an algorithm for detecting network motifs based on the statistical approach of Williams and Martinez (2000). They define network motifs as "recurring, significant patterns of interconnections." For a variety of biotic and abiotic networks, they identified and counted all possible configurations of three- and four-node subgraphs, and then compared the frequency of different subgraphs in empirical networks to their frequency in comparable randomized networks. Particular subgraph types are considered *motifs* when they occur significantly more often than expected for random webs. Milo et al. (2002) analyzed the seven food webs from Williams and Martinez (2000) and found that five of seven share a three-node motif referred to as three-chain or a three-species food chain. Additionally, all seven share a four-node motif referred to as bi-parallel in which two species share a common predator and prey. The food webs do not share the three-node motif with other types of networks, but do share their four-node motif with the *C. elegans* neuronal network and all five electronic circuit networks considered. The *presence* of significant motifs in food webs appears to be independent of network size (S), but the *frequency* of motifs in food web and other networks appears to grow linearly with size, unlike the frequency in randomized networks. Milo et al. (2002) speculate that motifs may be interpreted as "structures that arise because of the special constraints under which the network has evolved."

Food webs have also been investigated as a type of transportation network, using methods applied to river basins and vascular systems (see Cartozo et al. Chapter 3 for detailed discussion). Garlaschelli et al. (2003) decompose seven food webs with 42 to 123 trophic species into *minimal spanning trees*. A minimal spanning tree is a simplified version of a network that is created by removing links to minimize the distance between nodes and some destination. One way to create a minimal spanning tree of a food web is to add an additional environment node to which all basal species link, and then to trace the shortest food chain from each species to the environment node. Links that are not a necessary part of any of these shortest food chains (e.g., cannibalism, and other links in loops) are excluded from the tree. Garlaschelli et al. (2003) analyzed the allometric scaling of these trees, or how their branching properties change with network size. They found that all seven food webs display a power-law, scale-free relationship with an exponent of 1.13, although the three smallest webs display marginally larger exponents. This suggests that minimal spanning trees may successfully characterize a universal core structure in food-web networks. However, the "universality" of the exponent for food-web minimal spanning trees has been called into question by an analysis of a broader set of 17 food webs, which display exponents ranging from 1.09 to 1.26 (Camacho and Arenas 2005). The short range of the exponent values is attributed to the relatively small mean trophic level of most food webs. The particular exponents for food-web minimal spanning trees suggests that they can transport resources more efficiently across the whole network than river or vascular systems, which display higher exponents. Within a reasonable range of connectance (0.05 to 0.3), the niche model (Williams and Martinez 2000) underestimates empirically observed exponents, returning values of 1.06 to 1.08 (Garlaschelli et al. 2003).

Other approaches for identifying universal network structure have yet to be applied to food webs. For example, Song et al. (2005) found that the network structure of a variety of real-world networks obeys power-law scaling, as if they are fractal shapes. This self-similarity was found to apply to the relationship between the number of boxes (i.e., sub-groups of connected nodes) needed to cover a network and the size of the box. Song et al. (2005) also used a renormalization procedure to coarse-grain the networks by sequentially collapsing boxes into single nodes and then creating new boxes. These aggregated networks fit the same power-law scaling as individual-node networks. Given that food webs differ in some basic ways from many other empirical networks, it will be interesting to see if this self-similarity applies to ecological networks. While analysis of food-web data may be inconclusive, given how small they are relative to most other networks studied, food-web models such as the niche model could be used to explore higher diversity webs as well as any sensitivity to changes in diversity and connectance.

4.6   NETWORK STRUCTURE AND ROBUSTNESS OF FOOD WEBS

As the previous sections highlight, research on general or universal aspects of food-web network structure underwent a renaissance at the beginning of the 2000s as a result of improved data, new topological models, and concepts and approaches borrowed from general network theory. Network approaches also brought back into focus the question of how food-web structure might relate to issues of core ecological interest, such as ecosystem stability. Within ecology during the 1990s, questions about stability were increasingly transformed into questions about ecosystem responses to perturbations and the relationship between ecosystem complexity, especially diversity, and ecosystem function (McCann 2000). Classic research into connectance and how it relates to May's stability criterion (May 1973) was largely abandoned, as foreseen by Paine (1988). However, in a new introduction to the 2001 Princeton Landmarks in Biology Edition of his 1973 book, May wrote, "...the theme of the relationship between the network structure of food webs and their ability to handle perturbation is central in ecology, as in many other subjects. ...The reorientation of this question to what kinds of connectance patterns are likely to be most resistant to specific kinds of disturbance is of continuing relevance in ecology, as elsewhere."

May's (2001) comments were inspired partly by emerging research in the broader arena of network theory. About that time, a series of papers examined the response of a variety of networks including the Internet and WWW pages (Albert et al. 2000) and metabolic and protein networks (Jeong et al. 2000, 2001) to the simulated loss of nodes. In each case, the networks, all of which display highly skewed power-law degree distributions, appear very sensitive to the targeted loss of highly connected nodes but relatively robust to random loss of nodes. When highly connected nodes are removed from scale-free networks, the average path length tends to increase rapidly, and the networks also quickly fragment into isolated clusters. In essence, paths of information-flow in highly skewed networks are easily disrupted by the loss of nodes that are directly connected to an unusually large number of other nodes. In contrast, random networks with much less skewed Poisson degree distributions display similar responses to the targeted loss of highly connected nodes versus random node loss (Strogatz 2001).

Within ecology, species deletions on small ($S < 14$) hypothetical food-web networks as well as a subset of the 113 web catalog were used to examine the reliability of network flow, or the probability that sources (producers) are connected to sinks (consumers) in food webs (Jordán and Molnár 1999). They concluded that the structure of the empirical webs appeared to conform to reliable flow patterns identified using the hypothetical webs, but that result was based on the early poorly resolved data. Following Albert et al. (2000), Solé and Montoya (2001) used three improved, diverse food webs to conduct species knockout simulations. Instead of path length, Solé and Montoya looked at the level of secondary extinctions potentially triggered by different patterns of primary species loss. This is easily measured in binary food-web networks using the simple algorithm that if primary extinctions cause a consumer to lose all of its resources, it too goes extinct. In all three food webs, removal of highly connected species resulted in much higher rates of secondary extinctions than random loss of species, and also fragmented the webs more rapidly, similar to results seen for other types of networks (e.g., Albert et al. 2000). Solé and Montoya (2001) attributed this to highly skewed power-law degree distributions (Montoya and Solé 2002). However, most food webs do not have power-law link distributions (Camacho et al. 2002b; Dunne et al. 2002a), so the generality of those results was unclear. Also, the web that showed extreme fragility to the loss of highly connected species was a detailed source web based on only one basal species, Scotch broom (Memmott et al. 2000). In this web, Scotch broom has a large number of species linking to it and, as a result, is removed very early in the non-random deletion sequence, leading to the necessary collapse of the entire web. That collapse is attributable to the particular and peculiar characteristics of the Scotch broom dataset, and is not good evidence for a general trend in community food webs (Jordán 2002; Dunne et al. 2002b).

To address these issues, a set of 16 non-marine food webs (Dunne et al. 2002b) and 3 marine food webs (Dunne et al. 2004) was used for similar bio-diversity loss simulations. It was found that even without highly skewed degree distributions, food webs are much more robust to random loss of species than to loss of highly connected species. These results suggest that any substantial skewness in degree distribution will tend to alter the response of a network to different kinds of node loss. Similarly, the order of pollinator loss was found to have an effect on potential plant extinction patterns in two detailed, speciose plant-pollinator networks (Memmott et al. 2004). Loss of plant diversity associated with targeted removal of highly connected pollinators was not as extreme as comparable secondary extinctions in food webs, which Memmott et al. (2004) attribute to pollinator redundancy and the nested topology of the networks.

The previous studies all point to the trend that sequential loss of highly connected species has a greater impact than random losses. However, the "knock-out highly connected species" approach is not necessarily useful for identifying particular species likely to have the greatest impact: loss of a particular highly connected species may or may not result in a large number of secondary extinctions. To address this, Allesina and Bodini (2004) used a dominator tree approach to reduce the topological structure of 13 empirical food webs into linear pathways that define the essential chains of energy delivery in the network. A particular node *dominates* another node if it passes energy to it along a chain in the dominator tree. In addition to corroborating prior findings of higher secondary extinctions with targeted loss of species (in this case, the loss of species that dominate many other species) versus random losses, Allesina and Bodini (2004) showed that the higher the number of species that a particular species dominates, the greater the secondary extinctions that result from its removal.

Dunne et al. (2002b, 2004) provide yet another answer to May's (2001) question about patterns of connectance and ecosystem responses to perturbation. They found that food-web *robustness*, defined as the fraction of primary species loss that induces at least 50% total species loss (primary + secondary extinctions) for a particular trophic-species web, increases with increasing connectance, L/S^2. This holds both for random species loss and targeted removal of highly connected species. While systematic removal of species with few links generally leads to low levels of secondary extinctions, there are exceptions, and there is no obvious correlation of those exceptional responses with global food-web properties such as S or C (Dunne et al. 2002b). Such apparently idiosyncratic effects of the loss of least-connected species probably have to do with finer-level patterns of link distribution within particular webs. One possibility is that such species "dominate" many other species within a dominator-tree context (Allesina and Bodini 2004), even though they have few direct links to other species. Or such effects may simply be hard to assess *a priori* using standard measures of network topology. Unpredictability of species' likely importance for extinction dynamics was reported in a dynamical food-web modeling study on the resistance of communities to non-random extinctions, based on species' sensitivity to a theoretical stressor (Ives and Cardinale 2004).

While the previous species removal studies (Jordán and Molnár 1999; Solé and Montoya 2001; Dunne et al. 2002b, 2004; Memmott et al. 2004; Allesina and Bodini 2004) are informed by a purely network-structure perspective that ignores dynamics, it has been demonstrated repeatedly that constraints imposed by structure can have a significant role in the outcomes of dynamics (e.g., Pimm and Lawton 1978; McCann and Hastings 1997; Jordán et al. 2002; Martinez et al. Chapter 6; see also the review by Jordán and Scheuring 2004). Ideally, such *in silico* biodiversity loss and related species invasion experiments will be conducted using approaches that integrate structure and dynamics (see Integrating Structure and Dynamics section). Even without explicit dynamical analysis, food-web topology research shows that more than 95% of species are within three links of each other, and that species draw ever closer as connectance and species richness increase (Williams et al. 2002). This suggests that the dynamics of species in complex ecosystems are more tightly connected than conventionally thought, which has profound implications for the impact and spread of perturbations.

5 RELATED TOPICS AND FUTURE DIRECTIONS

As the previous sections highlight, a great deal of interesting research on general aspects of network structure of complex food webs has occurred since the first 30-web catalog was analyzed and published (Cohen 1977a,b, 1978). Such research is now thriving and has found a broader context in interdisciplinary network research, following a lull during the 1990s when much of the focus was on systematically exploring the impacts of methodology on patterns of food-web properties. By limiting this review to research explicitly focused on potential generalities in complex food-web network structure, I have neglected a number of exciting topics that relate to food-web topology, many of which provide more of an ecological context for such research. It is from these additional areas that many of the future directions of structural food-web research are likely to emerge. I wrap up the chapter with a brief discussion of a few of many possible related topics.

5.1 VARIABLE STRUCTURE ACROSS ECOSYSTEMS AND/OR ENVIRONMENT

Since the beginning of food-web structure research, some studies have looked for systematic differences among types of webs. For example, early data suggested that food webs from fluctuating environments have lower connectance than those from more constant environments (Briand 1983), and that three-dimensional habitats have longer food chains than more two-dimensional habitats such as forest canopy or grassland (Briand and Cohen 1987). Differences in food webs across broad categories of habitat have also been considered, for example aquatic versus terrestrial webs (Chase 2000), and marine versus "continental" webs (Cohen 1994; Link 2002; Dunne et al. 2004). Unfortunately, there is currently not enough high-quality data, representing multiple webs from a variety of ecosystems, to be more than suggestive in any assessment of systematic changes in structure across environment or habitat (Dunne et al. 2004).

In addition, any comparisons of the details of food-web structure need to be mediated by the understanding that species richness and connectance vary across webs in ways that systematically impact structure. Unless food webs are constructed with the same methodology, which is almost impossible to standardize across different habitats, such methodological variation may result in inconsistent levels of sampling effort and thresholds for link inclusion. This can make it difficult to discern systematic differences in food-web structure attributable to ecological or environmental processes. The niche model or other models that use S and C as inputs can help with this problem by acting as "benchmarks" for assessing similarities and differences between food webs with varying S and C (Dunne et al. 2004). The benchmark approach works by helping to factor out variation due to species and/or links being differentially observable among habitat types. S and C may also vary due to ecological or other non-methodological reasons. How and why they vary beyond methodology has important implications for food-web structure and dynamics and is a basic question for all of ecology. For example, a source of variation in the relationship between S and C has been suggested by the model of Solé et al. (2002), where the immigration probability of new species into the system appears to be critical. This suggests that the degree of openness of the system will influence the relationship in systematic ways (Pascual et al. Chapter 15).

5.2 COMPARTMENTS

The presence or lack of compartmentalization may have implications for the transmission of both beneficial and harmful effects throughout food webs, and thus can affect the stability and robustness of ecosystems. Early food-web research suggested that there is little detectable compartmentalization in food webs (Pimm and Lawton 1980) although other studies using early data suggested otherwise (Yodzis 1982; Raffaelli and Hall 1992). High connectance in food webs, compared to many other types of networks studied, should tend to obscure compartmentalization, as indicated by low clustering coefficients (Amaral et al. 2002b; Dunne et al. 2002a) and short path lengths (Williams et al. 2002). However, some recent studies suggest that there are ways to identify compartments in detailed food-web data. For example, Girvan and Newman (2002) borrowed a "centrality" or "betweenness" index common to social network research, applied it to links, and successfully detected pelagic and benthic subcommunities of species within the Chesapeake Bay food web (Baird and Ulanowicz 1989) using only network structure data. Also borrowing from social network analysis, Krause et al. (2003) used a clustering algorithm ("KliqueFinder") which identifies subgroups with concentrated interactions (Frank 1995). They found compartmentalization in three of five complex food webs, but the ability to find compartments depended on inclusion of flow data. They generally did not detect compartments in less complex food webs with or without flow data. Allesina et al. (in press) identified multiple strongly connected components (SCCs) in 17 food webs, where an SSC is a set of species interconnected by cycles or loops. However, they found that the number of SSCs identified is very sensitive to removal of weak links (Winemiller 1990).

A promising algorithm for systematically identifying a type of food-web compartment in any size web using only structural data was developed by Melián and Bascompte (2004). A k subweb is defined as a subset of species that are connected to at least k predator and/or prey species within that subset. Each species belongs to only one subweb, the subweb where each species has the highest k value. In effect, k is the measure of links per species, or linkage density, within the context of the subweb. Any particular species can have a higher absolute number of links if connected to species in other subwebs. Melián and Bascompte (2004) assessed the frequency distribution of subweb size in five speciose webs with S of 134 to 237 and found that all follow a power-law distribution, with zero or few links per species in most subwebs and a unique subweb with the most links per species. The scaling exponents increase with web size and vary from -1.87 to -0.65. Melián and Bascompte (2004) also calculated the connectance of the most dense subweb for each web. They compared those values to means for the most dense subweb of networks (with the same levels of S) generated by various topological models including the niche model (Williams and Martinez 2000). No model was highly successful at generating webs with most dense subweb connectance closely comparable to that of the five empirical webs. However, to be a more appropriate test of the niche model, the analysis should be redone for trophic species versions of the food webs.

An issue for all studies considering compartmentalization or sub-web structure is whether such structures play any functional role within ecosystems. There are likely innumerable ways for chopping up networks into clusters, but that does not mean that such clusters are necessarily meaningful for ecological function or dynamics. Ideally, future compartmentalization studies will be more directly linked to issues such as ecosystem function, robustness, or resilience to perturbations.

5.3 CHAIN LENGTHS AND TROPHIC LEVELS

While from one perspective, food-chain length and trophic-level properties are just a few of many possible metrics of food-web structure (Williams and Martinez 2000), there is a very rich body of research specifically on chain lengths and trophic levels that spans theoretical, observational, experimental, and applied approaches dating back at least to Elton (1927). This research reflects the understanding that food-chain length, or "the number of transfers of energy or nutrients from the base to the top of a food web," is a central characteristic of ecological communities due to its impact on ecosystem functioning, such as nutrient and carbon cycling, contaminant concentration, and trophic cascades (Post 2002). A great deal of research has gone into characterizing food-chain length, exploring its implications for ecosystem functioning, and determining whether and how it is constrained (Post 2002). Similar questions have been explored for the closely related concept of trophic level, or "the number of times chemical energy is transformed from a consumer's diet into a consumer's biomass along the food chains that lead to the species" (Williams and Martinez 2004b), although some researchers have claimed that the concept lacks scientific utility (Polis and Strong 1996).

Food-chain length and trophic level have been explored in detail using structural food-web data and models (e.g., Cohen and Newman 1991), but much of the research on these properties falls outside strictly structural approaches (Post 2002). Constraints on food-chain length or trophic levels in ecosystems have often been attributed to dynamical stability or resource availability, although this is increasingly questioned (e.g., Sterner et al. 1997) in favor of ecosystem size (Post et al. 2000). Post (2002) suggests that the debate is shifting from the search for singular explanations to "a complex and contingent framework of interacting constraints that includes the history of community organization, resource availability, the type of predator-prey interactions, disturbance and ecosystem size." Although the nuances regarding such interacting constraints would seem to require the inclusion of data on relative flows along links to accurately characterize food-chain lengths and trophic levels, complex binary food webs without such flow data appear to provide a successful and simple quantitative framework for analysis. For example, a binary link-based measure called *short-weighted trophic level* yields surprisingly accurate estimates of species' trophic and omnivory levels as compared to flow-weighted data (Williams and Martinez 2004b).

5.4 INTERACTION STRENGTH

Most early structural food-web research concerning May's stability criterion (1973) generally focused on diversity (S) and connectance (C), the two parameters readily computed from structural data. Those analyses were based on the implicit assumption that the third parameter, interaction strength (i), was roughly constant. May's analysis actually assigned interaction strengths randomly. From very early on, researchers noted that these assumptions are flawed since interaction strengths are likely to vary non-randomly in real communities, with ramifications for ecosystem structure and stability (e.g., Paine 1969, 1980). One of the earliest analyses of this issue using structural food-web data was by Yodzis (1981), who used 40 food webs (Briand 1983) to construct community matrices based on structured interaction patterns. He found that stability, in May's sense, was far more likely to occur if interaction strengths are assigned non-randomly based on ecological understanding. More recent research suggests that "non-random patterning of strong and weak links can be critical for the stability or persistence of theoretical and empirically observed complex communities" (Berlow et al. 2004; de Ruiter et al. 1995; Kokkoris et al. 1999; Neutel et al. 2002). There are many opportunities and pitfalls in interaction strength research, as discussed in an excellent review by Berlow et al. (2004). Done thoughtfully, such research can provide linkages between ecological network structure and dynamics, and can facilitate future links between theoretical work and experimental and other empirical research.

5.5 INTEGRATING STRUCTURE AND DYNAMICS

Most models of food-web dynamics have focused on small modules with only a few species (see review by Dunne et al. 2005). While topology is generally thought to constrain ecological dynamics, most studies have explored this issue using very simplified network structures that can be imposed on species-poor dynamical models (see review by Jordán and Scheuring 2004). However, researchers are increasingly integrating dynamics with complex food-web structure in modeling studies that move beyond modules. The Lotka-Volterra cascade model (Cohen et al. 1990b; Chen and Cohen 2001a,b) was probably the first incarnation of this type of integration. As its name suggests, the Lotka-Volterra cascade model runs classic L-V dynamics, including a non-saturating linear Type I functional response, on sets of species interactions structured according to the cascade model (Cohen and Newman 1985b). The cascade model was also used to generate the structural framework for a recent dynamical food-web model with a Type I functional response (Kondoh 2003) that was used to study the effects of prey-switching on ecosystem stability. Improving on aspects of biological realism of both dynamics and structure, Yodzis used a bioenergetic dynamical model with nonlinear functional responses (Yodzis and Innes 1992), in conjunction with empirically-defined trophic network structure among 29 species, to simulate the biomass dynamics of a marine fisheries food web (Yodzis 1998, 2000). This bioenergetic dynamical modeling approach has been integrated with cascade and niche model network structure by other researchers (Brose et al. 2003; Williams and Martinez 2004a; Martinez et al. Chapter 6).

All of the approaches mentioned, as well as others, have been used to examine a variety of aspects of food-web complexity and stability including persistence, and are proving to be a valuable way to explore structural constraints on dynamics in complex ecosystems (Martinez et al. Chapter 6). An alternative approach is to develop assembly models that include ecological and/or evolutionary dynamics and see if they generate plausible ecological diversity and network structure (see review by McKane and Drossel Chapter 9), or to use methods such as genetic algorithms to explore the space of dynamically possible or probable structures (Ruiz-Moreno et al. Chapter 7). Integration of plausible ecological and evolutionary dynamics and network structure is a grand challenge for ecological modeling, with potential benefits for conservation and management of ecosystems (Yodzis 1998, 2000), as well as for fundamental scientific understanding of complex adaptive systems.

5.6 QUANTIFYING SPECIES AND LINKS

There have been many calls to go beyond species presence/absence data and binary link designations to include quantitative aspects of species and/or links during documentation and analysis of food webs (Cohen et al. 1993; Borer et al. 2002). Obvious ways to quantify "species" include characterizing their population abundance, population biomass, average individual biomass, and average body size. Links can be quantified through the amount of flow, usually biomass, attributable to the link (e.g., Baird and Ulanowicz 1989; Winemiller 1990), or the frequency of the occurrence of the links (Martinez et al. 1999). While researchers have compiled "quantitative" data for food-web components for as long as predator-prey relationships and food webs have been objects of study, it is only recently that some studies are attempting to integrate quantitative estimates with comprehensive descriptions of complex food-web network structure.

For example, Bersier et al. (2002) introduced a set of quantitative descriptors corresponding to several commonly studied food-web structure metrics (e.g., link and species proportions, L/S, C, chain properties, omnivory, generality, vulnerability). They used indices based on incoming and outgoing biomass flow for each species, as inspired by prior ecological network analysis applications of information theory (Ulanowicz 1986; Ulanowicz and Wulff 1991). Binary link metrics and quantitative link metrics do not return the same values for the Chesapeake Bay web (Baird and Ulanowicz 1989), and Bersier et al. (2002) suggest that their combined use will provide the most insight into food-web structure and function. A follow-up study looking at ∼10 empirical webs with flow data suggests that quantitative indices are more robust to variable sampling effort than metrics based on binary links, although precision decreased (Banašek-Richter et al. 2004). A different approach focusing on species traits was introduced by Cohen et al. (2003), who document numerical abundance (varying across 10 orders of magnitude) and average body size (varying across ∼ 12 orders of magnitude) for each species in the detailed pelagic food web of Tuesday Lake, Michigan (Jonsson et al. in press). They characterize this approach as a new way to describe ecological communities and they identify new community patterns for analysis. The approach of Cohen et al. (2003) was used to describe the detrital soil food-web in agricultural grasslands in the Netherlands, where it may be helpful for assessing land-use quality (Mulder et al. 2005).

While these and other "quantitative" approaches are intriguing, it remains to be seen whether they are primarily a tool for richer description of particular ecological communities, or whether they also give rise to generalities, novel models and predictions, or theory. Such approaches have much greater data requirements than just trying to characterize network structure. In many cases it will be difficult to document the necessary link and species characteristics given time and monetary constraints. Also, many of the concerns about "binary" data apply to "quantitative" data, and may be even more complicated to resolve. For example, biomass flow along trophic links and levels of species abundances vary spatiotemporally. A snapshot of a set of flows or abundances at a particular time and in a particular space ignores variability that may result in a very misleading picture of typical flow or abundance levels. Indeed, it may be difficult to characterize "typical" or average levels that are ecologically meaningful. Additional quantitative detail may be unnecessary for some kinds of investigations, since indices based on binary link data can be a good proxy for, or improvement upon, those based on richer quantitative data. For example, as previously mentioned, calculation of mean trophic level for food webs using detailed flow data is well approximated by a binary link-based structural index (Williams and Martinez 2004b). Also, species similarity measures based on binary link data outperform flow-based similarity measures (Yodzis and Winemiller 1999).

5.7   ECOLOGICAL NETWORKS

The food webs that are the object of study of most of the research reviewed in this chapter generally have as their focus classic predator-herbivore-primary producer feeding interactions. However, the basic concept of food webs can be extended to a broader framework of *ecological networks* that is more inclusive of different components of ecosystem biomass flow, and that takes into consideration different kinds of species interactions that are not strictly trophic. I give three of many possible examples here. First, parasites have typically been given short shrift in traditional food webs, although exceptions exist (Huxham et al. 1996; Memmott et al. 1999). Almost a decade after "a plea for parasites" in food webs (Marcogliese and Cone 1997), there are still few food-web studies that systematically incorporate or focus on parasites. Dobson et al. (Chapter 4) take up this issue and explore it further. A second issue that has yet to be resolved adequately for structural or dynamical food-web studies is the role of detritus, or dead organic matter, in ecosystems. Detritus has been explicitly included as one or several separate nodes in many binary-link and flow-weighted food webs. In some cases, it is treated as an additional primary producer, while in other cases both primary producers and detritivores connect to it. Researchers must think much more carefully about how to include detritus in all kinds of ecological studies (Moore et al. 2004), given that it plays a fundamental role in most ecosystems and has particular characteristics that differ from other food-web nodes: it is dead and not living organic matter, it is heterogeneous, and it has an ambiguous trophic role. The third example concerns the analysis of ecological networks focused on other interactions besides strictly predator-prey relationships. Plant-animal mutualistic networks, particularly pollination and seed dispersal networks, have received the most attention thus far, and their network structure is discussed by Bascompte and Jordano (Chapter 5). While pollination and seed dispersal each involve a trophic interaction and can be portrayed as two trophic level networks; unlike in a classic predator-prey relationship, a strong positive benefit is conferred upon both partners in the interaction. The evolutionary and ecological dynamics of such mutualistic relationships may place unique constraints on the network structure of such interactions and the dynamical stability of such networks (Jordano 1987; Memmott 1999; Jordano et al. 2003; Bascompte et al. 2003; Vásquez and Aizen 2004).

5.8   ECOINFORMATICS

Future food-web research will depend on the collection of and access to increasingly high-quality data from systems spanning the globe. To address these needs, and to help identify gaps in knowledge that can be filled with strategic sampling, recent and historical food-web datasets are being compiled for a WWW-based publicly accessible trophic interaction database. This database will be integrated with structural and dynamical modeling tools as well as three-dimensional food-web graphics and animation tools. This project, called *Webs on the Web* (WoW), in effect updates and expands Cohen's (1989) *Ecologists Co-operative Web Bank* (ECOWeB), a set of over 200 "machine readable" food-web datasets, and adds many layers of additional functionality and database capabilities. Initially, the WoW knowledgebase will include hundreds of food-web datasets, thousands of instances of consumer-resource relationships, and associated quantitative species and link information where available. Database tools will facilitate the decentralized addition of data to the knowledgebase, as well as annotation of existing data. Ideally, WoW will increase the ability of scientists, managers, and students to exchange and analyze information regarding the structure, function, and dynamics of ecological networks and the species within them.

Webs on the Web is just one example of an ever-expanding array of biodiversity-related databases that are, or will be, available on the WWW (Graham et al. 2004). In turn, these types of databases are just one of many kinds of ecoinformatics tools, which include technologies and practices for gathering, analyzing, visualizing, storing, retrieving, and otherwise managing ecological knowledge and information. There is a great deal of ecological information potentially available on the WWW, but it is widely dispersed and comes in a large variety of formats. Current search tools are very limited in their ability to effectively mine the data. To address these issues, efforts are underway to develop languages and tools for a *Semantic Web* that will allow for more sophisticated, content-based access to dispersed data and information on the WWW (Hendler 2003). Some of these tools are being developed in an ecological network research context through the Semantic Prototypes in Research Ecoinformatics project (spire.umbc.edu) in conjunction with Webs on the Web (www.foodwebs.org). The emerging Semantic Web technologies, if properly developed and widely implemented, have the potential to transform the scope, effectiveness, and efficiency of ecological research across spatial and temporal scales (Green et al. 2005).

6 FINAL THOUGHTS

It is a vibrant time for research on the structure and dynamics of complex systems. Food webs, and more broadly, ecological networks, are a paradigmatic example of such systems. Research on complex food webs can not only benefit from research on other types of networks, but can provide novel insights that have implications for network research beyond ecology. However, as a cautionary note, what Lawton stated in 1989 is completely relevant today. Our data are still limited and of highly variable quality, and we may come to realize that some of the apparent generalities or universalities we see in that data are merely artifacts of poor information. Fortunately, researchers have not succumbed to "hand-wringing paralysis" given these limitations. However, the emergence of novel ways of analyzing and modeling ecological network data need to be accompanied by meticulous collection of detailed, comprehensive field data that seeks to address or overcome the previous empirical limitations. In addition, scientists must make careful decisions about what data to use, and perhaps more importantly, what data *not* to use in analyses. Beyond the data issues, researchers need to think continually about how to connect food-web properties, both structural and dynamical, back to issues of fundamental ecological interest, to avoid chasing research down dead-end alleys of little scientific interest. While some hand-wringers may continue to assert that research on food-web structure is one of those unfortunate alleys, the rich history outlined here combined with the recent tonic of interdisciplinary network research where it is well understood that structure always affects function (Strogatz 2001), shows that ecological network research is an important avenue for future advances. Such advances are likely to occur both with regard to more applied concerns such as the response of ecosystems to perturbations including biodiversity loss and species invasions (Memmott et al. Chapter 14), as well as more basic research questions such as identifying the processes or constraints that give rise to general structural patterns. Thoughtful and careful development of ecological network data, analyses, and models can provide the backbone for robust and general ecological theory on the complex networks of interactions among species across many scales, especially if integrated with other macroecological theory on allometric scaling and species-area relationships (Brose et al. 2004).

ACKNOWLEDGMENTS

This work was supported by grants from the National Science Foundation (DBI-0234980l, ITR-0326460). I thank the Santa Fe Institute and the Rocky Mountain Biological Laboratory for hospitality and administrative support, and appreciation goes to Brooke Ray Smith and Neo Martinez for helpful edits and comments made on a draft of the manuscript.

REFERENCES

Abrams, P. A. 1996. "Dynamics and Interactions in Food Webs with Adaptive Foragers." In *Food Webs: Integration of Patterns and Dynamics*, ed. G. A. Polis and K. O. Winemiller, 113–122. New York: Chapman and Hall.

Albert, R., H. Jeong, and A. L. Barabási. 2000. "Error and Attack Tolerance of Complex Networks." *Nature* 406:378–382.

Albert, R., and A. L. Barabási. 2002. "Statistical Mechanics of Complex Networks." *Rev. Mod. Phys.* 74:47–97.

Allesina, S., and A. Bodini. 2004. "Who Dominates Whom in the Ecosystem? Energy Flow Bottlenecks and Cascading Extinctions." *J. Theor. Biol.* 230:351–358.

Allesina, S., A. Bodini, and C. Bondavalli. 2004 (in press). "Ecological Subsystems via Graph Theory: The Role of Strongly Connected Components." *Oikos.*

Amaral, L. A. N., A. Scala, M. Barthélémy, and H. E. Stanley. 2000. "Classes of Small-World Networks." *Proc. Natl. Acad. Sci. USA* 97:11149–11152.

Arditi, R., and J. Michalski. 1996. "Nonlinear Food Webs Models and Their Responses to Increased Basal Productivity." In *Food Webs: Integration of Patterns and Dynamics*, ed. G. A. Polis and K. O. Winemiller, 122–133. New York: Chapman and Hall.

Arnold, S. J. 1972. "Species Densities of Predators and Their Prey." *Amer. Natur.* 106:220–236.

Arii, K., and L. Parrott. 2004. "Emergence of Non-Random Structure in Local Food Webs Generated from Randomly Structured Regional Webs." *J. Theor. Biol.* 227:327–333.

Auerbach, M. J. 1984. "Stability, Probability and the Topology of Food Webs." In *Ecological Communities: Conceptual Issues and the Evidence*, edited by D. R. Strong, D. Simberloff, L. G. Abele, and A. B. Thistle, 413–436. Princeton, NJ: Princeton University Press.

Baird, D., and R. E. Ulanowicz. 1989. "The Seasonal Dynamics of the Chesapeake Bay Ecosystem." *Ecol. Monogr.* 59:329–364.

Banašek-Richter, C., M-F., Cattin, and L-F. Bersier. 2004. "Sampling Effects and the Robustness of Quantitative and Qualitative Food-Web Descriptors." *J. Theor. Biol.* 7:23–32.

Bascompte, J., P. Jordano, C. J. Melián, and J. M. Olesen. 2003. "The Nested Assembly of Plant-Animal Mutualistic Networks." *Proc. Natl. Acad. Sci. USA* 100:9383–9387.

Berlow, E. L., A-M. Neutel, J. E. Cohen, P. De Ruiter, B. Ebenman, M. Emmerson, J. W. Fox, V. A. A. Jansen, J. I. Jones, G. D. Kokkoris, D. O. Logofet, A. J. McKane, J. Montoya, and O. L. Petchey. 2004. "Interaction Strengths in Food Webs: Issues and Opportunities." *J. Animal Ecol.* 73:585–598.

Bersier, L.-F., C. Banašek-Richter, and M-F. Cattin. 2002. "Quantitative Descriptors of Food Web Matrices." *Ecology* 83:2394–2407.

Bersier, L.-F., P. Dixon, and G. Sugihara. 1999. "Scale-Invariant or Scale-Dependent Behavior of the Link Density Property in Food Webs: A Matter of Sampling Effort?" *Amer. Natur.* 153:676–682.

Briand, F. 1983. "Environmental Control of Food Web Structure." *Ecology* 64:253–263.

Briand, F., and J. E. Cohen. 1984. "Community Food Webs Have Scale-Invariant Structure." *Nature* 398:330–334.

Briand, F., and J. E. Cohen. 1987. "Environmental Correlates of Food Chain Length." *Science* 238:956–960.

Brose, U., R. J. Williams, and N. D. Martinez. 2003. "Comment on 'Foraging Adaptation and the Relationship between Food-Web Complexity and Stability.'" *Science* 301:918b.

Brose, U., A. Ostling, K. Harrison, and N. D. Martinez. 2004. "Unified Spatial Scaling of Species and Their Trophic Interactions." *Nature* 428:167–171.

Camacho, J., and A. Arenas. 2005. "Universal Scaling in Food-Web Strucutre?" *Nature* 435:E3–E4.

Camacho, J., R. Guimera, D. B. Stouffer, and L. A. N. Amaral. 2005. "Quantitative Patterns in the Structure of Model and Empirical Food Webs." *Ecology* 86:1301–1311.

Camacho, J., R. Guimerà, and L. A. N. Amaral. 2002a. "Analytical Solution of A Model for Complex Food Webs." *Phys. Rev. Lett. E* 65:030901.

Camacho, J., R. Guimerà, and L. A. N. Amaral. 2002b. "Robust Patterns in Food Web Structure." *Phys. Rev. Lett.* 88:228102.

Camerano, L. 1880. "Dell'equilibrio dei viventi mercè la reciproca distruzione." *Atti della Reale Accademia della Scienze di Torino* 15:393–414.

Cameron, G. N. 1972. "Analysis of Insect Trophic Diversity in Two Salt Marsh Communities." *Ecology* 53:58–73.

Cattin, M.-F., L.-F. Bersier, C. Banašek-Richter, M. Baltensperger, and J-P. Gabriel. 2004. "Phylogenetic Constraints and Adaptation Explain Food-Web Structure." *Nature* 427:835–839.

Chase, J. M. 2000. "Are There Real Differences Among Aquatic and Terrestrial Food Webs?" *TREE* 15:408–411.

Chen, X., and J. E. Cohen. 2001a. "Global Stability, Local Stability and Permanence in Model Food Webs." *J. Theor. Biol.* 212:223–235.

Chen, X., and J. E. Cohen. 2001b. "Transient Dynamics and Food Web Complexity in the Lotka-Volterra Cascade Model." *Proc. Roy. Soc. Lond. B* 268:869–877.

Christensen, V., and D. Pauly. 1993. "Trophic Models of Aquatic Ecosystems." ICLARM, Manila.

Christian, R. R., and J. J. Luczkovich. 1999. "Organizing and Understanding a Winter's Seagrass Foodweb Network through Effective Trophic Levels." *Ecol. Model.* 117:99–124.

Cohen, J. E. 1977a. "Ratio of Prey to Predators in Community Food Webs." *Nature* 270:165–167.

Cohen, J. E. 1977b. "Food Webs and the Dimensionality of Trophic Niche Space." *Proc. Natl. Acad. Sci. USA* 74:4533–4563.

Cohen, J. E. 1978. *Food Webs and Niche Space.* Princeton, NJ: Princeton University Press.

Cohen, J. E. 1989. "Ecologists Co-operative Web Bank (ECOWeBTM)." Version 1.0. Machine Readable Data Base of Food Webs. Rockefeller University, NY.

Cohen, J. E. 1990. "A Stochastic Theory of Community Food Webs: VI. Heterogeneous Alternatives to the Cascade Model." *Theor. Pop. Biol.* 37:55–90.

Cohen, J. E. 1994. "Marine and Continental Food Webs: Three Paradoxes?" *Phil. Trans. Roy. Soc. Lond. B* 343:57–69.

Cohen, J. E., and F. Briand. 1984. "Trophic Links of Community Food Webs." *Proc. Natl. Acad. Sci. USA* 81:4105–4109.

Cohen, J. E., and C. M. Newman. 1985a. "When Will A Large Complex System be Stable?" *J. Theor. Biol.* 113:153–156.

Cohen, J. E., and C. M. Newman. 1985b. "A Stochastic Theory of Community Food Webs: I. Models and Aggregated Data." *Proc. Roy. Soc. Lond. B* 224:421–448.

Cohen, J. E., and C. M. Newman. 1991. "Community Area and Food-Chain Length: Theoretical Predictions." *Amer. Natur.* 138:1542–1554.

Cohen, J. E., and Z. J. Palka. 1990. "A Stochastic Theory of Community Food Webs: V. Intervality and Triangulation in the Trophic Niche Overlap Graph." *Amer. Natur.* 135:435–463.

Cohen, J. E., F. Briand, and C. M. Newman. 1986. "A Stochastic Theory of Community Food Webs: III. Predicted and Observed Length of Food Chains." *Proc. Roy. Soc. Lond. B* 228:317–353.

Cohen, J. E., F. Briand, and C. M. Newman. 1990a. *Community Food Webs: Data and Theory*. New York: Springer-Verlag.

Cohen, J. E., C. M. Newman, and F. Briand. 1985. "A Stochastic Theory of Community Food Webs: II. Individual Webs." *Proc. Roy. Soc. Lond. B* 224:449–461.

Cohen, J. E., T. Jonsson, and S. R. Carpenter. 2003. "Ecological Community Description Using the Food Web, Species Abundance, and Body Size." *Proc. Natl. Acad. Sci. USA* 100:1781–1786.

Cohen, J. E., T. Luczak, C. M. Newman, and Z-M. Zhou. 1990b. "Stochastic Structure and Nonlinear Dynamics of Food Webs: Qualitative Stability in a Lotka-Volterra Cascade Model." *Proc. Roy. Soc. Lond. B* 240:607–627.

Cohen, J. E., R. A. Beaver, S. H. Cousins, D. L. De Angelis, L. Goldwasser, K. L. H oeng, R. D. Holt, A. J. Kohn, J. H. Lawton, N. D. Martinez, R. O'Malley, L. M. Page, B. C. Patten, S. L. Pimm, R. M. Polis, T. W. Schoener, K. Schoenly, W. G. Sprules, J. M. Teal, R. E. Ulanowicz, P. H. Warren, H. M. Wilbur, and P. Yodzis. 1993. "Improving Food Webs." *Ecology* 74:252–258.

de Ruiter, P. C., A-M. Neutel, and J. C. Moore. 1995. "Energetics, Patterns of Interaction Strengths, and Stability in Real Ecosystems." *Science* 269:1257–1260.

Deb, D. 1995. "Scale-Dependence of Food Web Structures: Tropical Ponds as Paradigm." *Oikos* 72:245–262.

Drossel, B., and A. J. McKane. 2003. "Modelling Food Webs." In *Handbook of Graphs and Networks: From the Genome to the Internet*, ed. S. Bornholt and H. G. Schuster, 218–247. Berlin: Wiley-VCH.

Dunne, J. A., R. J. Williams, and N. D. Martinez. 2002a. "Food-Web Structure and Network Theory: The Role of Connectance and Size." *Proc. Nat. Acad. Sci. USA* 99:12917–12922.

Dunne, J. A., R. J. Williams, and N. D. Martinez. 2002b. "Network Structure and Biodiversity Loss in Food Webs: Robustness Increases with Connectance." *Ecol. Lett.* 5:558–567.

Dunne, J. A., R. J. Williams, and N. D. Martinez. 2004. "Network Structure and Robustness of Marine Food Webs." *Mar. Ecol. Prog. Ser.* 273:291–302.

Dunne, J. A., U. Brose, R. J. Williams, and N. D. Martinez. 2005. "Modeling Food-Web Dynamics: Complexity-Stability Implications." In *Aquatic Food Webs: An Ecosystem Approach*, ed. A. Belgrano, U. Scharler, J. A. Dunne, and R. E. Ulanowicz. New York: Oxford University Press.

Elton, C. S. 1927. *Animal Ecology*. London: Sidgwick and Jackson.

Elton, C. S. 1958. *Ecology of Invasions by Animals and Plants.* London: Chapman & Hall.

Emmerson, M. C., and D. G. Raffaelli. 2004. "Predator-Prey Body Size, Interaction Strength and the Stability of a Real Ecosystem." *J. Animal Ecol.* 73:399–409.

Erdös, P., and A. Rényi. 1960. "On the Evolution of Random Graphs." *Publ. Math. Inst. Hung. Acad. Sci.* 5:17–61.

Evans, F. C., and W. W. Murdoch. 1968. "Taxonomic Composition, Trophic Structure and Seasonal Occurrence in a Grassland Insect Community." *J. Animal Ecol.* 37:259–273.

Forbes, S. A. 1977. *Ecological Investigations of Stephen Alfred Forbes.* New York: Arno Press.

Frank, K. 1995. "Identifying Cohesive Subgroups." *Soc. Networks* 17:27–56.

Gardner, M. R., and W. R. Ashby. 1970. "Connectance of Large Dynamic (Cybernetic) Systems: Critical Values for Stability." *Nature* 228:784.

Garlaschelli, D., G. Caldarelli, and L. Pietronero. 2003. "Universal Scaling Relations in Food Webs." *Nature* 423:165–168.

Girvan, M., and M. E. J. Newman. 2002. "Community Structure in Social and Biological Networks." *Proc. Nat. Acad. Sci. USA* 99:8271–8276.

Goldwasser, L., and J. A. Roughgarden. 1993. "Construction of a Large Caribbean Food Web." *Ecology* 74:1216–1233.

Graham, C. H., S. Ferrier, F. Huettman, C. Moritz, and A. T. Peterson. 2004. "New Developments in Museum-Based Informatics and Applications in Biodiversity Analysis." *Trends Ecol. & Evol.* 19:497–503.

Green, J. L., A. Hastings, P. Arzberger, F. Ayala, K. L. Cottingham, K. Cuddington, F. Davis, J. A. Dunne, M-J. Fortin, L. Gerber, and M. Neubert. 2005. "Complexity in Ecology and Conservation: Mathematical, Statistical, and Computational Challenges." *Bioscience* 55:501–510.

Hall, S. J., and D. Raffaelli. 1991. "Food-Web Patterns: Lessons from a Species-Rich Web." *J. Animal Ecol.* 60:823–842.

Hall, S. J., and D. Raffaelli. 1993. "Food Webs: Theory and Reality." *Ad. Ecol. Res.* 24:187–239.

Hall, S. J., and D. Raffaelli. 1997. "Food-Web Patterns: What Do We Really Know." In *Multi-Trophic Interactions in Terrestrial Systems*, ed. A. C. Gange and V. K. Brown, 395–417. Oxford: Blackwell Scientific.

Hardy, A. C. 1924. "The Herring in Relation to Its Animate Environment. Part 1. The Food and Feeding Habits of the Herring with Special Reference to the East Coast of England." *Fisheries Investigations Series II* 7:1–53.

Hastings, A. 1996. "What Equilibrium Behavior of Lotka-Volterra Models Does Not Tell Us About Food Webs." In *Food Webs: Integration of Patterns and*

Dynamics, ed. G. A. Polis and K. O. Winemiller, 211–217. New York: Chapman and Hall.

Havens, K. 1992. "Scale and Structure in Natural Food Webs." *Science* 257:1107–1109.

Hawkins, B. A., N. D. Martinez, and F. Gilbert. 1997. "Source Food Webs as Estimators of Community Web Structure." *Acta Oecologia* 18:575–586.

Haydon, D. 1994. "Pivotal Assumptions Determining the Relationship between Stability and Complexity: An Analytical Synthesis of the Stability-Complexity Debate." *Amer. Natur.* 144:14–29.

Hendler, J. 2003. "Science and the Semantic Web." *Science* 299:520–521.

Hildrew, A. G., C. R. Townsend, and A. Hasham. 1985. "The Predatory Chironimidae of an Iron-Rich Stream: Feeding Ecology and Food Web Structure." *Ecol. Entomol.* 10:403–413.

Hutchinson, G. E. 1959. "Homage to Santa Rosalia, or Why are There so Many Kinds of Animals?" *Amer. Natur.* 93:145–159.

Huxham, M., S. Beany, and D. Raffaelli. 1996. "Do Parasites Reduce the Chances of Triangulation in a Real Food Web?" *Oikos* 76:284–300.

Ives, A. R, and B. J. Cardinale. 2004. "Food-Web Interactions Govern the Resistance of Communities after Non-Random Extinctions." *Nature* 429:174–177.

Jaccard, P. 1900. "Contribution au problème de l'immigration post-glaciaire de la flore alpine." *Bull. Soc. Vaudoise Sci. Nat.* 36:87–130.

Jeong, H., S. P. Mason, A-L. Barabási, and Z. N. Oltvai. 2001. "Lethality and Centrality in Protein Networks." *Nature* 411:41.

Jeong, H., B. Tombor, R. Albert, Z. N. Oltvia, and A-L. Barabási. 2000. "The Large-Scale Organization of Metabolic Networks." *Nature* 407:651–654.

Jonsson, T., J. E. Cohen, and S. R. Carpenter. 2003 (in press). "Food Webs, Body Size and Species Abundance in Ecological Community Description." *Adv. Ecol. Res.* 36.

Jordán, F., and I. Molnár. 1999. "Reliable Flows and Preferred Patterns in Food Webs." *Ecol. Res.* 1:591–609.

Jordán, F., and I. Scheuring. 2004. "Network Ecology: Topological Constraints on Ecosystem Dynamics." *Phys. Life Rev.* 1:139–229.

Jordán, F., I. Scheuring, and G. Vida. 2002. "Species Positions and Extinction Dynamics in Simple Food Webs." *J. Theor. Biol.* 215:441–448.

Jordano, P. 1987. "Patterns of Mutualistic Interactions in Pollination and Seed Dispersal: Connectance, Dependence Asymmetries, and Coevolution." *Amer. Natur.* 129:657–677.

Jordano, P., J. Bascompte, and J. M. Olesen. 2003. "Invariant Properties in Coevolutionary Networks of Plant-Animal Interactions." *Ecol. Lett.* 6:69–81.

Kenny, D., and C. Loehle. 1991. "Are Food Webs Randomly Connected?" *Ecology* 72:17940–1799.

Kokkoris, G. D., A.Y. Troumbis, and J. H. Lawton. 1999. "Patterns of Species Interaction Strength in Assembled Theoretical Competition Communities. *Ecol. Lett.* 2:70–74.

Kondoh, M. 2003. "Foraging Adaptation and the Relationship Between Food-Web Complexity and Stability." *Science* 299:1388–1391.

Krause, A. E., K. A. Frank, D. M. Mason, R. E. Ulanowicz, and W. W. Taylor. 2003. "Compartments Revealed in Food-Web Structure." *Nature* 426:282–285.

Law, R. M., and J. C. Blackford. 1992. "Self-Assembling Food Webs: A Global Viewpoint of Coexistence of Species in Lotka-Volterra Communities." *Ecology* 73:567–579.

Lawlor, L. R. 1978. "A Comment on Randomly Constructed Model Ecosystems." *Amer. Natur.* 112:445–447.

Lawton, J. H. 1989. "Food Webs." In *Ecological Concepts*, ed. J. M. Cherett, 43–78. Oxford: Blackwell Scientific.

Lawton, J. H. 1995. "Webbing and WIWACS." *Oikos* 72:305–306.

Levin, S. A. 1992. "The Problem of Pattern and Scale in Ecology." *Ecology* 73:1943–1967.

Levine, S. 1980. "Several Measures of Trophic Structure Applicable to Complex Food Webs." *J. Theor. Biol.* 83:195–207.

Link, J. 2002. "Does Food Web Theory Work for Marine Ecosystems?" *Mar. Ecol. Prog. Ser.* 230:1–9.

MacArthur, R. H. 1955. "Fluctuation of Animal Populations and a Measure of Community Stability." *Ecology* 36:533–536.

MacDonald, N. 1979. "Simple Aspects of Foodweb Complexity." *J. Theor. Biol.* 80:577–588.

Marcogliese, D. J., and D. K. Cone. 1997. "Food-Webs: A Plea for Parasites." *Trends Ecol. & Evol.* 12:320–325.

Martinez, N. D. 1988. "Artifacts or Attributes? Effects of Resolution on the Food-Web Patterns in Little Rock Lake, Wisconsin." Masters Thesis, University of Wisconsin, Madison.

Martinez, N. D. 1991. "Artifacts or Attributes? Effects of Resolution on the Little Rock Lake Food Web." *Ecol. Monogr.* 61:367–392.

Martinez, N. D. 1992. "Constant Connectance in Community Food Webs." *Amer. Natur.* 139:1208–1218.

Martinez, N. D. 1993a. "Effect of Scale on Food Web Structure." *Science* 260:242–243.

Martinez, N. D. 1993b. "Effects of Resolution on Food Web Structure." *Oikos* 66:403–412.

Martinez, N. D. 1994. "Scale-Dependent Constraints on Food-Web Structure." *Amer. Natur.* 144:935–953.

Martinez, N. D. 1995. "Unifying Ecological Subdisciplines with Ecosystem Food Webs." In *Linking Species and Ecosystems*, ed. C. G. Jones and J. H. Lawton, 166–175. London: Chapman and Hall.

Martinez, N. D., and J. A. Dunne. 1998. "Time, Space, and Beyond: Scale Issues in Food-Web Research." In *Ecological Scale: Theory and Applications*, ed. D. Peterson and V. T. Parker, 207–226. New York: Columbia University Press.

Martinez, N. D., and J. H. Lawton. 1995. "Scale and Food-Web Structure—From Local to Global." *Oikos* 73:148–154.

Martinez, N. D., B. A. Hawkins, H. A. Dawah, and B. P. Feifarek. 1999. "Effects of Sampling Effort on Characterization of Food-Web Structure." *Ecology* 80:1044–1055.

May, R. M. 1972. "Will a Large Complex System be Stable?" *Nature* 238:413–414.

May, R. M. 1973. *Stability and Complexity in Model Ecosystems*. Princeton, NJ: Princeton University Press.

May, R. M. 1983. "The Structure of Food Webs." *Nature* 301:566–568.

May, R. M. 1986. "The Search for Patterns in the Balance of Nature: Advances and Retreats." *Ecology* 67:1115–1126.

May, R. M. 2001. *Stability and Complexity in Model Ecosystems*. Princeton Landmarks in Biology Edition. Princeton, NJ: Princeton University Press.

McCann, K. S. 2000. "The Diversity-Stability Debate." *Nature* 405:228–233.

McCann, K., and A. Hastings. 1997. "Re-evaluating the Omnivory-Stability Relationship in Food Webs." *Proc. Roy. Soc. Lond. B* 264:1249–1254.

Melián, C. J., and J. Bascompte. 2004. "Food Web Cohesion." *Ecology* 85:352–358.

Memmott, J. 1999. "The Structure of a Plant-Pollinator Network." *Ecol. Lett.* 2:276–280.

Memmott, J., N. D. Martinez, and J. E. Cohen. 2000. "Predators, Parasitoids and Pathogens: Species Richness, Trophic Generality and Body Sizes in a Natural Food Web." *J. Animal Ecol.* 69:1–15.

Memmott, J., N. M. Waser, and M. V. Price. 2004. "Tolerance of Pollination Networks to Species Extinctions." *Proc. Roy. Soc. Lond. Ser. B* 271:2605–2611.

Milo, R., S. Shen-Orr, S. Itzkovitz, N. Kashtan, D. Chklovskii, and U. Alon. 2002. "Network Motifs: Simple Building Blocks of Complex Networks." *Science* 298:824–827.

Montoya, J. M., and R. V. Solé. 2002. "Small World Patterns in Food Webs." *J. Theor. Biol.* 214:405–412.

Montoya, J. M., and R. V. Solé. 2003. "Topological Properties of Food Webs: From Real Data to Community Assembly Models." *Oikos* 102:614–622.

Moore, J. C., E. L. Berlow, D. C. Coleman, P. C. de Ruiter, Q. Dong, A, Hastings, N. Collin Johnson, K. S. McCann, K. Melville, P. J. Morin, K. Nadelhoffer, A. D. Rosemond, D. M. Post, J. L. Sabo, K. M. Scow, M. J. Vanni, D. H. Wall. 2004. "Detritus, Trophic Dynamics and Biodiversity." *Ecol. Lett.* 7:584–600.

Moore, J. C., and H. W. Hunt. 1988. "Resource Compartmentation and the Stability of Real Ecosystems." *Nature* 333:261–263.

Mulder, C., J. E. Cohen, H. Setälä, J. Bloem, and J. M. Breure. 2005. "Bacterial Traits, Organism Mass, and Numerical Abundance in the Detrital Soil Food Web of Dutch Agricultural Grasslands." *Ecol. Lett.* 8:80–90.

Murtaugh, P. A., and D. R. Derryberry. 1998. "Models of Connectance in Food Webs." *Biometrics* 54:754–761.

Murtaugh, P. A., and J. P. Kollath. 1997. "Variation of Trophic Fractions and Connectance in Food Webs." *Ecology* 78:1382–1387.

Neutel, A. M., J. A. P. Heesterbeek, and P. C. de Ruiter. 2002. "Stability in Real Food Webs: Weak Links in Long Loops." *Science* 296:1120–1123.

Newman, C. M., and J. E. Cohen. 1986. "A Stochastic Theory of Community Food Webs: IV. Theory of Food Chain Lengths in Large Webs." *Proc. Roy. Soc. Lond. B* 228:355–377.

Newman, M. E. J. 2002. "Assortative Mixing in Networks." *Phys. Rev. Lett.* 89:208701.

Odum, E. 1953. *Fundamentals of Ecology.* Philadelphia, PA: Saunders.

Opitz, S. 1996. "Trophic Interactions in Caribbean Coral Reefs." Technical Report 43, ICLARM, Manila.

Paine, R. T. 1969. "A Note on Trophic Complexity and Community Stability." *Amer. Natur.* 103:91–93.

Paine, R. T. 1980. "Food Webs, Linkage Interaction Strength, and Community Infrastructure." *J. Animal Ecol.* 49:667–685.

Paine, R. T. 1988. "Food Webs: Road Maps of Interactions or Grist for Theoretical Development?" *Ecology* 69:1648–1654.

Pierce, W. D., R. A. Cushman, and C. E. Hood. 1912. "The Insect Enemies of the Cotton Boll Weevil." *USDA Bur. Entomol. Bull.* 100:9–99.

Pimm, S. L. 1980. "Bounds on Food Web Connectance." *Nature* 284:591.

Pimm, S. L. 1982. *Food Webs.* London: Chapman and Hall.

Pimm, S. L. 1984. "The Complexity and Stability of Ecosystems." *Nature* 307:321–326.

Pimm, S. L., and J. H. Lawton. 1977. "Number of Trophic Levels in Ecological Communities." *Nature* 268:329–331.

Pimm, S. L., and J. H. Lawton. 1978. "On Feeding on More than One Trophic Level." *Nature* 275:542–544.

Pimm, S. L., and J. H. Lawton. 1980. "Are Food Webs Divided into Compartments?" *J. Animal Ecol.* 49:879–898.

Pimm, S. L., J. H. Lawton, and J. E. Cohen. 1991. "Food Web Patterns and Their Consequences." *Nature* 350:669–674.

Polis, G. A. 1991. "Complex Desert Food Webs: An Empirical Critique of Food Web Theory." *Amer. Natur.* 138:123–155.

Polis, G. A., and D. R. Strong. 1996. "Food Web Complexity and Community Dynamics." *Amer. Natur.* 147:813–846.

Post, D. M. 2002. "The Long and Short of Food-Chain Length." *Trends Ecol. & Evol.* 17:269–277.

Post, D. M., M. L. Pace, and N. G. Hairston, Jr. 2000. "Ecosystem Size Determines Food-Chain Length in Lakes." *Nature* 405:1047–1049.

Rafaelli, D., and S. J. Hall. 1992. "Compartments and Predation in an Estuarine Food Web." *J. Animal Ecol.* 61:551–560.

Rejmánek, M., and P. Starý. 1979. "Connectance in Real Biotic Communities and Critical Values for Stability of Model Ecosystems." *Nature* 280:311–313.

Richards, O. W. 1926. "Studies on the Ecology of English Heaths. III. Animal Communities of the Felling and Burn Successions at Oxshott Heath, Surrey." *J. Ecol.* 14:244–281.

Schoener, T. W. 1989. "Food Webs from the Small to the Large." *Ecology* 70:1559–1589.

Schoenly, K., R. Beaver, and T. Heumier. 1991. "On the Trophic Relations of Insects: A Food Web Approach." *Amer. Natur.* 137:597–638.

Schoenly, K., and J. E. Cohen. 1991. "Temporal Variation in Food Web Structure: 16 Empirical Cases." *Ecol. Monogr.* 61:267–298.

Schmid-Araya, J. M., P. E. Schmid, A. Robertson, J. Winterbottom, C. Gjerløv, and A. G. Hildrew. 2002. "Connectance in Stream Food Webs." *J. Animal Ecol.* 71:1056–1062.

Shelford, V. E. 1913. *Animal Communities in Temperate America as Illustrated in the Chicago Region: A Study in Animal Ecology.* Chicago, IL: University of Chicago Press.

Solé, R. V., and J. M. Montoya. 2001. "Complexity and Fragility in Ecological Networks." *Proc. Roy. Soc. Lond. B* 268:2039–2045.

Solé, R. V., D. Alonso, and A. McKane. 2002. "Self-Organized Instability in Complex Ecosystems." *Phil. Trans. Roy. Soc.* 357:667–668.

Solow, A. R. 1996. "On the Goodness of Fit of the Cascade Model." *Ecology* 77:1294–1297.

Solow, A. R., and A. R. Beet. 1998. "On Lumping Species in Food Webs." *Ecology* 79:2013–2018.

Song, C., S. Havlin, and H. A. Makse. 2005. "Self-Similarity of Complex Networks." *Nature* 433:392–395.

Sprules, W. G., and J. E. Bowerman. 1988. "Omnivory and Food Chain Length in Zooplankton Food Webs." *Ecology* 69:418–426.

Sterner, R. W., A. Bajpai, and T. Adams. 1997. "The Enigma of Food Chain Length: Absence of Theoretical Evidence for Dynamic Constraints." *Ecology* 78:2258–2262.

Stouffer, D. B., J. Camacho, R. Guimerà, C. A. Ng, and L. A. Nunes Amaral. 2005. "Quantitative Patterns in the Structure of Model and Empirical Food Webs." *Ecology* 86:1301–1311.

Strogatz, S. H. 2001. "Exploring Complex Networks." *Nature* 410:268–275.

Sugihara, G. 1982. "Niche Hierarchy: Structure, Organization, and Assembly in Natural Communities." Ph.D. diss., Department of Biology, Princeton University.

Sugihara, G., L-F. Bersier, and K. Schoenly. 1997. "Effects of Taxonomic and Trophic Aggregation on Food Web Properties." *Oecologia* 112:272–284.

Sugihara, G., K. Schoenly, and A. Trombla. 1989. "Scale Invariance in Food Web Properties." *Science* 245:48–52.

Summerhayes, V. S., and C. S. Elton. 1923. "Contributions to the Ecology of Spitzbergen and Bear Island." *J. Ecol.* 11:214–286.

Summerhayes, V. S., and C. S. Elton. 1928. "Further Contributions to the Ecology of Spitzbergen and Bear Island." *J. Ecol.* 16:193–268.

Tavares-Cromar, A. F., and D. D. Williams. 1996. "The Importance of Temporal Resolution in Food Web Analysis: Evidence from a Detritus-Based Stream." *Ecol. Monogr.* 66:91–113.

Taylor, P. J. 1988. "Consistent Scaling and Parameter Choice for Linear and Generalized Lotka-Volterra Models Used in Community Ecology." *J. Theor. Biol.* 135:18–23.

Thompson, R. M., and C. R. Townsend. 1999. "The Effect of Seasonal Variation on the Community Structure and Food-Web Attributes of Two Streams: Implications for Food-Web Science." *Oikos* 87:75–88.

Townsend, C. R., R. M. Thompson, A. R. McIntosh, C. Kilroy, E. Edwards, and M. R. Scarsbrook. 1998. "Disturbance, Resource Supply, and Food-Web Architecture in Streams." *Ecol. Lett.* 1 200–209.

Ulanowicz, R. E. 1986. *Growth and Development: Ecosystem Phenomenology.* New York: Springer-Verlag.

Ulanowicz, R. E., and W. F. Wulff. 1991. "Ecosystem Flow Networks: Loaded Dice?" *Math. Biosci.* 103:45–68.

Vásquez, D. P., and M. A. Aizen. 2004. "Asymmetric Specialization: A Pervasive Feature of Plant-Pollinator Interactions." *Ecology* 85:1251–1257.

Waide, R. B., and W. B. Reagan, eds. 1996. *The Food Web of a Tropical Rainforest.* Chicago, IL: University of Chicago Press.

Warren, P. H. 1988. "The Structure and Dynamics of a Fresh-Water Benthic Food Web." D. Phil. Thesis, University of York, England.

Warren, P. H. 1989. "Spatial and Temporal Variation in the Structure of a Freshwater Food Web." *Oikos* 55:299–311.

Warren, P. H. 1990. "Variation in Food-Web Structure—The Determinants of Connectance." *Amer. Natur.* 136:689–700.

Warren, P. H., and J. H. Lawton. 1987. "Invertebrate Predator-Prey Body Size Relationships: An Explanation for Upper Triangularity Food Webs and Patterns in Food Web Structure?" *Oecologia* 74:231–235.

Watts, D. J., and S. H. Strogatz. 1998. "Collective Dynamics of 'Small-World' Networks." *Nature* 393:440–442.

Williams, R. J., and N. D. Martinez. 2000. "Simple Rules Yield Complex Food Webs." *Nature* 404:180–183.

Williams, R. J., and N. D. Martinez. 2004a. "Diversity, Complexity, and Persistence in Large Model Ecosystems." Working Paper 04-07-022, Santa Fe Institute, Santa Fe, NM.

Williams, R. J., and N. D. Martinez. 2004b. "Trophic Levels in Complex Food Webs: Theory and Data." *Amer. Natur.* 163:458–468.

Williams, R. J., E. L. Berlow, J. A. Dunne, A. L. Barabási, and N. D. Martinez. 2002. "Two Degrees of Separation in Complex Food Webs." *Proc. Natl. Acad. Sci. USA* 99:12913–12916.

Winemiller, K. O. 1989. "Must Connectance Decrease with Species Richness?" *Amer. Natur.* 134:960–968.

Winemiller, K. O. 1990. "Spatial and Temporal Variation in Tropical Fish Trophic Networks." *Ecol. Monogr.* 60:331–367.

Winemiller, K. O., E. R. Pianka, L. J. Vitt, and A. Joern. 2001. "Food Web Laws or Niche Theory? Six Independent Empirical Tests." *Amer. Natur.* 158:193–199.

Winemiller, K. O., and G. A. Polis. 1996. "Food Webs: What Can They Tell Us about the World?" In *Food Webs: Integration of Patterns and Dynamics*, ed. G. A. Polis and K. O. Winemiller, 1–22. New York: Chapman and Hall.

Wulff, F., J. G. Field, and K. H. Mann, eds. 1989. *Network Analysis in Marine Ecology. Coastal and Estuarine Studies Series.* Berlin: Springer-Verlag.

Yodzis, P. 1980. "The Connectance of Real Ecosystems." *Nature* 284:544–545.

Yodzis, P. 1981. "The Stability of Real Ecosystems." *Nature* 289:674–676.

Yodzis, P. 1982. "The Compartmentation of Real and Assembled Ecosystems." *Amer. Natur.* 120:551–570.

Yodzis, P. 1984. "The Structure of Assembled Communities II." *J. Theor. Biol.* 107:115–126.

Yodzis, P. 1996. "Food Webs and Perturbation Experiments: Theory and Practice." In *Food Webs: Integration of Patterns and Dynamics*, ed. G. A. Polis and K. O. Winemiller, 192–200. New York: Chapman and Hall.

Yodzis, P. 1998. "Local Trophodynamics and the Interaction of Marine Mammals and Fisheries in the Benguela Ecosystem." *J. Animal Ecol.* 67:635–658.

Yodzis, P. 2000. "Diffuse Effects in Food Webs." *Ecology* 81:261–266.

Yodzis, P., and K. O. Winemiller. 1999. "In Search of Operational Trophospecies in a Tropical Aquatic Food Web." *Oikos* 87:327–340.

Box A
Additional Model Complexity Reduces Fit to Complex Food-Web Structure

Neo D. Martinez
Lara J. Cushing

Cattin et al. (2004) assert that the simple "niche model" (N) of food-web structure (Williams and Martinez 2000) "fail[s] to adequately describe recent and high-quality data" and that, in contrast, their more complex version of the niche model called the "nested-hierarchy model" (NH) yields "food webs whose structure is very close to real data" and "better reflects the complexity" of ecosystems. However, we suggest that Cattin et al.'s results actually support a different conclusion: that N, the simpler model, more accurately predicts the structure of complex food webs than NH.

Cattin et al. compare both models against 14 network properties of 7 food webs (Williams and Martinez 2000). Degree distribution and two small-world properties that N predicts well (Camacho et al. 2002b; Williams et al. 2002; Dunne et al. 2004) were not included. Cattin et al. assert both models perform equally for three properties, N better predicts five properties, and NH better predicts six properties. Even without further analysis, it is already apparent that their claim that N fails

continued on next page

Ecological Networks: Linking Structure to Dynamics in Food Webs, edited by Mercedes Pascual and Jennifer A. Dunne, Oxford University Press.

87

continued

to adequately describe recent food-web data is a mischaracterization, since there is little difference in this coarse-grained assessment of overall performance of the two models. A closer look shows that two of the six properties that NH predicts better were chosen to characterize related aspects of intervality, an assumption of N that was acknowledged to not match empirical data in the original study (Williams and Martinez 2000). In sum, the link distribution rules of N create webs that are interval, while the rules of NH (like the cascade model before it) do not. It has long been known that most large webs ($S > 30$) are not interval (Cohen and Palka 1990). Cattin et al. have clearly made contributions by quantifying intervality, creating a model that does not generate intervality, and including intervality in a quantitative comparison of models. However, by introducing two measures related to intervality, and by ignoring other common properties of food webs that N is known to fit well, Cattin et al. bias their results in a way that overemphasizes the non-intervality fit to data of NH.

Going beyond the coarse-grained assessment, a careful look at Cattin et al.'s results paint a picture that differs from their characterization of their results (table 1). Both models perform similarly for mean chain length. N better predicts the fraction of species at the top, intermediate, and basal trophic levels, looping, and the variability of food-chain length, while NH better predicts the fraction of omnivores. More importantly, N fits three properties, I, Lo, and Ch_{sd}, significantly better than NH and NH fails to fit any property significantly better than N ($P <$ 0.05, Wilcoxian paired-sample test). N and NH are more frequently closer to the reference webs for 8 and 5 of the 14 properties, respectively, which become 8 and 4 of 13 properties, respectively, when 1 of the 2 intervality-related properties is excluded. Considering each individual prediction for properties of each reference web, N is closer to 47 observed properties while NH is closer to 43. Excluding Cy_4 (an intervality property that N predicts more poorly relative to NH), N and NH more closely predict 47 and 39 properties, respectively. Overall, this means that N performs slightly better than NH even when analyses are biased toward intervality measures. When only one intervality measure is considered, NH underperforms compared to N. The increased complexity of NH, while allowing a better fit to intervality measures, decreases its performance on other properties compared to the niche model.

continued on next page

continued

Further exploration should study more properties and employ the more rigorous statistical methodology (Williams and Martinez 2002; Dunne et al. 2004) that considers variability of stochastic network models in addition to the models' means discussed above. Based on the results of Cattin et al. (2004), it is difficult to judge the nested-hierarchy model's fit to particular food-web properties because the variability of the model is not specified, important information that has been provided for the niche and cascade models (Williams and Martinez 2000; Dunne et al. 2004). We note that although natural food webs are generally not interval, they are close enough to interval that the niche model is very successful in its overall characterization of food-web structure. While Cattin et al.'s new measures of intervality do highlight a limitation of the niche model, the empirical evidence presented by Cattin et al. (2004) suggests that the more complicated nested-hierarchy model provides a worse fit to empirical data, rather than a better fit, as was claimed.

TABLE 1 Summary of table 1 in Cattin et al. (2004) indicating a better fit of the niche (N) or nested-hierarchy (NH) model.

	Skipwith Pond	Little Rock Lake	Bridge Brook Lake	Chesa-peake Bay	Ythan Estuary	Coachella Desert	St. Martin Island
T	N	NH	N	NH	N	N	N
I*	N	NH	N	N	N	N	N
B	N	N	N	NH	NH	N	NH
Gen_{sd}	NH	NH	NH	N	N	N	N
Vul_{sd}	N	NH	N	NH	NH	NH	NH
M_{sim}	N	N	N	NH	NH	N	NH
Ch_{mean}	NH	x	NH	N	N	N	NH
Ch_{sd}*	N	x	N	N	N	N	N
Ch_{log}	NH	x	NH	NH	NH	N	NH
Lo*	N	x	N	NH	N	N	N
Can_{sp}	N	N	NH	NH	N	N	NH
O	NH	x	NH	NH	N	NH	N
Cy_4	=	NH	=	=	NH	NH	NH
D_{diet}	NH	NH	N	NH	NH	NH	NH

Entries indicate whether the means (medians for Cy_4) of the niche (N) or nested-hierarchy (NH) models are closer or equally close (=) to the observed properties of the empirical food webs (x indicates properties not calculated due to computational limits). Three properties* were better fit by one model (N) at $P < 0.05$. Properties and food webs are fully described elsewhere (Williams and Martinez 2000, Cattin et al. 2004). Briefly, the properties are the fractions of trophic species that are omnivores (O), cannibals (Can_{sp}), or are found in loops (Lo) or at top (T), intermediate (I), or basal (B) trophic levels; the variability of generality (Gen_{sd}), vulnerability (Vul_{sd}), and food-chain length (Ch_{sd}); and the mean length (Ch_{mean}) and log number (Ch_{log}) of food chains. Cy_4 and D_{diet} indicate the degree to which food webs are "interval" in that each species' diet can be represented as an unbroken segment of a fixed sequence of species.

Box B
Reply to Martinez and Cushing

Louis-Félix Bersier
Marie-France Cattin
Carolin Banašek-Richter
Richard Baltensperger
Jean-Pierre Gabriel

There is no doubt that the niche model (N) is excellent at describing standard food web properties, and we do not contest that N can perform better than our nested-hierarchy model (NH) for some properties in some observed food webs. There are however two important points that must be considered. First, observed values for standard properties always fall within the range predicted by NH; N may be better than NH in some instances, but NH always provides an adequate fit. Second and more fundamentally, N completely fails in the description of real food webs in that it generates only contiguous diets. This structural feature, captured by the property Ddiet, is never observed in recent and highly resolved food webs (Ddiet is fundamental because it is closely related to the organization of the food web). This failure leads us to reject N and points out that its basic assumption is wrong. **continued on next page**

Ecological Networks: Linking Structure to Dynamics in Food Webs,
edited by Mercedes Pascual and Jennifer A. Dunne, Oxford University Press.

continued

By constraining all consumers to be ordered along a single niche dimension, N may more aptly describe taxonomically defined sub-systems. Only our niche-hierarchy model, which postulates that the evolutionary history of the species comprising the food web underlies its organization, fully captures the complexity and variability of real food webs.

Graph Theory and Food Webs

Cecile Caretta Cartozo
Diego Garlaschelli
Guido Caldarelli

Recently the study of complex networks has received great attention. One of the most interesting applications of these concepts is found in the study of food webs. Food webs provide fascinating examples of biological organization in ecological communities and display characteristic and unexpected statistical properties. In particular, comparison to other complex networks shows that food webs lack the scale-free properties observed in almost all other artificial and natural networks. That is, the frequency distribution for the degree (number of different predators per species) does not display scale-free behavior. Nevertheless, we show here that self-similar and universal behavior are still present. By considering food webs as transportation networks (for the flow of resources between species), we can recover scaling properties typical of other transportation systems, such as vascular and river networks. The importance of these properties for models of structure in food webs is discussed.

Ecological Networks: Linking Structure to Dynamics in Food Webs, edited by Mercedes Pascual and Jennifer A. Dunne, Oxford University Press.

93

1   INTRODUCTION

We are presently witnessing an explosion of studies in the field of complex networks (Albert and Barabási 2001; Watts and Strogatz 1998). In particular, networks as different as the Internet (Faloutsos et al. 1999; Caldarelli et al. 2000; Pastor-Satorras and Vespignani 2004), the World Wide Web (Adamic and Huberman 1999; Albert et al. 1999), social and financial networks (Garlaschelli et al. 2003; Bonanno et al. 2003), protein interaction networks (Vazquez et al. 2003), and food webs (Montoya and Solé 2002; Camacho et al. 2002; Williams et al. 2002) can be represented and studied by means of the so-called *graph theory* (Erdös and Rényi 1959; Bollobás 1985). This theory provides a unified framework for different systems and insights into their topological properties. Graphs are mathematical objects composed of *vertices* representing the basic units of the various systems. Every connection or relationship between a pair of units is represented by an *edge* between the corresponding vertices.

Interestingly, in all the different physical systems mentioned above, data analysis show a common statistical behavior. That is, regardless of the different kinds of networks, the frequency distribution of the *degree* per vertex (defined as the number of edges per vertex) follows in most cases a power law similar to

$$P(k) \propto k^{-\gamma},\qquad(1)$$

(where the value of the exponent is always $2 \leq \gamma \leq 3$). The analysis also shows a small value of the *average distance* (minimum number of intermediate links) between pairs of vertices and a high level of *clustering* (the presence of many links between neighbors of a given vertex).

These properties are very characteristic. They are not observed either in simple network structures, such as regular lattices, or in the random graphs. For that reason, they are often considered as the signature of some underlying mechanism shaping the observed complex topology of real networks (Albert and Barabási 2001; Strogatz 2001). This unexpected finding of structural similarities across different networks has stimulated fascinating interdisciplinary research. In various scientific domains the tools of network theory have been used to characterize traditional systems from a novel point of view.

Here, we shall focus on a specific case study in which networks are formed by predation relationships in ecological communities, commonly defined as *food webs* (Lawton 1989; Pimm 1982; Cohen et al. 1990). After a brief introduction to the subject, we clarify the importance of looking for universal features across different food webs. However, by reviewing some recent results (Montoya and Solé 2002; Williams et al. 2002; Dunne et al. 2002a; Camacho et al. 2002) regarding the topological organization of food webs, we show that these systems are actually different from almost all other networks. Furthermore, no clear universal pattern appears to rule their topological properties. We then introduce the scenario proposed by some of the authors (Garlaschelli et al. 2003) suggesting that the controversial behavior of food webs can be explained in terms of their specific functional role, namely the transfer of resources in an ecosystem.

2   NETWORKS AND FOOD WEBS

2.1   EMPIRICAL RESULTS ON FOOD WEBS

The first generalization of the simple concept of *food chain* (a set of species feeding sequentially on each other in a resulting linear structure, see fig. 1(left)) due to C. S. Elton. In his pioneering work, Elton (1927) suggested the idea of representing the predation relationships among a set of species in the form of a network. He introduced his definition of *food cycle* in order to provide a more complete and realistic description of real predator-prey interactions among species. In such a description, which is now referred to as a *food web*, each species observed in a limited geographic area is represented by a vertex, and a directed link is drawn from each species to each of its predators (Lawton 1989; Pimm 1982; Cohen et al. 1990) (see fig. 1(right)). This description defines an *oriented network* reporting the trophic organization of ecological communities. The understanding of these networks is not only interesting and important from a theoretical point of view, but also for practical ecological reasons such as environmental policy and the preservation of biodiversity. More recently (Martinez 1991; Williams and Martinez 2000), it has been suggested that a less biased description can be achieved when each group of functionally equivalent species (which means species sharing the same set of predators and the same set of prey) is aggregated in the same *trophic species* and treated as a single vertex in the web. In the following, when addressing the properties of a food web, we shall always refer to the aggregated version, also called a *trophic web*.

In the ecological literature there are several quantities introduced in order to characterize food web structure. Such quantities include the fractions B, T, and I (respectively, *basal* [with no prey], *top* [with no predators], and *intermediate* [with both predators and prey] species in a web); the fractions of links between basal and top, intermediate and basal, top and intermediate species; and the number of *trophic levels* (the length of the shortest chain separating each species from the environment, commonly defined as the *first trophic level*) (Lawton 1989; Pimm 1982; Cohen et al. 1990). Whether these properties are scale-invariant or instead display any trend with system size is a debated issue (Martinez 1991; Sugihara et al. 1989; Pimm et al. 1991; Bersier et al. 1999). Moreover, patterns observed in smaller webs (Cohen et al. 1990) do not seem to persist when more recent and larger webs are considered (Martinez 1991; Warren 1989; Polis 1991; Goldwasser and Roughgarden 1993; Christian and Luczkovich 1999; Martinez et al. 1999; Memmott et al. 2000; Hall and Raffaelli 1991; Huxham et al. 1996). The only stable result seems to be the small value of the maximum trophic level $l_{\max}$ (typically $l_{\max} \leq 4$) even when the total number of species is large (Lawton 1989, Pimm 1982, Cohen et al. 1990).

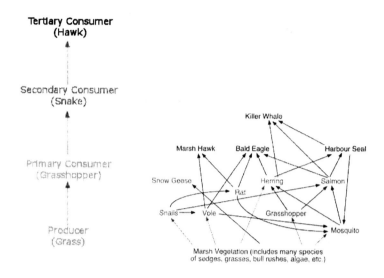

FIGURE 1   Left: an example of a food chain. Right: an example of a food web.

We shall present a description of the topology of food webs by means of graph theory. Indeed, once it has been reduced to a series of predation relationships, a food web is actually a collection of species that can be represented as vertices connected by various edges in a graph. Moreover, the specific nature of this kind of system allows the introduction of some new quantities, leading to a nice extension of the traditional theory.

2.2   TOPOLOGY OF FOOD WEBS

Food webs can be described by means of graphs. Readers not familiar with such a concept can find a brief list of definitions and properties in the following boxes. Box 1 contains the fundamental definitions in graph theory. Box 2 introduces the quantifier of interest. One particular kind of graph that will be very useful in the following discussion is a tree. In Box 3 we present some properties holding for trees.

Box 1: Definitions.

- A *graph* is a mathematical object composed of *vertices* connected by *edges*.

- Edges can have arrows; that is, they can be crossed in one direction only (if one species preys on another, the opposite is not necessarily true). In this case the graph is *oriented*. When there are no arrows, the graph is *non-oriented* (fig. 2).

- A graph is *connected* when starting from each vertex (eventually following the crossing directions defined on each edge), it is possible to reach all the other vertices in the graph. Otherwise (as in the case of isolated vertices), the graph is *not connected*.

- A graph is *minimally connected* when it is connected without loops or double-linked vertices (as we shall see later, this is the case of tree graphs).

Box 2: Graph Quantities. Among the various quantities that we could define, the following three are the most important for our purposes.

- The *degree* of a vertex in a graph is the number of edges that connect it to other vertices. In the case of oriented graphs this quantity splits into *in-degree* and *out-degree* for edges pointing in and out respectively. Networks with a power-law distribution for the degree frequency are called *scale-free networks*.

- The *distance* d_{ij} between two vertices i, j is the shortest number of edges one needs to cross to go from vertex i to vertex j. The neighbors of a vertex are all the vertices which are connected to that vertex by a single edge. In the case of oriented graphs one has to follow the direction of the edges, so that the distances are generally larger than in the homologous non-oriented graphs. Networks with a distribution for the distance frequency peaked around very little values as 5 or 6 show *small-world properties*.

- The *clustering coefficient* C is a basic characterization of clustering. C is given by the average fraction of a pair of neighbors of a node that are also neighbors of each other (the three vertices will then form a triangle). Almost any real network shows a surprisingly high value of the cluster coefficient (with respect to theoretical models).

Let us now consider in detail the graph properties of food webs. As we have already suggested, food webs are quite different from most complex systems. They lack the regularities and universal patterns found in most topological properties. However, this should not be considered as evidence of a lack of structure. The variability of these properties can actually depend on some specific biological parameter (see section 2.2.4 for an example on connectance variance according to the number of parasitic links), thus differing from random patterns. The real challenge is to identify all possible regularities, within both universal and variable properties.

Box 3: Large Trees. A *tree graph* is a connected graph with no loops (called *cycles* in the language of graph theory). As in the case of ordinary connected graphs, if the tree is undirected there is always a path between all couples of vertices. For oriented trees instead, it is possible that some of the vertices are isolated from the others. In this case we consider the tree connected in the sense that starting from a particular vertex named *source* it is possible to reach all other vertices in the graph following the oriented edges. A couple of measures that are relative to this particular kind of graph will be very important in the following.

- For each node i in a tree graph, it is possible to compute the value of a topological quantity named *area* A_i. It is a quantity first introduced in the theory of river networks and represents the set of nodes that are uphill from the node i. In this case, the area is also called a *drained area* and it actually represents the amount of water collected at point i. In other networks with tree-like structures such as social communities webs or taxonomic trees, the area represents the dimension (number of vertices) of the nested subtree up to the node i (see fig. 3 (left)). An interesting topological property of tree graphs is the *Density Probability Function* $P(A)$ of the size A for the various nested subtrees. In most real networks, for example those previously introduced, this quantity $P(A)$ notably behaves as a power law, i.e., $P(A) \propto A^{-\tau}$ (see fig. 3 (right)). As we show better in what follows, this behavior is not necessarily linked to the presence of some self-organizing mechanism that influences the properties of scale invariance of the topology.

- Once we have computed the value of the area for each node in the graph, it is possible to go further and to compute the integral C of the above quantity A on the tree (fig. 3(left)). As we will see in the next sections, the functional relationship between C and A is able to be established if the topology of the tree graph is closer to a chain-like structure or to a star-like structure. These properties are useful for the study of the efficiency of the structure. Furthermore, we can also study the *density probability function* $P(C)$ of the integral C on the tree (see fig. 3 (right)).

2.2.1 CLUSTERING COEFFICIENT. The *clustering coefficient* C can also be computed in the case of food webs. The clustering properties are indeed relevant in the study of the local wiring properties of networks. Remember that C is defined as the fraction of observed links between neighbors of a vertex (out of the total possible neighbor pairs) averaged over all vertices. In any real network this quantity is always found to be larger than expected in a random graph (Albert and Barabási 2001; Watts and Strogatz 1998). As we said in the previous section, if a network displays a small average distance between all couples of vertices, it is said to show a *small-world* behavior. If the network simultaneously displays a large clustering coefficient *and* a small average distance it is said to display a *small-world* behavior in a "strong" sense (Watts and Strogatz 1998).

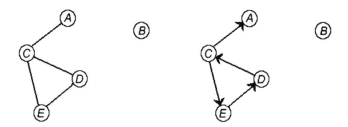

FIGURE 2   (a) Graph $G(5,4)$. The *degree* of vertex A is 1, the *degree* of vertex C is 3. (b) An oriented graph of the same size. In this case the *in-degree* of vertex A is 1, the *in-degree* of vertex C is 1 and its *out-degree* is 2.

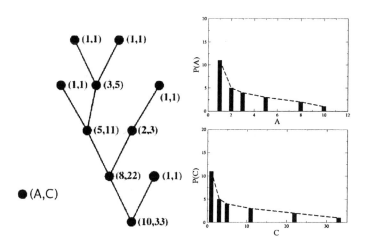

FIGURE 3   Left: values of A and C on a simple tree. Right: corresponding P(A) and P(C) distributions.

The clustering coefficient of real food webs has been independently addressed by Montoya and Solé (2002) and by Dunne, Williams, and Martinez (2002a). Table 1 reports the values of C for a set of webs. While in some cases the value of C is unambiguously larger than the one corresponding to random graphs (Montoya and Solé 2002; Dunne et al. 2002a), in others it is even smaller (Dunne et al. 2002a). In their analysis, Dunne, Williams, and Martinez suggest that the reason for this ambiguous behavior is again the small size of food webs. They show that the ratio of observed to random clustering coefficient increases roughly linearly with network size in a large number of real networks:

$$C_{\text{observed}}/C_{\text{random}} \propto N \, . \tag{2}$$

However the value of the intercept is such that, when N is sufficiently small, such as in food webs, $C_{\text{observed}}/C_{\text{random}}$ can display values smaller than one (Dunne et al. 2002a). As a consequence, food webs cannot be considered as small-world networks in the aforementioned "strong" sense. That is why we have to speak of "weak" small-world behavior. Interestingly, as for the average distance, a recent model (Williams and Martinez 2000) reproduces the empirical values of C in the observed range of size and predicts (Camacho et al. 2002) that indeed $C \propto 1/N$ for large food webs.

2.2.2 AVERAGE DISTANCE. The *average distance* (D) of an undirected graph is defined as the minimum number of links separating two vertices, averaged over all vertex pairs. This quantity is always found to be small in real networks, with a typical logarithmic dependence on the number of vertices N (Albert and Barabási 2001):

$$D \propto \log N \, . \tag{3}$$

Note that in a regular d-dimensional lattice $D \propto N^{1/d}$. The average distance of food webs has been studied independently by Montoya and Solé (2002) and by Williams et al. (2002), with similar results showing that its value is always $D \leq 3$. In both cases food webs were treated as undirected graphs, since from an ecological point of view, one is mainly interested in the propagation of perturbations in the webs, which are likely to "travel" in both directions along each link. The values of D computed on a set of real webs are reported in table 1. Due to the small size ($N < 200$) of the recorded food webs, it is, however, difficult to determine whether the observed values of D actually show a logarithmic dependence on N or not. Some indications of this come from a recent food web model (Williams and Martinez 2000) that reproduces well the empirical values of D in the observed range of size and is found to predict (Camacho et al. 2002) a logarithmic dependence of the form (3) for large web sizes. In any case, the small value of the average distance is one of the clear properties of real food webs together with the aforementioned (and obviously related) small value of l_{max} (maximum trophic level).

2.2.3 DEGREE DISTRIBUTION. We now turn to the degree distribution $P(k)$, which is probably considered as the most important topological property characterizing a network. In almost all real-world networks studied, the functional form of the degree distribution is a power law (Albert and Barabási 2001):

$$P(k) \propto k^{-\gamma} \qquad 2 \leq \gamma \leq 3 \, . \tag{4}$$

Note that when the exponent γ is lower than 3 in absolute value, then the standard deviation diverges.

In the case of food webs, however, the behavior is quite irregular (Montoya and Solé 2001; Dunne et al. 2002a; Camacho et al. 2002). First note that, since food webs are directed networks, one can distinguish between the *in-degree* k^{in} (number of incoming links) and the *out-degree* k^{out} (number of outgoing links) of a vertex. The degree distribution was studied both in its "undirected" version $P(k)$, where $k = k^{in} + k^{out}$ (Montoya and Solé 2001; Dunne et al. 2002a), and in the two possible "directed" forms $P^{in}(k^{in})$ and $P^{out}(k^{out})$ (Camacho et al. 2002). Montoya and Solé (2002) suggested that in a few isolated cases $P(k)$ can be fitted by a power law, while in others it has an irregular behavior. However, their study was based on the analysis of the probability density $P(k)$, and not of the cumulative one $P_>(k) \equiv \int_k^\infty P(k')dk'$. Since the size of food webs is very small (generally less than 200 species), the resulting data are very noisy and the analysis cannot be considered as conclusive.

A more statistically reliable analysis by Camacho, Guimerà, and Amaral (2002) showed that for the food webs in their study the cumulative distributions $P_>^{in}(k^{in})$ and $P_>^{out}(k^{out})$ have distinct functional forms, which are, however, universal across different webs, and that none of them is scale-free.

Finally, Dunne, Williams, and Martinez (2002a) studied the behavior of $P_>(k)$ on more food webs and showed that the functional form of the distribution is not universal, and that it seems to depend on the connectance of the web. In a couple of webs with extremely low connectance $P(k)$ is consistent with a power-law distribution (although it can be fitted by an exponential distribution as well), while as the connectance increases, the webs tend to display exponential and then uniform degree distributions.

2.2.4 CONNECTANCE. In the specific case of food webs it is possible to introduce some other quantities in order to obtain a very simple description of this kind of network. The first example is the *connectance*, defined as the fraction of observed links (L) out of the N^2 possible ones (Martinez 1991, 1992):

$$c \equiv \frac{L}{N^2}. \qquad (5)$$

Some early studies (Sugihara et al. 1989; Pimm et al. 1991) suggest that a linear or nonlinear scaling of connectance versus system size should hold: $c \propto N^\alpha$. In contrast, Martinez observed that, when trophic webs are considered, the connectance appears to display the almost constant value $c \approx 0.1$ (Martinez 1991). While this *constant connectance hypothesis* is confirmed by a range of empirical webs (Martinez 1991; Goldwater and Roughgarden 1993; Christian and Luczkovich 1999), other data deviate from this expectation and display either a larger (Warren 1989; Polis 1991) ($c \approx 0.3$) or a smaller (Martinez et al. 1999; Memmott et al. 2000; Hall and Raffaelli 1991; Huxham et al. 1996) ($0.02 \leq c \leq 0.06$) value (see table 1). While a value smaller than the expected one can be a consequence of the small size of some webs (Warren 1989; Polis 1991), a larger one appears to be a genuine property, which is increasingly evi-

TABLE 1 Properties of nine empirical food webs, in order of increasing number of trophic species N, and references to the original papers.

Food web [and reference]	N	c	D	C	η
Skipwith Pond (Warren 1982)	25	0.31	1.33	0.33	1.13*
Coachella Valley (Polis 1991)	29	0.31	1.42	0.43	1.13*
St Martin Island (Goldwater and Roughgarden 1993)	42	0.12	1.88	0.14	1.16
St Marks Seagrass (Christian and Luczkovich 1999)	48	0.10	2.04	0.14	1.16
Grassland (Martinez et al. 1999)	63	0.02	3.74	0.11	1.15
Silwood Park (Memmott et al. 2000)	81	0.03	3.11	0.12	1.13
Ythan Estuary 1 (Hall and Raffaelli 1991)	81	0.06	2.20	0.16	1.13
Little Rock Lake (Martinez 1991)	93	0.12	1.89	0.25	1.13
Ythan Estuary 2 (Huxham et al. 1996)	123	0.04	2.34	0.15	1.13

The Ythan Estuary food web is present in two versions: with (2) and without (1) parasites. The values of η are obtained by plotting C_i versus A_i for each vertex in the corresponding web, except those marked with (*) which are inferred by plotting the value of C_0 versus A_0 for all webs in the table together (these values are therefore interpreted as "expected" figures that could in principle differ from those computed directly on the individual webs, which on the other hand would be less reliable due to the small size of the webs; see Garlaschelli et al. 2003).

dent when the number of parasites in the web increases. In any case, there is no clear monotonous dependence of c on the number of species N. Therefore, like some other fundamental quantities in food webs, the connectance does not show any clear trend across different webs. As a consequence, fundamental functional properties which depend on the connectance (such as food web stability under species removal (Dunne et al. 2002a)) also vary across different webs. We shall return to this point in the following.

2.2.5 ALLOMETRIC RELATIONS. The real meaning of these quantities is intimately related to the study of river networks and biological and social systems where these quantities appear (like, for example, taxonomic classification trees of plant species or community networks). As we will see in the next chapters, all real river networks are characterized by the same $P(A)$. On the other hand, biological and natural systems are characterized by the same $C(A)$. In the following chapter, the occurrence of a power-law-like frequency distribution of the degree $P(k)$ or of the probability density function $P(A)$, in general, is considered as a sign of the presence of some self-organizing mechanism ruling the structure of the observed system. In the particular case of the probability density function $P(A)$ of a tree graph, this is not always true. In a recent work (Caldarelli et al. 2004), the authors show that the inverse square distribution of the size of sub-trees of a tree graph is not the consequence of some kind of self-organization of the system under consideration. They have studied the distribution of sub-communities within communities in social networks and of sub-species within species in taxonomic trees, and they have observed that both self-organized and truly random treelike systems display similar statistical behaviors when considering the size of their sub-branches. Although self-organization is surely present in social and ecological systems, the above-mentioned inverse square relationship of the density distribution to the size of sub-trees is an inescapable universal consequence of the treelike nature of the building algorithm of these systems.

2.2.6 OPTIMIZATION. In this paper, whenever we describe optimization or optimal states, we are referring to a optimization measure as it relates to the transfer of resources. In this respect the concept of optimization is simply related to the minimization of some cost function. In the case of river networks, for example, it has been conjectured that those structures have been shaped in order to reduce the total amount of the dissipation of gravitational energy. By using this as a working hypothesis, one can start with a random network and accept evolution of the patterns only if this potential is minimized. This is the case of the *optimal channel networks* model (Rodriguez-Iturbe and Rinaldo 1996; Rodriguez-Iturbe et al. 1992). Interestingly, the steady state of this iteration produces networks with statistical properties similar to those of the real rivers. It is even more interesting, that in some simple models of river evolution based on cellular automata reproducing erosion and reshaping of the river basins, an almost monotonical decrease of this cost function is still observed (Rodriguez-Iturbe and Rinaldo 1996; Caldarelli 2001). In the case of food webs, it is not possible to conjecture the existence of such a cost function. From our analysis we find that real networks seem to arrange themselves in order to maximize the efficiency of transportation provided by the presence of species competition. The meaning of competition in this sense, is access to the resources as in the case of the dynamical model explained in section 4.2.

3   RESOURCE DISTRIBUTION NETWORKS AND SPANNING TREES

In the previous chapter we have seen that food webs display a weak small-world effect, and do not display a clear scale-free degree distribution or a large clustering coefficient. Nevertheless they still present some kind of complex network behavior. We shall see that by considering food webs as transportation networks of resources from basal to intermediate and top species, we can discover new universal scaling properties that also characterize other transportation systems.

There are two possible configurations for a transportation system. In the first case we have a *source* and a set of N points to be reached. The most obvious biological example of this kind of structure is the vascular system, which delivers blood from the heart to the various parts of an organism. This is also the case in food webs. The second configuration is given by the inverse problem in which there are N sources draining into one final point or *sink* and it is

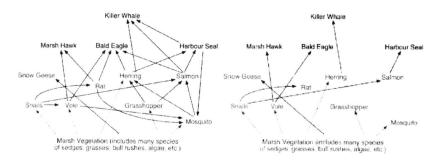

FIGURE 4   Left: example of a simple food web. Right: example of *spanning tree* of the previous food web; loop-less subset in which each vertex can be reached from the source.

simply obtained by reversing the direction of the flow. The most common and well-studied example of this situation is the river network (Rodriguez-Iturbe and Rinaldo 1996), where the rain collected throughout the basin is transferred through channels to a final outlet where the main stream of the river originates. Despite the physical difference between the two cases we shall see in the following that results from river networks can be used for food webs as well.

In nature, all species living in an ecosystem need resources to survive. These resources are obtained by feeding on other species, or, in the case of primary producers, by directly exploiting the abiotic environmental resources such as water, light, and chemicals. Food webs can, therefore, be treated as ecological transportation networks. In general, in a food web, the set of all these abiotic environmental resources can be considered as one of the trophic species in the transportation network and represented as the *environment vertex* at the lowest trophic level of the food web. This kind of structure is connected, in the sense that considering this environment vertex as the source of the network, and starting from it, it is possible to reach every species in the food web by following the direction of the links (see fig. 4(left)). This means that every food web includes a *spanning tree*, defined as a loop-less subset of the network such that each vertex can be reached from the source (see fig. 4(right)). This particular interpretation of a food web can be very interesting because the topological properties of suitably chosen spanning trees are closely related to the efficiency of a transportation network.

Before proceeding further, we note that in a transportation system each of the N sites in the web needs to receive (or to deliver, in the case of river networks) a certain amount of resources per unit time. In this hypothetical transportation system the entire amount of these resources is delivered from each site to its neighbors. But in real transportation networks things are different because every species, represented by a vertex, actually *dissipates energy*, which means that it transfers only a part of its resources to other vertices. In a food web, the transportation mechanism goes through prey-predator relationships. Each species (except top species that have no predators) transfers part of the energy that it receives to its predators, but a non-vanishing part of these resources needs to be kept by the species in order to maintain the equilibrium population size of the individuals (food webs are always assumed to be snapshots of an equilibrium state of population dynamics). Otherwise there will be no individuals left and the species will become extinct.

3.1 FREE AND GEOMETRICALLY CONSTRAINED TOPOLOGY

The topology of transportation networks is shaped in a nontrivial way by the mechanism of resource transfer associated with some optimization criterion (Banavar et al. 1999; West et al. 1999). Resources can be delivered in a more or less efficient way, and if the system is subject to some evolutionary process its structure may change in order to reach an optimized configuration. This optimized state is usually a compromise obtained considering a limited set of possible configurations for the system and trying to maximize the transportation efficiency of these structures.

There are two extreme opposite cases in the set of possible configurations of a transportation network: the *star-like* (see fig. 5(b)) and the *chain-like* (see fig. 5(a)) structures. In the first case we have all points in the structure that are directly connected to the source (point in the middle of the star), while in the second one all vertices have only one incoming and one outgoing link, with exceptions made for the source and last vertices of the chain. Let us assume that all vertices in both structures require the same amount of resources and that each of them keeps a certain fraction of these resources for its own survival. If we double the number N of vertices in the structure, in the star-like network the amount of resources provided by the source per unit time has to be doubled, but in the chain-like case the consequences are even stronger because the amount of resources provided by the source will undergo a great number of dissipation steps before it reaches the last vertex. This argument can be simplified by saying that the star-like network is generally more efficient that the chain-like one.

Note that, in both cases if the number of links is increased in such a way that the shortest paths from the source to all vertices are unchanged (see for instance fig. 5(a) and (b)), the efficiency is not significantly increased. This means that the efficiency of the network is essentially determined by the topology of its *spanning tree*, obtained by minimizing the distance of each vertex from the source. Furthermore, in a transportation network the presence of a link is always associated to a "cost" for the system; in the vascular case, for example, the formation of unnecessary tissues such as additional blood vessels represents a too high cost for the system and is clearly discouraged. Therefore, the least expensive choice is a tree-like network, even if in real networks other factors can make loops necessary.

The star and the chain are actually the most and the least efficient tree-like transportation networks respectively (Banavar et al. 1999; West et al. 1999). If there are no particular constraints on the topology, each network can be decomposed in a certain number of possible spanning trees. The spanning tree can be chain-like, star-like or something in between (see fig. 5). If the star-like configuration is allowed then the system can reach the most efficient state. But if there are some constraints limiting the range of possibilities, the star-like configuration can be forbidden and there will be a different state of optimized efficiency for the topology of the system. As an example we can consider the case of a network embedded in a bi-dimensional space where only nearest-neighbor connections are allowed. The optimal transportation system can then be reached by minimizing the distances. It then looks like the one shown in figure 5(c), where every vertex is reached by one of the possible shortest chains originating at the source. In general, every spanning tree of a d-dimensional lattice obtained by minimizing the distance of each vertex from the source is an optimal (geometrically constrained) transportation network in dimension d. Note that the least efficient chain-like topology can also be realized in the presence of geometric constraints. In this case too, adding loops such that the shortest chains are unchanged does not significantly affect the efficiency of the network.

3.2   MEASURING THE ALLOMETRIC GROWTH

The transportation cost and efficiency of a transportation network can also be studied by exploiting the tools of river networks theory (Rodriguez-Iturbe and Rinaldo 1996) on the corresponding spanning trees. In a tree-like graph (just like the length-minimizing spanning tree of a transportation network) for each vertex i it is possible to define the number A_i of vertices in the subtree (or *branch*) $\gamma(i)$ rooted at i (hereafter we assume that such a branch also includes the vertex i itself). In transportation networks this quantity has a particular meaning. In a river network, for example, this counts the number of sites "uphill" from point i (the number of points that are in the basin of vertex i) and is also called a *drained area*. By assuming a unit rainfall rate at each site in the network, the drained area actually gives the total rate (expressed in time units) at which the site i transfers water downhill. In vascular systems A_i is instead the *metabolic rate* at point i, or the amount of blood needed per time unit by the part of the organism reached by the branch of vertex i.

In general, A_i can be viewed as the quantity of resources flowing through (or *weight* of) the only incoming (outgoing, for rivers) link of vertex i in a tree-like transportation network. Note that A_i is completely independent of the topology of the tree, being simply equal to the size of the branch, irrespective of its internal structure. However, the quantity of resources flowing through each link in the branch can change significantly depending on how links are arranged within each branch. The sum of the weights of all links within the branch $\gamma(i)$ rooted at i

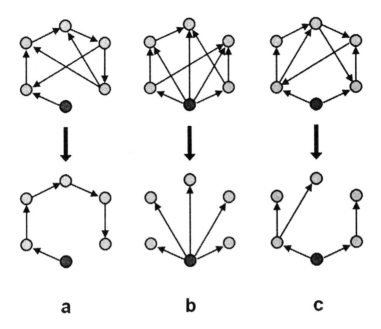

a   **b**   **c**

FIGURE 5   Possible transportation networks (top) in the case of no geometric constraints and corresponding spanning trees (bottom) using the method of chain length minimization. (a) Maximally inefficient network: the spanning tree is chain-like. (b) Maximally efficient network: the spanning tree is star-like. (c) Intermediate case with a nontrivial spanning tree.

can be computed as

$$C_i \equiv \sum_{j \in \gamma(i)} A_j . \tag{6}$$

The quantity C_i is regarded as the *transportation cost* at i (as we have seen in section 2.2.5, C is simply the integral of A on the tree, see fig. 3). If the source is labeled by $i = 0$, the quantities A_0 and C_0 represent the total transportation rate (which simply equals $N + 1$) and the total transportation cost (amount of resources flowing in the whole network per unit time) respectively. By plotting C_i versus A_i for each vertex i in the network, or by plotting C_0 versus A_0 for several networks of the same type, one obtains the so-called *allometric* scaling relations (Banavar et al. 1999; West et al. 1997, 1999)

$$C(A) \propto A^\eta \tag{7}$$

where the scaling exponent η quantifies the transportation efficiency. Clearly, the larger the value of η the less efficient the transportation system. It is easy to show that, for star-like networks, $\eta = 1$ and one recovers the expected linear scaling of cost with system size. For chain-like networks, $\eta = 2$ confirming a much worse efficiency. In the case of length-minimizing spanning trees in a d-dimensional space, it can be proved (Banavar et al. 1999; West et al. 1999) that

$$\eta_d = \frac{d+1}{d} \tag{8}$$

which clearly reduces to the previous cases $\eta_1 = 2$ (chain) and $\eta_\infty = 1$ (star), where $d = \infty$ formally indicates no geometric constraint. All the above trends are shown in figure 6.

Remarkably, real river basins always display the value $\eta = 3/2$ (Rodriguez-Iturbe and Rinaldo 1996), while for vascular systems the value $\eta = 4/3$ is observed (West et al. 1997, 1999) (see fig. 6). This means that the evolution of these networks shaped them in order to span their embedding space ($d = 2$ and $d = 3$ respectively) in an optimal (length-minimizing) way (Banavar et al. 1999; West et al. 1999), independently of other specific conditions of the system. Moreover, the value of η in both cases is different from what was expected by simple scaling arguments based on Euclidean geometry, such as quadratic scaling of the embedding area (or cubic scaling of embedding volume) with the fundamental length in the network (Banavar et al. 1999; West et al. 1999). In other words, the scaling is not isometric (hence *allometric*). By contrast, the observed values are those predicted by fractal geometry (West et al. 1999; Mandelbrot 1983), confirming that self-similar structures are often obtained as a result of optimization processes driving the evolution of complex systems (Mandelbrot 1983).

3.3   SPANNING TREES OF FOOD WEBS

We finally turn to food webs. In this case, there is clearly no geometric constraint and all the different spanning trees presented in figure 5 are possible. We assume that biological evolution drives ecological communities towards an optimization of the resource transfer. Therefore, the natural expectation is the tendency of the spanning trees to be closer to a star than to a chain. However, in principle each food web could display a spanning tree with specific properties, different from that of any other food web, due to particular environmental and evolutionary conditions. It is unclear whether selection at the level of populations can generate these optimization patterns.

However, a recent analysis (Garlaschelli et al. 2003) of the allometric scaling relation of eq. (7) in length-minimizing spanning trees of food webs shows that, remarkably, $\eta \approx 1.13$ on a large set of real webs, independently of the properties of the corresponding habitat and surrounding environment (see table 1). Due to the absence of an embedding geometric space, in food webs η is smaller than in river or vascular networks, as expected. However, it also deviates from the most efficient value $\eta = 1$ in a systematic way. The same value of the exponent is observed both within each analyzed web, highlighting self-similar branching

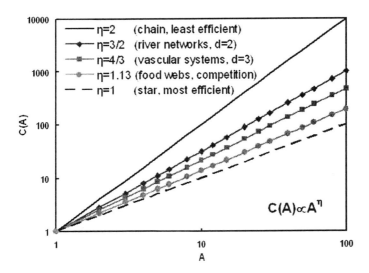

FIGURE 6   Plot of the transportation cost (C) versus system or subsystem size (A) in different networks. The optimal case corresponds to $\eta = 1$ (star-like spanning tree), and the least efficient to $\eta = 2$ (chain-like spanning tree). Real transportation systems display intermediate values of the exponent: in the optimized d-dimensional geometrically constrained case $\eta = (d+1)/d$, while in food webs the observed value is $\eta = 1.13$ (competition-like constraint).

properties, and across all webs, revealing an invariant scaling of total transportation cost C_0 with network size $A_0 = N + 1$ (Garlaschelli et al. 2003) (see table 1 and the corresponding caption). Marine, terrestrial, desert, freshwater, and island food webs fall, therefore, within the same "universality class."

We also note that the self-similarity of the spanning trees (or in other words the statistical equivalence of the whole tree and of its branches) is confirmed. Indeed a *source web* (that is a web reporting only species "sustained" by resources coming from a unique environmental element playing the role of environment vertex) like Silwood Park, centered on the Scotch Broom *Cytisus scoparium* (Memmott et al. 2000), displays the same value of the allometric exponent. This is consistent with the interpretation that the spanning tree of a source web, being a branch of the larger (undocumented) spanning tree of the whole community food web, is described by the same statistics as the whole web (self-similarity) and other webs (universality). As compared to the "irregular" behavior of food webs with respect to other important topological quantities (see section 2.2), this is really an encouraging result. The results reported so far suggest that, while other commonly used topological quantities do not capture robust properties of food webs, the allometric scaling relation of eq. (7) appears to highlight an invariant functional property, namely the resource transportation across the ecosystem. Such *function* seems to shape the *structure* of food webs in a universal fashion.

This is instructive in general, since the choice of the relevant topological quantities characterizing a network often ignores whether they capture a true underlying functional aspect. If they do not, they may display no clear behavior across different networks of the same type, especially if the boundary conditions change from case to case.

The above results have interesting evolutionary implications as well. Ecological communities are known to evolve due to immigration, speciation and extinction of species (Caldarelli et al. 1998; Drossel et al. 2001). This clearly changes the topology of food webs; however, different snapshots of such evolution at different places suggest that allometric scaling is quantitatively invariant in time and space. The resulting picture is that further evolution of the food webs would not result in a greater efficiency. In some sense, the observed property is already the "asymptotic" one. The deviation from the star-like case $\eta = 1$ is, therefore, not due to transient factors, and has to be interpreted as the result of an ecological mechanism which is constantly at work.

3.4 FOOD WEB STRUCTURE DECOMPOSITION

3.4.1 SPANNING TREES AND COMPETITION. A natural candidate for this ecological mechanism is *inter-specific competition* (Garlaschelli et al. 2003). A star-like spanning tree corresponds to maximum competition for the same resources, unless there is an infinite amount of the latter. With finite resources, if the number of competing species is large the gain associated to feeding directly on the environment may become less than the competitive effort it requires. When this occurs, species tend to differentiate their "diet" in such a way that some of them do not feed directly on the environment, causing the spanning tree to deviate from being star-like. The argument can then be generalized, so that each species realizes a trade-off between maximizing its resource input (by minimizing its trophic level) and minimizing its competitive effort (by maximizing its trophic level). This results in the universal structure of the spanning trees. Note that the finding is consistent with the observed small number of trophic levels mentioned in section 2.1, and at the same time it is far more general since it also reveals, as we pointed out, the self-similar and universal character of more detailed properties of food web topology. Note that these quantities are very difficult to be monitored from real data. A way to test these assumptions is by means of a statistical model such as the one presented in section 4.2.

3.4.2 ROBUSTNESS AND RECOVERING OF LOOPS. Besides resource transportation, there is another functional aspect of food webs which was recently shown to be tightly related to a structural property, namely the *robustness* under species removal. An analysis by Dunne, Williams, and Martinez (2002b) showed that the

fraction of species to be eliminated in order to induce the simulated secondary loss of half the initial number of species increases monotonically (with an approximately logarithmic dependence) with the connectance c defined in eq. (5). The connectance is, therefore, a simple topological quantity determining the stability of food webs, much like the scaling exponent η is a simple way to characterize their efficiency.

Now, note that c can also be regarded as a measure of the number of loops in the web. More specifically, if $L_{\text{tree}} = N$ denotes the number of links in the spanning tree of a food web, the remaining $L_{\text{loops}} = L - L_{\text{tree}}$ links are those responsible for the presence of loops in the original web. The connectance can then be written as

$$c = \frac{L}{N^2} = \frac{L_{\text{tree}} + L_{\text{loops}}}{N^2} = \frac{1}{N} + \frac{L_{\text{loops}}}{N^2}. \tag{9}$$

Therefore, the results in Dunne, Williams, and Martinez (2002b) show that, when the number of loop-forming links in a food web increases, the robustness of the web also increases.

Note that in the spanning tree analysis presented in sec. 3.3 all loops are clearly ignored. In other words, while, in order to characterize food web efficiency, loops are irrelevant, they are of fundamental importance in determining web stability. This observation allows us to recover the role of loops that was ignored so far in the spanning tree analysis.

The puzzling behavior of food webs now appears clearer. Food web structure can be decomposed in the suitably defined (length-minimizing) spanning tree and in the remaining loop-forming links. The former determine the transportation properties of the webs, in particular their efficiency, and can be simply characterized by the value of the allometric exponent η. The latter instead determine the robustness under species removal (stability), and are simply related to the connectance c. In this picture, spanning trees and loops are complementary ingredients of food web topology as well as of food web function. As we mentioned in section 2.2.4, the connectance is one of the properties that vary across food webs in an unclear fashion. As a consequence, food web stability is a highly varying property as well, whereas transportation efficiency appears universal. The reason for the non-universal behavior of connectance is not completely clear yet. However, an argument proposed by Cousins (1991) can be used to provide a simple possible explanation in terms of the different compositions of food webs. Cousins argued that when the energy flows from large to small organisms (as in host-parasite interactions) the consumers tend to feed on only one or a few species (Cousins 1991). By contrast, when the energy flows from small to large organisms (such as in most ordinary prey-predator interactions) the consumer has to face weaker constraints and can, therefore, have more different prey species. As a consequence, food webs with a large fraction of host-parasite interactions will tend to have smaller connectance than webs with few or no parasites. This expectation is rigorously verified in real food webs (Garlaschelli et al. 2003). For example, in the lowest-connectance food web (Grassland [Martinez et al. 1991]) almost all links are from large to small species. Similarly, adding parasites (Huxham et al. 1996) to the Ythan Estuary food web (Hall and Raffaelli 1991) decreases the connectance. The value of the scaling exponent η is, however, unchanged in all these cases (see table 1).

It is not clear whether the connectance varies due to either intrinsic or experimentally induced reasons. However, it is straightforward to note that the variability of c accounts for the non-universal behavior of related quantities such as the clustering coefficient C and, as mentioned in section 2.2.3, the degree distribution $P(k)$ (Dunne, Williams, and Martinez 2002a). By contrast, note that even if in principle the abundance of loops also affects the average distance D (see for instance the relatively large value of D in the lowest-connectance web in table 1), in food webs the effect is suppressed since, due to the efficient shape of the spanning tree, the distances from the environment (or in other words the number $l_{\max}$ of trophic levels) are always small independently of the connectance. This suggests that C and $P(k)$ vary due to the non-universal behavior of c, whereas D and $l_{\max}$ are always small due to the universal small value of η.

4 MODELS OF FOOD WEBS

We finally compare the empirical results to the outcomes of various models (Cohen et al. 1990; Caldarelli et al. 1998; Williams and Martinez 2000) and show how our interpretation can help to improve some aspects of food web modeling.

4.1 THE CASCADE AND NICHE MODELS

The *cascade model* (Cohen et al. 1990) and the more recent *niche model* (Williams and Martinez 2000) generate food webs in a static fashion, by assigning each species i a variable x_i and by drawing a link from species i to species j according to some rule depending on the variables x_i and x_j. Both models introduce a single tuning parameter which implicitly determines the connectance of the web (Cohen et al. 1990; Williams and Martinez 2000). Although the *Niche model* improves the predictions of the *cascade model* by successfully reproducing a wider range of observed properties (Williams and Martinez 2000) (see also the discussion regarding the average distance and clustering coefficient), both models do not reproduce the empirical value $\eta \approx 1.13$ of the scaling exponent of the allometric relations. The spanning trees of the model webs display values of η systematically smaller than the observed one (Garlaschelli et al. 2003). However, due to the static character of these models, the high degree of efficiency has not to be interpreted as the result of some optimization process, but simply as a particular property of the randomly generated webs (Cohen et al. 1990; Williams and Martinez 2000). This means that a static model like the cascade or the niche model cannot successfully reproduce the behavior of real food webs related to efficiency considerations.

4.2 THE WEBWORLD MODEL

Different from the previous examples, the *webworld model* (Caldarelli et al. 1998; Drossel et al. 2001) is an example of a dynamic model. In this model the simulated food web is generated at first in a random initial state and subsequently evolves through speciation and the extinction of species. This is accomplished by assigning each species a (time-dependent) set of features determining its trophic abilities and, therefore, influencing the final topology of the food web. The topological structure can now range from a fully connected web to the limit of a chain-like web as in figure 5. In any intermediate case, many quantities (including η and c) undergo an initial transient evolution and then reach a sort of equilibrium state that fluctuates around a stable "asymptotic" value (Caldarelli et al. 1998). Interestingly, during the first evolution period, the value of c increases and the value of η decreases from the corresponding initial values (Garlaschelli et al. 2003). This means that both efficiency and stability increase. For a suitable parameter choice, the asymptotic webs display the empirical value of η. However, this parameter choice yields an asymptotic value of the connectance $c \approx 0.12$ (Garlaschelli et al. 2003), therefore, only the webs consistent with the "constant connectance hypothesis" (Martinez 1992) mentioned in section 2.2.4 are correctly reproduced. This means that webs produced by the model are close to the empirical ones dominated by ordinary prey-predator interactions, but are significantly different from those including many host-parasite interactions. In summary, our observation of a possible "decoupling" of the spanning tree topology from the loop structure strongly suggests that, in order to reproduce both features, static and dynamic food web models need at least two independent parameters tuning network topology, each one accounting for the properties of one of these two subsets of the network.

5 CONCLUSIONS

We have focused our attention on the results obtained from the application of graph theory to the study of food webs. First of all we have observed that these ecological networks do not show the typical universal properties of complex networks that could be expected. Food webs that are simple networks of predator-prey relationships display anomalous behavior compared to the fundamental topological properties such as the degree distribution or the clustering coefficient. This lack of universality apparently denotes a problematic or irregular structural organization of these webs. Nevertheless, a food web can also be regarded as a resource transportation network. Our results reveal that, when these functional aspects are properly considered, interesting universal properties emerge. Our findings suggest that this new universality can be considered as a consequence of the influence of two kinds of mechanisms. On one side is the universal mechanism of the trade-off between maximizing resource input and minimizing inter-specific competition that shapes the topology of the spanning tree of the web. On the other side is the non-universal mechanism of development of more or fewer links in food web evolution, depending on the abundance of host-parasite interactions that determines the loops' structure. Both structures are necessary to justify simultaneously the variability of many quantities and the relative stability. We note here that it is very difficult to measure from data the degree of competition between species. We base our hypothesis on the property of the dynamical webworld model explained in section 4.2. Here we find that by varying the parameter η determining the access to resources of different species we can influence effectively the competition. This results in different topologies of food webs such as those shown in figure 5.

ACKNOWLEDGMENTS

This work has been partially supported by European Commission FET Open Project IST-2001-33555 COSIN (www.cosin.org).

REFERENCES

Adamic, L. A., and B. A. Huberman. 1999. "Growth Dynamics of the World Wide Web." *Nature* 401:131.

Albert, R., and A.-L. Barabási. 2001. "Statistical Mechanics of Complex Networks." *Rev. Modern Phys.* 74:47.

Albert, R., H. Jeong, and A.-L. Barabási. 1999. "Diameter of the World Wide Web" *Nature* 401:130–131.

Banavar, J., A. Maritan and A. Rinaldo. 1999. "Size and Form in Efficient Transportation Networks." *Nature* 399:130–132.

Barrat, A., M. Barthélémy, R. Pastor-Satorras, and A. Vespignani. 2003. "The Architecture of Complex Weighted Networks." Available at http://arxiv.org/abs/cond-mat/0311416 (accessed May 3, 2005).

Bersier, L.-F., P. Dixon, and G. Sugihara. 1999. "Scale Invariant or Scale Dependent Behavior of the Link Density Property in Food Webs: A Matter of Sampling Effort?" *Amer. Natur.* 153:676–682.

Bollobás, B. 1985. *Random Graphs*. London: Academic Press.

Bonanno, G., G. Caldarelli, F. Lillo, and R. Mantegna. 2003. "Topology of Correlation Based Minimal Spanning Trees in Real and Model Markets." *Phys. Rev. E* 68:046130.

Caldarelli, G. 2001. "Cellular Model for River Networks." *Phys. Rev. E* 63:021118.

Caldarelli, G., P. G. Higgs, and A. J. McKane. 1998. "Modelling Coevolution in Multispecies Communities." *J. Theor. Biol.* 193:345–358.

Caldarelli, G., R. Marchetti, and L. Pietronero. 2000. "The Fractal Properties of Internet." *Europhys. Lett.* 52:386–391.

Caldarelli, G., C. Caretta Cartozo, P. De Los Rios, and V. Servedio. 2004. "On the Widespread Occurrence of the Inverse Square Distribution in Social Sciences and Taxonomy." *Phys. Rev. E* 69:035101(R).

Camacho, J., R. Guimerà, and L. A. N. Amaral. 2002. "Analytical Solution of A Model for Complex Food Webs." *Phys. Rev. E* 65:030901-1–030901-4.

Camacho, J., R. Guimerà, and L. A. N. Amaral. 2002. "Robust Patterns in Food Web Structure." *Phys. Rev. Lett.* 88:228102.

Christian, R. R., and J. J. Luczkovich. 1999. "Organizing and Understanding a Winter's Seagrass Foodweb Network through Effective Trophic Levels." *Ecol. Model* 117:99–124.

Cohen, J. E., F. Briand, and C. M. Newman. 1990. *Community Food Webs: Data and Theory.* Berlin: Springer.

Cousins, S. H. 1991. "Species Diversity Measurement: Choosing the Right Index." *Trends Ecol. & Evol.* 6:190–192.

Drossel, B., P. G. Higgs, and A. J. McKane. 2001. "The Influence of Predator-Prey Population Dynamics on the Long-Term Evolution of Food Web Structure." *J. Theor. Biol.* 208:91–107.

Dunne, J. A., R. J. Williams, and N. D. Martinez. 2002. "Food-Web Structure and Network Theory: The Role of Connectance and Size." *Proc. Natl. Acad. Sci. USA* 99:12917–12922.

Dunne, J. A., R. J. Williams, and N. D. Martinez. 2002. "Network Structure and Biodiversity Loss in Food Webs: Robustness Increases with Connectance." *Ecol. Lett.* 5:558–567.

Elton, C. S. 1927. *Animal Ecology.* London: Sidgwick & Jackson.

Erdös, P., and A. Rényi. 1959. "On Random Graphs." *Publicationes Mathematicae Debrecen* 6:290–297.

Faloutsos, M., P. Faloutsos, and C. Faloutsos. 1999. "On Power-Law Relationships of the Internet Topology." *Proc. ACM SIGCOMM, Comput. Commun. Rev.* 29(4):251–262. Also available at http://powerlaws.media.mit.edu (accessed May 4, 2005).

Garlaschelli, D., G. Caldarelli, and L. Pietronero. 2003. "Universal Scaling Relations in Food Webs." *Nature* 423:165.

Garlaschelli, D., S. Battiston, M. Castri, V. D. P. Servedio, and G. Caldarelli. 2003. "The Scale-Free Topology of Market Investments." 2003. Submitted to *Europhys. Lett.* Available at http://arxiv.org/abs/cond-mat/0310503 (accessed May 4, 2005).

Goldwasser, L., and J. Roughgarden. 1993. "Construction and Analysis of a Large Caribbean Food Web." *Ecology* 74:1216–1233.

Hall, S. J., and D. G. Raffaelli. 1991. "Static Patterns in Food Webs: Lessons from a Large Web." *Animal Ecol.* 63:823—842.

Huxham, M., S. Beaney, and D. G. Raffaelli. 1996. "Do Parasites Reduce the Chances of Triangulation in a Real Food Web?" *Oikos* 76:284–300.

Lawton, J. H. 1989. *Ecological Concepts*, ed. J. M. Cherret, Blackwell Scientific, 43–78. Oxford University Press.

Mandelbrot, B. B. 1983. *The Fractal Geometry of Nature*. San Francisco: Freeman.

Martinez, N. D. 1991. "Artifacts or Attributes? Effects of Resolution on the Little Rock Lake Food Web." *Ecol. Monogr.* 61:367–392.

Martinez, N. D. 1992. "Constant Connectance in Community Food Webs." *Amer. Natur.* 139:1208–1218.

Martinez, N. D., B. A. Hawkins, H. A. Dawah, and B. P. Feifarek. 1999. "Effects of Sampling Effort on Characterization of Food-Web Structure." *Ecology* 80:1044–1055.

Memmott, J., N. D. Martinez, and J. E. Cohen. 2000. "Predators, Parasites and Pathogens: Species Richness, Trophic Generality, and Body Sizes in a Natural Food Web." *J. Animal Ecol.* 69:1–15.

Montoya, J. M., and R. V. Solé. 2002. "Small World Patterns in Food Webs." *J. Theor. Biol.* 214:405–412.

Pastor-Satorras, R., and A. Vespignani. 2004. *Evolution and Structure of the Internet. A Statistical Physics Approach*. Cambridge University Press.

Pimm, S. L. 1982. *Food Webs*. London: Chapman & Hall.

Pimm, S. L., J. H. Lawton, and J. E. Cohen. 1991. "Food Web Patterns and Their Consequences." *Nature* 350:669–674.

Polis, G. A. 1991. "Complex Trophic Interactions in Deserts: An Empirical Critique of Food-Web Theory." *Amer. Natur.* 138:123–155.

Rodriguez-Iturbe, I., and A. Rinaldo. 1996. *Fractal River Basins: Chance and Self-Organization*. Cambridge: Cambridge University Press.

Rodriguez-Iturbe, I., A. Rinaldo, R. Rigon, R. L. Bras, A. Marani, and E. J. Ijjasz-Vasquez. 1992. "Energy Dissipation, Runoff Production, and the Three-Dimensional Structure of River Basins." *Water Resour. Res.* 28(4):1095–1103.

Strogatz, S. H. 2001. "Exploring Complex Networks." *Nature* 410:268–276.

Sugihara, G., K. Schoenly, and A. Trombla. 1989. "Scale Invariance in Food-Web Properties." *Science* 245:48–52.

Vazquez, A., A. Flammini, A. Maritan, and A. Vespignani. 2003. "Global Protein Function Prediction from Protein-Protein Interaction Networks." *Nat. Biotechnol.* 21:697–700.

Warren, P. H. 1989. "Spatial and Temporal Variation in the Structure of a Freshwater Food Web." *Oikos* 55:299–311.

Watts, D. J., and S. H. Strogatz. 1998. "Collective Dynamics of Small-World Networks." *Nature* 393:440–442.

West, G. B., J. H. Brown, and B. J. Enquist. 1997. "A General Model for the Origin of Allometric Scaling Laws in Biology." *Science* 276:122–126.

West, G. B., J. H. Brown, and B. J. Enquist. 1999. "The Fourth Dimension of Life: Fractal Geometry and Allometric Scaling of Organisms." *Science* 284:1677–1679.

Williams, R. J., and N. D. Martinez. 2000. "Simple Rules Yield Complex Food Webs." *Nature* 404:180–183.

Williams, R. J., E. L. Berlow, J. A. Dunne, A.-L. Barabási, and N. D. Martinez. 2002. "Two Degrees of Separation in Complex Food Webs." *Proc. Nat. Acad. Sci.* 99:12913–12916.

Yook, S. H., H. Jeong, A.-L. Barabási, and Y. Tu. 2001. "Weighted Evolving Networks." *Phys. Rev. Lett.* 86:5835–5838.

Parasites and Food Webs

Andy P. Dobson
Kevin D. Lafferty
Armand M. Kuris

Every free-living species is host to a large diversity of parasitic species, the presence and influence of which have only recently received attention in studies of food webs. Initial studies of the St. Martin and Ythan food webs suggest that the consideration of parasites significantly changes the proportions of species at higher trophic levels; by definition the addition of parasites of top predators will lengthen the maximum lengths of food chains. However, until recently, the addition of parasites seems not to have had a major influence on levels of connectedness, nor on the scaling properties of food webs. Empirical long-term studies of the Serengeti illustrate the dramatic impact that a single pathogen may have on the long-term dynamics of food webs; ongoing comparative studies of salt marshes along the California coast illustrate that parasites may form a significant component of food-web biomass. Furthermore, parasites with complex (heteroxenic) life cycles have life-history stages that occupy a variety of different trophic positions in webs, these lead to complex long-loops in interactions between species that may

Ecological Networks: Linking Structure to Dynamics in Food Webs, edited by Mercedes Pascual and Jennifer A. Dunne, Oxford University Press.

considerably enhance the stability of food webs. We will conclude by examining how to include parasites into current theoretical models for food webs. In particular, we will examine whether the "niche model" may be scaled by metabolic rate, rather than body size, and examine how this affects our ability to organize species along a logical trophic-dynamic interaction spectrum.

1 INTRODUCTION

The world looks very different to a parasitologist or a pathologist (and not just because *Scientific American* listed parasitic worm research as the second worst career in science). The diversity of free-living organisms that compete with and consume each other is but a subset of biodiversity. To a parasitologist, free-living species represent a metapopulation of resource patches that exist for the benefit of an equal diversity of free-living species. Conceivably the diversity of parasitic species exceeds that of free-living species (Price 1980; Toft 1986). Thus, food webs that ignore parasites are seeing less than half of the species in an ecosystem; if parasites are as promiscuous in their interactions as free-living species, then at least three quarters of the links of the web are also missed. The main point of this chapter is to illustrate the major role that pathogens play in determining the structure and stability of food webs (Marcogliese and Cone 1997).

The last 20 years have seen a significant increase in our ecological understanding of the role that parasites and pathogens play in regulating host abundance, in modifying host behavior, and in mediating interactions between free-living species. In this chapter we discuss the role that parasites play in food webs. This discussion will be empirical, theoretical and speculative. Except for a few excellent exceptions, this field of study is wide open. If we take the recent significant increase in ecological understanding of pathogen ecology as an example (Grenfell and Dobson 1995; Hudson et al. 2002), there is clearly a huge potential for parasites to play a major role in structuring food webs. Yet given the diversity of parasites and pathogens, and our limited understanding of these taxa, then a full understanding of their ecological role is a long-term goal. We shall, therefore, focus on providing a handful of examples in which dramatic roles have been illustrated. We then review the current literature on the role of parasites in food webs, and conclude by examining how further consideration of parasites would influence current exciting discussions about the structure and dynamics of food webs.

Although there has been a historic absence of food web studies that contain information about parasites, a number of recent studies have reversed this. Long-term studies on the evolution of Anolis lizards on Carribean islands (Roughgarden 1995) included information on the gut and blood parasites of the lizards (Dobson et al. 1992; Schall 1992). When a food web for the island of St. Marten was constructed that included the helminth parasites of the lizards, it had two main effects on the web: (1) it added species to the top trophic level; (2) it emphasized that a significant component of the web may be missing. Only lizards were examined for parasites, so there is likely a significant diversity of parasites in other animal and plant species that have been left out of the web. Of particular importance here will be parasites of the top carnivores; these pathogens will add a further trophic level to the food web.

The food webs developed for the Ythan estuary and Loch Leven in Scotland represent comprehensive efforts to include all of the parasitic helminths into the food webs for these two aquatic systems. Many of these species have complex (heteroxenic) life cycles that require them to sequentially utilize two or more hosts, each of which lives on a different trophic level. For example, the trematode, *Diplostomum spathaceum*, lives in the alimentary canals of fish-eating birds; this stage of its life cycle adds an additional trophic level above the fish-eating birds that would traditionally appear at the top of the food chain. The adult worms produce eggs that pass with the birds faeces into the water. On hatching, the miracidial stages actively search for snails which they penetrate and infect. This parasitic stage reproduces asexually, eventually taking over most of the reproductive tissue of the snail. Other digenetic trematodes are capable of both castrating their snail hosts and of prolonging their lives and increasing their body sizes. All of this increases the parasites' ability to produce the cercarial stages that are released from the snail back into the water, where they attempt to locate a fish to penetrate and parasitize. They then migrate to the eyes of the fish, where their presence causes the fish to view its world through increasingly opaque vision. This causes the fish to spend more time feeding in the better illuminated surface waters where they are more susceptible to predation by the birds that act as final hosts. Many trematodes, cestodes, nematodes, and acanthocephalans have life cycles of this complexity. Only the nematodes have species which can exhibit simple life cycles involving a single species of host.

Inclusion of parasitic helminthes instantly creates several problems when classifying their position in food webs; should we include *Diplostomum* as a single species, or should we include each of its lifecycle stages as a separate trophic species? If we choose the latter, then *Diplostomum* counts as five species, three parasitic, and the two free-living stages that make a substantial contribution to the benthos. The parasitized Ythan and Loch Leven webs include each parasitic helminth only as a single species—but this leads to more than a fifty percent increase in species numbers in the Ythan case and a doubling in the case of Loch Leven. These numbers would increase still further if we included each of the parasitic species' life history stages as different trophic species. Inclusion of the parasites also significantly increases the length of the food chains within the web and the average position of species in the web (fig. 2). Here we should note that these webs will still underestimate the net diversity of parasitic species; none of the microparasitic bacterial, viral, fungal or protoan parasites has been included. Studies of marine sediment from an estuarine habitat in southern California (which is broadly comparable to the Ythan estuary) have found over a thousand

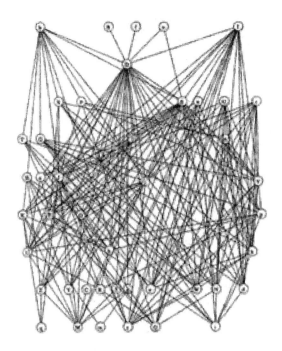

FIGURE 1   St. Martin food web—species b, f, and o are parasitic helminthes (after Goldwasser and Roughgarden 1993).

distinct viral genotypes, only 292 of which are currently represented in GenBank (Breitbart et al. 2004).

1.1   SILWOOD AND COSTA RICAN PARASITOID WEBS

A recent food web for an estuary in southern California (Carpinteria Salt Marsh) uses the concept of sub-webs to integrate parasites (Lafferty et al. in press). This reveals four sub-webs in the estuary. The predator-prey sub-web is what constitutes most published food webs and contains 81 species. The connectance of this sub-web is consistent with predictions of food-web theory. The addition of 39 parasite species provides a rich parasite-host web with many parallels to the Ythan and Loch Levan webs. The sub-web concept, when envisioned as a 2 × 2 matrix of parasites and free-living species, forced the authors to consider two previously unconceptualized sub webs. The first, predator-parasite, is the most richly linked of all. It includes all the parasites consumed when predators eat infected prey. This has dramatic evolutionary consequences as it indicates that parasites suffer great risks of predation. Many have adapted, and one third of the

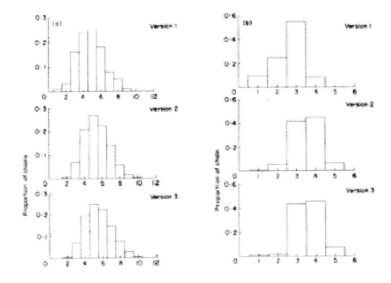

FIGURE 2   Ythan estuary and Loch Leven food webs; an increasing amount of information about parasites is included as we go from version 1, to 2, to 3. (After Huxham, Raffaelli, and Pike 1995).

TABLE 1   Nine parameters from four versions of the Ythan and Loch Leven webs. Version 4 of the Ythan web is exceptional in having tropho-, rather than biological, species.

	Food-chain lengths			Species proportions			Omnivory		Species
	Mode	Mean	$(L/S)^d$	% top	% middle	% basal	Degree	%	(S)
Ythan									
Ver. 1	5	5·00	4·42	0·28	0·68	0·04	7·65	33·7	94
Ver. 2	5	5·38	3·82	0·45	0·53	0·02	3·90	35·0	135
Ver. 3	5	5·54	4·37	0·45	0·53	0·02	5·06	43·0	135
Ver. 4	5	5·54	3·51	0·56	0·42	0·02	2·12	18·5	168
L. Levin									
Ver. 1	3	2·61	1·45	0·41	0·41	0·18	0·11	4·5	22
Ver. 2	4	3·56	1·45	0·67	0·25	0·08	2·80	19·6	52
Ver. 3	4	3·52	1·65	0·67	0·25	0·08	2·92	22·0	52
Ver. 4	3	3·57	1·91	0·65	0·29	0·06	3·02	55·0	65

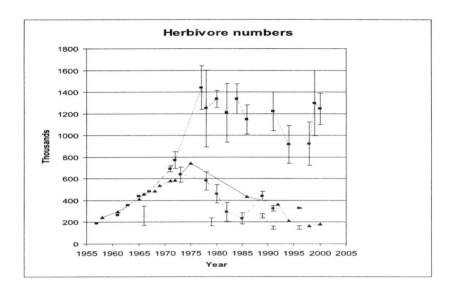

FIGURE 3   Abundance of different ungulate species in the Serengeti (data from Sinclair et al. 2005). The squares are wildebeest numbers, the triangles are Cape buffalo numbers (× 10), the circles are Thompsons gazelles, and the crosses are zebra numbers.

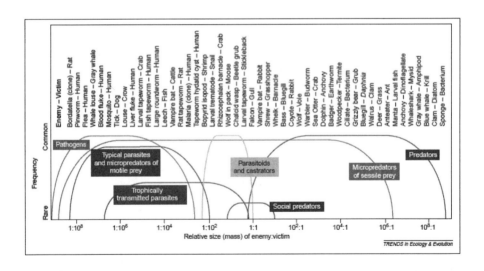

FIGURE 4   After Lafferty and Kuris, TREE (2002).

predator-parasite links lead to host-parasite links. In other words, parasites frequently adapt to the predation of their hosts by parasitizing the host's predators. The final sub web in the Carpinteria web is a parasite-parasite web which, in this estuary, is dominated by intraguild predation among the larval trematodes that parasitize the first intermediate host snail (Sousa 1992; Esch and Fernandez 1994). The parasite sub webs add greatly to connectance, such that including parasites in the Carpinteria food web leads to a threefold increase in connectance, greater than expected from food web theory. Three quarters of the nearly 2000 links in this web involve parasites, even though the authors acknowledge that they have left out many parasitic species that are likely to be common.

Interestingly, the one other web that makes a comprehensive attempt to include parasites is the grassland web for Scotch Broom at Silwood Park (Memmott et al. 2000). Here the authors find that the inclusion of parasites and parasitoids again leads to an almost doubling of species: of the 154 taxa, 69 were parasitoids or pathogens (suggesting pathogens were undersampled!). In this web, resource species had a higher vulnerability to consumers than has been recorded in any other web. Each species was preyed upon by about 13 consumers, and around 50% of these were parasitoids. Similar features are also observed in the Carpinteria web, where the consumers at lower trophic levels are predominantly predators and are at higher trophic levels predominantly pathogens (Lafferty et al. in press). In the Scotch broom web, parasitoids were significantly more specific than predators. This is the opposite to what occurs in the Carpinteria, Loch Leven, and Ythan food webs, where the large numbers of parasitic helminthes utilize a diversity of host species, often at different trophic levels within the same life cycle. All of these webs that include parasites and parasitoids create problems for organizing principles that attempt to organize species along an axis based simply on body size: while predators are usually larger than the prey they consume (93% of species on Scotch broom), the parasitoids and pathogens are significantly smaller (79% on broom). We'll return to this point in the final discussion.

2   WHY PARASITES ARE IMPORTANT

We suspect that most food webs are like the one described for the Carpinteria salt marsh. Some will exhibit a lower diversity of parasites, (such as the communities of deep-sea vents that appear almost devoid of parasitic species (Kuris personal communication), but other communities will have higher parasite diversity. Here it's less easy to give an example, as only a few habitats have been extensively examined for parasites. As is often the case with protozoan and other invertebrates, the taxonomic capacity to undertake these studies is simply unavailable. Hopefully, developments in molecular taxonomy may allow us to obtain estimates of the parasitic diversity we are undersampling (Breitbart et al. 2004), but even this will be restricted to what we have already sampled. So we conclude this section with a plea to other workers to extend their studies of other well documented webs to include parasites and pathogens in their collation of food web data.

3   TWO EMPIRICAL EXAMPLES

Do parasites play any significant role in food web structure? There is a handful of examples that suggest their introduction, or removal, may have very dramatic effects. The two examples described below show how parasites cause ecosystem-wide cascading effects on the abundance of nearly all the free-living species in two well-studied communities at very different spatial scales. Anecdotal evidence exists for other similar ecosystem-wide outbreaks, for example, pathogen-driven die-offs of Diadema urchins in the Caribbean (Lessios et al. 1984; Lessios 1988). Around California's Channel Islands, food webs affect disease and disease, in turn, affects food webs. Fishing for lobsters leads to increases in the density of sea urchin populations and at high densities, bacterial epidemics are more common (Lafferty 2004). Reductions in sea urchin populations following an epidemic lead to a shift in the state of the rocky reef community from sea urchin barrens to kelp forests (Berhens and Lafferty 2004). It would be interesting to know if there are other examples. Disease outbreaks certainly occur and disrupt long-term studies, but these are too often dismissed as unfortunate accidents, rather than accepted as natural events.

3.1   RINDERPEST IN THE SERENGETI

Perhaps the best example of a pathogen completely modifying the structure of a food web would be the introduction of the rinderpest virus into sub-Saharan Africa in the 1890s. Rinderpest is a morbillivirus that infects hoofed animals: cattle, wild buffalo, wildebeest, giraffe, and other large antelope. It is closely related to both canine distemper (CDV) and measles, two of the commonest diseases of humans and their domestic dogs. The recent evolution of these three pathogens is intimately entwined with the domestication of dogs and cattle, which created the opportunities for the pathogen to establish itself in new host species, where a few mutations allowed it to differentiate itself from rinderpest which the ancestral main trunk of the morbillivirus tree (Norrby et al. 1985; Barrett 1987). The split between the three pathogens is so recent (<5000 years) that there is still strong cross immunity between them; inoculation of dogs with rinderpest vaccine will protect them against distemper. This again raises interesting questions about how we classify pathogens in food webs where they may fail to establish a dependence upon a host, but stimulate an immunological response which allows the host to protect itself against invasion by a potentially lethal natural enemy.

Rinderpest caused one of the largest pandemics in recorded history. It took ten years to spread from the Horn of Africa to the Cape of Good Hope, during which time it reduced the abundance of many ungulate/artiodactyl species by as much as 80% (Branagan and Hammond 1965). This, in turn, produced a temporary glut of food for decomposers and scavengers, such as vultures and jackals. However, this quickly led to a massive reduction in food supply for the predators that relied on wildebeest and other game for food. The removal of the ungulates changed the grazing intensity on both shrubs and grasses. This seems to have allowed some tree species to undergo a pulse of recruitment; thus, many of the individual fever trees that create woodlands in damper areas of the savanna are now just over a hundred years old. In contrast, reduced levels of grass grazing led to an increased fire frequency, which prevented the reestablishment of Miombo bushland that had previously covered the savanna. This, in turn, modified the habitat for many of the predators that require thicker bush coverage to successfully attack their prey.

The development of a vaccine for rinderpest in the 1950s allowed these processes to be reversed. There is an instructive irony here: the presence of rinderpest in wildlife was blamed as the major reason why it had proved almost impossible to establish large-scale cattle ranches in East Africa. The rinderpest vaccine was largely developed to help the cattle industry, and although it was only ever applied to cattle, this, in turn, led to its disappearance from wildlife. Thus, cattle had been the reservoir and the repeated epidemics observed in wildlife were in response to constant spillovers from cattle (Plowright 1982). Rinderpest vaccination has successfully eradicated the disease from most parts of Africa, except in times of civil unrest, when declines in vaccination coverage allow it to resurge. The impact on wildlife has been spectacular. In the Serengeti, wildebeest numbers have grown from around 250,000 to over 1.5 million, buffalo have appeared in areas where they were previously unrecorded, and lion and hyena numbers have increased dramatically in response to the enhanced food supply (Sinclair 1979; Sinclair and Arcese 1995). Ironically, this strengthens our contention that predators are less effective than pathogens in regulating host abundance. The numbers of some species have declined; for example, there are fewer Thompson's gazelles, perhaps because of more competition for grassland forage, but more likely because of increased predation pressure from the more numerous hyenas (Sinclair et al. 1985; Dublin et al. 1990). African hunting dogs have also declined, since wide-scale rinderpest vaccination allowed their prey to increase in abundance, this is at first surprising. However, the decline may be due primarily to increase competition with more abundant hyenas (Creel and Creel 1996). There may also be increased risk of infectious disease, particularly distemper. In the absence of rinderpest, it may be that hunting dogs (and other carnivores such as lions), no longer acquire cross-immunity to distemper. This pathogens may have caused CDV outbreaks that led to the extinction of several wild dog packs and a 30% decline in lion abundance. The increased wildebeest abundance has in turn reduced the excess of dried grass during the dry season, this has, in turn, led to a reduction in fire frequency, which has allowed the miombo bushland to return to some areas of the Serengeti (Sinclair et al. in review).

The key point here is that while the biomass of the rinderpest virus in the Serengeti was always less than 10kg, the virus had an impact on the abundance and dynamics of nearly all of the dominant plant and animal species in the system—even on those which it did not infect. It strengthens the notion that indirect effects may be as important in shaping species abundance and web linkages as are direct interactions. It also reinforces the earlier comments that pathogens may be as powerful as predators in regulating abundance and distribution.

3.2 PARASITIC EYE-FLUKES IN SLAPTON LEY, DORSET

A second empirical example comes from a long-term study of a small lake adjacent to the south coast of the United Kingdom. Clive Kennedy has been studying the fish and parasites of Slapton Ley since the early 1970s (Kennedy 1981; Kennedy and Burroughs 1981; Kennedy 1987; Kennedy et al. 1994). Although no detailed data are available on the invertebrates that form the major component of the diet of the fish, long-term data are available on the relative abundance of the main fish species, their principle parasitic helminthes, and the birds that act as definitive hosts for several key species of trematodes that exhibit the complex (multi-host) life cycles described above. In the early years of the study, the fish populations and their parasites appeared relatively constant. However, nutrient runoff from neighboring agricultural land was leading to increased eutrophication. This favored one species of fish over its two potential competitors. By the late 1970s, rudd completely dominated the fishery and the high densities of this species meant that most individuals were stunted and reached reproductive maturity at late age and small body size. Sometime in the early 1980s, the lake was colonized by Great Crested Grebes (*Podiceps major*). This bird acts as the definitive host for two helminth species, the cestode *Ligula intestinalis*, and the digenetic trematode, *Tylodelphys clavata*. Both of these parasite species were introduced into the fish community when the birds recolonized, they both have complex life cycles which sequentially involve crustacea, fish, and birds in the case of the cestode, and snails, fish and birds in the case of the trematode. Both species have dramatic impacts on their fish hosts: the trematode lives in the eyes of the fish, occluding its vision and increasing its susceptibility to predation by the birds that act as the final host; the cestode lives in the body cavity of the fish and can grow to a body size that may exceed that of an uninfected fish. This significantly increases the energetic demands of the fish, which in turn causes its host to grow to a significantly bloated size, increasing its susceptibility to predation by the piscivorous birds that act as predators on the fish and final hosts for the parasite. The introduction of the two parasite species quickly had a dramatic impact on the rudd population, whose numbers rapidly declined over the following few years. While this reduced the level of intraspecific competition and alleviated the stunting (Burroughs and Kennedy 1979), the intensity of interspecific competition increased as the numbers of roach increased. A final twist to the complexity of the system is provided by the continued presence of a second species of trematode, which also parasitizes the eyes of fish. This species competes directly with the introduced trematode as both use exactly the same niche for the same part of their life cycle. The increase in the abundance of the introduced eye-fluke was matched by a significant decline in the abundance of the original eye-fluke. After a time, a third species of eye-fluke colonized the lake. This also competes with the other two species of eye-fluke. The important point here is that the direct impact of the piscivorous bird species was relatively minor, their abundance was always restricted to one or two breeding pairs and their young. However, the presence of the birds has a profound indirect effect on the fish species in the system because of their ability to complete the life cycle of the parasites that impact nearly every fish in the lake.

4 THEORETICAL IMPLICATIONS

Consideration of the two examples provided above and the brief historical review of food web studies that incorporate parasites suggests a number of important insights that point to new directions for future studies of parasites, food webs, and ecosystem function. First, the inclusion of parasites and pathogens significantly increases the diversity of species in food webs. In its simplest form, diversity is increased because most free-living species harbor "several specific pathogens" that are specialized to use only one host species. This will always lead to somewhere between a doubling and a quadrupling of the number of species in the web, depending on how accurately we can quantify "several species of specific pathogens." An important consideration here is that these pathogens may play a key role in regulating the abundance of individual host species, which will magnify the regulatory action of other intraspecific regulatory factors and thus help maintain the stability of the web as a whole. Where direct life-cycle pathogens are shared between species, such as occurs with rinderpest in the Serengeti, then pathogens can again play an important role in allowing species with similar resource requirements to coexist stably (Hudson and Greenman 1998; Dobson 2004). This occurs due to the strong frequency-dependent regulation of species abundance when rates of within-species pathogen transmission exceed those of between-host species transmission; this effectively prevents any one competitor from becoming too common as it then receives the majority of the impact of the pathogen. Such asymmetries in transmission are likely to arise when hosts are aggregated into conspecific herds or groups, or when small degrees of spatial niche separation ensure that opportunities for transmission occur more often within than between species. In contrast, where asymmetries occur in either the hosts' susceptibility to infection, or in the duration of time for which they are infectious, then this may allow one species to reverse a competitive disadvantage with a more susceptible species and potentially drive it extinct.

Thus, a healthy ecosystem (or food web) is not one that is devoid of parasites and pathogens (Lafferty 1997; Huspeni and Lafferty 2004), on the contrary, the absence of these species may allow one or two species to significantly increase in abundance and outcompete species that may have to battle both their own parasites and an increased intensity of competition from a released competitor. This situation may occur frequently when alien species are introduced into an area through accidental or deliberate anthropogenic introduction. Surveys that compare the parasite burdens of host species in their native ranges with those of species in areas where they been introduced and have successfully established show less than half of the parasite species diversity in the introduced area (Mitchell and Power 2003; Torchin et al. 2003)

4.1 COMPLEX LIFE CYCLES CREATE LONG LOOPS

The second major impact of parasites on food webs relates to recent work on the potential stabilizing role of species that introduce weak interactions into long loops (Neutel et al. 2002). In the absence of such weak links, long loops would be destabilizing. Many of the parasitic helminthes with complex life cycles fulfill these criteria. Although they may have strong impacts on some species in their life cycle (witness the worms that impact the fish at Slapton Ley), these species are less specific in their "choice" of birds that act as definitive hosts, and have only a weak, and experimentally non-detectable impact on the fitness of their avian hosts. When they parasitize the snails in their life cycle, they often only infect a small proportion of the snail population (<1–10%), their impact on these hosts is usually in the form of castration, which maintains the host within the same functional role in the food web. Castration of the snail by the worm may considerably increase snail body mass and its survival, which is considerably advantageous to the parasite, as most of the snail's tissue is producing parasite infective stages. Although only a small proportion of these will successfully locate and infect fish, the rest are likely to form an important addition to the food supply of freshwater invertebrates and even the fish species they are attempting to parasitize. The impact of this on overall web dynamics requires further exploration; in particular, food web studies need to more sharply consider interactions between species, where the interaction between the species has impacts on the fecundity component of fitness, rather than the survival term. However, the net effect of including complex life cycle parasites into food webs is that we have added a significant number of species which have complex, but relatively weak, links with a significant number of free-living species on two or three different trophic levels. This again could theoretically have an important stabilizing impact on food structure and long-term persistence.

4.2 PARASITES AND THE CASCADE MODEL

The third theoretical area in which the inclusion of parasites may be important is in consideration of their role in modifying the cascade and niche models which have recently deepened our understanding of the ways food webs are organized. These models are discussed in more detail elsewhere in this volume (Dunne Chapter 2; Cohen et al. 2003). Here, we will simply note that they seek to arrange species along an axis in a way that reflects a cascade of energy up the food chain. At first it is hard to imagine how parasites might fit into this framework. In particular, species with complex life cycles may need to be split into "meta-morpho-species," which appear at multiple positions along the trophic cascade. Obviously, we would also have this problem with aphids, mosquitoes, and other insects that occupy less dramatically different trophic locations at different points in their complex life cycles. Alternatively, consideration of parasites may allow us to consider the major ecological determinants of the species we wish to organize in a trophic cascade along a niche axis. While initial attention has focused on body size as a way of organizing the niche/cascade model, the small size of parasites and their position at the top of the trophic pyramid almost completely falsifies the notion of organizing trophic species along lines of body size. However, the limited amount of physiological evidence available suggests that parasites (and insect parasitoids) are highly efficient at energy assimilation and have relatively high metabolic rates for their body sizes (Bailey 1975; Calow 1979, 1983). This raises the intriguing prospect that perhaps the niche/cascade models reflect an underlying organization along an axis that is driven by either assimilation efficiency, or by basal metabolic rate. There is a certain "poetic justice" to this as it would shift food web studies closer to their origins in the work of Lindeman (1942). Simultaneously, it could allow us to bring many of the developments in food web ecology closer to work on ecosystem function; an exciting goal for the future of ecology.

5 CONCLUSIONS

Developing a deeper understanding of the role that parasites and pathogens play in food webs is perhaps the greatest empirical challenge in food web biology. As we mentioned at the outset, they may double the number of species in some systems and quadruple the number of links (Lafferty et al. in press). A variety of theoretical studies suggest that parasites have properties that will allow them to play major roles in stabilizing the long-term dynamics of food webs. Particularly important here is the strong potential that specific, directly transmitted pathogens have to regulate the abundance of their hosts. Complementary to this are the properties of parasites with complex life cycles which may also be stabilizing through the long loops of weak interactions they create with many hosts species on two or more trophic levels. In both these respects, parasites are likely to be as important, if not more important, than the predator-prey links that form most of the structure of food webs. Here it is important to recognize that simply dividing natural enemies into predators and parasites is perhaps too great a simplification. Lafferty and Kuris (2001) and Dobson and Hudson (1986) have proposed a more comprehensive classification of resource consumer interactions

TABLE 2   Predators versus Pathogens

Predators	Pathogens
Usually same size as victim	Much smaller than victims
Similar, or slower, rate of population increase	Much faster rate of population increase
Tend to satiate	Insatiable, unless vectored, or STD's
Indirect impact when change victim's behavior	Subtle changes of host behavior
Energetically fairly efficient	Energetically very efficient

based on the ratio of the body mass of the consumer to its resources (fig. 5). The ratios span at least 16 orders of magnitude, with microbial pathogens of vertebrates covering several orders of magnitude at one end of the spectrum, and vertebrate herbivores and their prey at the other. All other forms of resource consumer relationship occupy overlapping ranges at other points along this spectrum. In many ways this attempt to classify the major form of interactions between species in food webs takes up a challenge first proposed by T. R. E. Southwood in his Presidential Address to the British Ecological Society (Southwood 1977); it provides a way of organizing species interactions to form one potential axis of a Periodic Table of Ecological Interactions, distantly equivalent to the table that appears on the walls of chemistry and science laboratories in schools. Ultimately, studies of food webs will need a similar organizing principle and some form of this body-size-dependent classification of interactions between species and their natural enemies is likely to be one major organizing axis.

These ideas echo one of the major organizing principles in current studies of food web organization, the cascade model and the niche model (Cohen et al. 1990; Williams and Martinez 2000). These models provide a framework for describing the properties of interactions between species in a food web when parasites are ignored. Species are organized along an axis for which a potential explanation is body size. However, once parasites, parasitoids, and pathogens are included in this classification, the system would begin to break down, mainly because these natural enemies are often significantly smaller than those of the species they exploit.

When the food web is organized as a matrix of interactions, the inclusion of parasites and pathogens creates significant numbers of interactions below as well as above the diagonal—this will initially violate the cascade model assumptions. Nevertheless, this should only be seen as a challenge for food web ecologists to search for a modified organizing principle. It may be that species are actually ordered based on energetic efficiency or nitrogen assimilation rates rather than on body size. This would recapture many of the properties observed when the cascade model and niche models are applied to webs that ignore parasites. Ultimately, parasites and pathogens present a variety of challenges to food web ecologists, none are unassailable, and many may provide important new insights into the complex properties of ecological webs.

REFERENCES

Bailey, G. N. A. 1975. "Energetics of a Host Parasite System: A Preliminary Report." *Intl. J. Parasit.* 5:609–613.

Barrett, T. 1987. "The Molecular Biology of the Morbillivirus (Measles) Group." *Biochem. Soc. Symp.* 53:25–37.

Branagan, D., and J. A. Hammond. 1965. "Rinderpest in Tanganyika: A Review." *Bull. Dis. Afr.* 13:225–246.

Burroughs, R. J., and C. R. Kennedy. 1979. "The Occurrence and Natural Alleviation of Stunting in a Population of Roach, *Rutilus rutilus*." *J. Fish Biol.* 15:93–109.

Calow, P. 1979. "Costs of Reproduction—A Physiological Approach." *Biol. Rev.* 54:23–40.

Calow, P. 1983. "Pattern and Paradox in Parasite Reproduction." *Parasitology* 86:197–207.

Cohen, J. E., F. Briand, and C. M. Newman. 1990. *Community Food Webs: Data and Theory.* Biomathematics Series 20. New York: Springer-Verlag.

Cohen, J. E., T. Jonsson, and S. R. Carpenter. 2003. "Ecological Community Description using the Food Web, Species Abundance, and Body Size." *PNAS* 100(4):1781–1786.

Creel, S., and N. M. Creel. 1996. "Limitation of African Wild Dogs by Competition with Larger Carnivores." *Conserv. Biol.* 10(2):526–538.

Dobson, A. P. 2004. "Population Dynamics of Pathogens with Multiple Hosts." *Amer. Natur.* 164:S64–S78.

Dobson, A. P., and P. J. Hudson. 1986. "Parasites, Disease and the Structure of Ecological Communities." *Trends Ecol. & Evol.* 1:11–15.

Dobson, A. P., S. W. Pacala, J. Roughgarden, E. Carper, and H. Harris. 1992. "The Parasites of Anolis Lizards in the Northern Lesser Antilles.1. Patterns of Distribution and Abundance." *Oecologia* 91:110–117.

Dublin, H. T., A. R. E. Sinclair, S. Boutin, E. Anderson, M. J. Jago, and P. Arcese. 1990. "Does Competition Regulate Ungulate Populations? Further Evidence from Serengeti, Tanzania." *Oecologia* 82:283–288.

Esch, G. W., and J. C. Fernandez. 1994. "Snail-Trematode Interactions and Parasite Community Dynamics in Aquatic Systems: A Review." *Am. Midl. Nat.* 131:209–237.

Grenfell, B. T., and A. P. Dobson. 1995. *Ecology of Infectious Diseases in Natural Populations.* Cambridge, IL: Cambridge University Press.

Hudson, P., and J. Greenman. 1998. "Competition Mediated by Parasites: Biological and Theoretical Progress." *Trends Ecol. & Evol.* 13:387–390.

Hudson, P. J., A. Rizzoli, B. Grenfell, H. Heesterbeek, and A. Dobson. 2002. *The Ecology of Wildlife Diseases.* Oxford: Oxford University Press.

Huxham, M., D. Raffaelli, and A. Pike. 1995. "Parasites and Food Web Patterns." *J. Animal Ecol.* 64:168–176.

Kennedy, C. R. 1981. "The Establishment and Population Biology of the Eye-Fluke *Tylodelphys podicipina* (Digenea: Diplostomatidae) in Perch." *Parasitology* 82:245–255.

Kennedy, C. R. 1987. "Long-Term Stability in the Population Levels of the Eye-fluke *Tylodelphys podicipina* (Digenea: Diplostomatidae) in Perch." *J. Fish Biol.* 31:571–581.

Kennedy, C. R., and R. Burroughs. 1981. "The Population Biology of Two Species of Eyefluke, *Diplostomum gasterostei* and *Tylodelphys clavata* in Perch." *J. Fish Biol.* 11:619–633.

Kennedy, C. R., R. J. Watt, and K. Starr. 1994. "The Decline and Natural Recovery of an Unmanaged Coarse Fishery in Relation to Changes in Land Use and Attendant Eutrophication." In *Rehabilitation of Freshwater Fisheries,* ed. I. G. Cowx, 366–375. Oxford: Blackwell Scientific.

Lafferty, K. D., K. Whitney, J. Shaw, R. F. Hechinger, and A. M. Kuris. 2005. "Food Webs and Parasites in a Salt Marsh Ecosystem." In *Disease Ecology: Community Structure and Pathogen Dynamics,* ed. S. Collinge and C. Ray. Oxford: Oxford University Press.

Lafferty, K. D., and A. M. Kuris. 2002. "Trophic Strategies, Animal Diversity and Body Size." *Trends Ecol. & Evol.* 17:507–513.

Lessios, H. A. 1988. "Mass Mortality of *Diadema antillarum* in the Caribbean: What Have We Learned?" *Ann. Rev. Ecol. & Sys.* 19:371–393.

Lessios, H. A., J. D. Cubit, D. R. Robertson, M. J. Shulman, M. R. Parker, S. D. Garrity, and S. C. Levings. 1984. "Mass Mortality of *Diadema antillarum* on the Caribbean Coast of Panama." *Coral Reefs* 3:173–182.

Lindeman, R. L. 1942. "The Trophic-Dynamic Aspect of Ecology." *Ecology* 23:399–418.

Marcogliese, D. J., and D. K. Cone. 1997. "Food Webs: A Plea for Parasites." *Trends Ecol. & Evol.* 12(8):320–325.

Memmott, J., N. D. Martinez, and J. E. Cohen. 2000. "Predators, Parasitoids and Pathogens: Species Richness, Trophic Generality, and Body Sizes in a Natural Food Web." *J. Animal Ecol.* 69:1–15.

Mitchell, C. E., and A. G. Power. 2003. "Release of Invasive Plants from Fungal and Viral Pathogens." *Nature* 421:625–627.

Neutel, A.-M., J. A. P. Heesterbeek, and P. C. de Rieter. 2002. "Stability in Real Food Webs: Weak Links in Long Loops." *Science* 296:1120–1123.

Norrby, E., H. Sheshberadaran, K. C. McCullogh, W. C. Carpenter, and C. Örvell. 1985. "Is Rinderpest Virus the Archevirus of the Morbillivirus Genus?" *Intervirology* 23:228–232.

Plowright, W. 1982. "The Effects of Rinderpest and Rinderpest Control on Wildlife in Africa." *Symp. Zool. Soc. Lond.* 50:1–28.

Price, P. W. 1980. *Evolutionary Biology of Parasites.* Princeton, IL: Princeton University Press.

Roughgarden, J. D. 1995. *Anolis Lizards of the Caribbean: Ecology, Evolution, and Plate Tectonics.* Oxford, Oxford University Press.

Schall, J. J. 1992. "Parasite-Mediated Competition in *Anolis* Lizards." *Oecologia* 92:58–64.

Sinclair, A. R. E. 1979. *The Eruption of the Ruminants. Serengeti: Dynamics of an Ecosystem,* ed. A. R. E. Sinclair and M. Norton-Griffiths, 82–103. Chicago, IL: University of Chicago Press.

Sinclair, A. R. E., and P. Arcese. 1995. *Serengeti II. Dynamics, Management, and Conservation of an Ecosystem.* Chicago, IL: Chicago University Press.

Sinclair, A. R. E., H. Dublin, and M. Morner. 1985. "Population Regulation of the Serengeti Wildebeest: A Test of the Food Hypothesis." *Oecologia* 65:266–268.

Sousa, W. P. 1992. "Interspecific Interactions among Larval Trematode Parasites of Freshwater and Marine Snails." *Am. Zool.* 32:583–592.

Southwood, T. R. E. 1977. "Habitat, the Templet for Ecological Strategies." *J. Animal Ecol.* 46:337–365.

Toft, C. A. 1986. "Coexistence in Organisms with Parasitic Lifestyles." In *Community Ecology,* ed. J. M. Diamond and T. J. Case, 445–463. New York: Harper and Row.

Torchin, M. E., K. D. Lafferty, A. P. Dobson, V. M. McKenzie, and A. M. Kuris. 2003. "Introduced Species and their Missing Parasites." *Nature* 421:628–630.

Williams, R. J., and N. D. Martinez. 2000. "Simple Rules Yield Complex Food Webs." *Nature (London)* 404:180–183.

Box C
Sea Lampreys in Great Lakes Food Webs

Sarah Cobey

The introduction of sea lampreys (*Petromyzon marinus*) to the Great Lakes demonstrates how invasive parasites can affect community structure. Sea lampreys had invaded Lake Ontario via canals from the Atlantic Ocean by 1835, but they were unable to advance farther until the canal between Lake Ontario and Lake Erie was improved in 1919. Within twenty years they were reported in the remaining lakes. They fed preferentially on the native piscivores, lake trout (*Salvelinus namaycush*) and burbot (*Lota lota*), and on large planktivores such as the lake whitefish (*Coregonus clupeaformis*).

continued on next page

Ecological Networks: Linking Structure to Dynamics in Food Webs, edited by Mercedes Pascual and Jennifer A. Dunne, Oxford University Press.

continued

Sea lampreys played a major, if not deciding, role in the extirpation of these three species mid-century (Smith 1969; Coble et al. 1990; Hansen 1999). Some stocks began to decline while sea lamprey populations were still small, suggesting initial perturbation from overfishing. These losses may have intensified pressure from commercial fisheries and sea lampreys on the remaining host populations. Lett et al. (1975) and Jensen (1994) modeled dynamics between sea lampreys and lake trout, concluding that extirpation of prey was inevitable even in the absence of fishing.

Following the suppression of predators, populations of prey species exploded. But the smaller native ciscoes (*Coregonus* spp.) in turn were preyed upon by sea lampreys, and two invasives, the alewife (*Alosa pseudoharengus*) and rainbow smelt (*Osmerus mordax*), came to dominate the planktivorous community (Eshenroder and Burnham-Curtis 1999). Alewife comprised 90% of the fish biomass in Lake Michigan in 1966 and also preyed on the larvae of piscivores and other planktivores (Tody and Tanner 1966). To resuscitate commercial and recreational fishing, managers began restocking lake trout in the 1950s and introduced Pacific salmonines, including chinook (*Oncorhyncus tshawytscha*) and coho (*O. kisutch*) in 1966–1967 (fig. 1).

Efforts to eradicate sea lampreys began in the early 1950s, when mechanical weirs were constructed in tributaries of Lakes Huron, Michigan, and Superior. In 1958, the larval lampricide TFM (3-trifluoromethyl-4-nitrophenol) was tested in Lake Superior and quickly depressed spawner abundance to 8% of pre-treatment levels (Smith and Tibbles 1980). Treatment was gradually extended to the other lakes and continues today in selected streams. Since 1982, alternative control methods have supplemented TFM treatment. The aim is no longer eradication but optimizing ecological, economic, and social benefits with the costs of suppression (Christie and Goddard 2003).

Analysis of interactions between sea lampreys and hosts is troubled by uncertainty over numerical and functional responses (reviewed by Bence et al. 2003). Stock-recruitment relationships suggest that sea lamprey populations may exhibit moderate compensation that is overshadowed by large density independent variability in several life stages (Haeseker et al. 2003). Estimates of attack rates and host mortality from laboratory experiments and measures of wounding marks on dead fish hint at a type II functional response (Farmer 1980; Schneider et al. 1996). Kitchell and Breck (1980) suggest that sea lampreys act as facultative parasites, with their behavior along a parasite-predator continuum determined by the frequency with which they encounter hosts. When hosts are relatively abundant, lampreys maximize blood yield by feeding on multiple hosts for short periods; when hosts are rare, lampreys attach for long periods to individual hosts, and function more as predators.

continued on next page

continued

Continual suppression of sea lampreys may be required to maintain a stable community dominated by desirable piscivores (Eshenroder and Burnham-Curtis 1999). Ironically, managers have recently been concerned by possible overpredation of alewives, which reduce pressure on native planktivores. Untangling the ecology of sea lampreys will contribute to effective management of Great Lakes food webs.

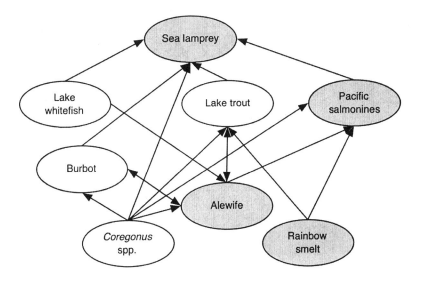

FIGURE 1   A subsection of a typical food web of the Great Lakes, showing predatory and parasitic relationships among sea lampreys and major piscivores and planktivores. Arrows indicate the direction of energy flow. Non-native species are shaded.

REFERENCES

Bence, J. R., R. A. Bergstedt, G. C. Christie, P. A. Cochran, M. P. Ebener, J. F. Koonce, M. A. Rutter, and W. D. Swink. 2003. "Sea Lamprey (*Petromyxon marinus*) Parasite-Host Interactions in the Great Lakes." *J. Great Lakes Resh.* 29(sup 1):253–282.

Christie, G. C. and C. I. Goddard. 2003. "Sea Lamprey International Symposium (SLIS II): Advances in the Integrated Management of Sea Lampreys in the Great Lakes." *J. Great Lakes Resh.* 29(sup 1):1–14.

Coble, D. W., R. E. Bruesewitz, T. W. Fratt, and J. W. Scheirer. 1990. "Lake Trout, Sea Lampreys, and Overfishing in the Upper Great Lakes—A Review and Reanalysis." *Trans. Am. Fish. Soc.* 119(6):985–995.

Eshenroder, R. L., and M. K. Burnham-Curtis. 1999. "Species Succession and Sustainability of the Great Lakes Fish Community." In *Great Lakes Fisheries Policy and Management: A Binational Perspective*, ed. W. W. Taylor and C. P. Ferreri. East Lansing: Michigan State University Press.

Farmer, G. J. 1980. "Biology and Physiology of Feeding in Adult Lampreys." *Can. J. Fish. Aquat. Sci.* 37(11):1751–1779.

Haeseker, S. L., M. L. Jones, and J. R. Bence. 2003. "Estimating Uncertainty in the Stock-Recruitment Relationship for St. Marys River Sea Lampreys." *J. Great Lakes Resh.* 29(sup 1):728–741.

Hansen, M. J. 1999. "Lake Trout in the Great Lakes: Basinwide Stock Collapse and Binational Restoration." In *Great Lakes Fisheries Policy and Management: A Binational Perspective*, ed. W. W. Taylor and C. P. Ferreri. East Lansing: Michigan State University Press.

Jensen, A. L. 1994. "Larkin's Predation Model of Lake Trout (*Salvelinus namaycush*) Extinction with Harvesting and Sea Lamprey (*Petromyzon marinus*) Predation: A Qualitative Analysis." *Can. J. Fish. Aquat. Sci.* 51(4):942–945.

Kitchell, J. F. and J. E. Breck. 1980. "Bioenergetics Model and Foraging Hypothesis for Sea Lamprey (*Petromyzon marinus*)." *Can. J. Fish. Aquat. Sci.* 37(11):2159–2168.

Lett, P. F., F. W. H. Beamish, and G. J. Farmer. 1975. "System Simulation of the Predatory Activity of Sea Lampreys (*Petromyzon marinus*) on Lake Trout (*Salvelinus namaycush*)." *J. Fish. Res. Board Can.* 32:623–631.

Schneider, C. P., R. W. Owens, R. A. Bergstedt, and R. O'Gorman. 1996. "Predation by Sea Lamprey (*Petromyzon marinus*) on Lake Trout (*Salvelinus namaycush*) in southern Lake Ontario, 1982–1992." *Can. J. Fish. Aquat. Sci.* 53:1921–1932.

Smith, S. H. 1968. "Species Succession and Fishery Exploitation in the Great Lakes." *J. Fish. Res. Board Can.* 25:667–693.

Smith, B. R. and J. J. Tibbles. 1980. "Sea Lamprey (*Petromyzon marinus*) in Lakes Huron, Michigan, and Superior: History of Invasion and Control, 1936–1978." *Can. J. Fish. Aquat. Sci.* 37:780–1801.

Tody, W. H. and H. A. Tanner. 1966. "Coho Salmon for the Great Lakes." Fish Management Report No. 1. Michigan Department of Conservation, Lansing, Michigan.

The Structure of Plant-Animal Mutualistic Networks

Jordi Bascompte
Pedro Jordano

Plant-animal mutualistic networks can be described as bipartite graphs depicting the interactions between two distinct sets: plants and animals. These mutualistic networks have been found to be highly structured. Specifically, they show a nested pattern in which specialists interact with proper subsets of the set of species with which generalists interact. This pattern is important for understanding coevolution in species-rich communities, which can be reduced neither to pairs of coevolving species nor to diffuse, randomly interacting assemblages. We discuss the dynamic implications of network structure from the points of view of coevolution, community ecology, and conservation biology.

1 INTRODUCTION

Studies on community organization have largely focused on a small subset of species interactions: trophic relationships depicted in traditional food webs. In order to improve our understanding of community organization, we need to

Ecological Networks: Linking Structure to Dynamics in Food Webs, edited by Mercedes Pascual and Jennifer A. Dunne, Oxford University Press. 143

explore how other types of interactions shape communities. In this chapter we consider a very important and largely unexplored interaction at the community level: plant-animal mutualisms. It is well accepted that the interactions between plants and the animals that pollinate them or disperse their fruits have played a major role in the generation of terrestrial biodiversity (Ehrlich and Raven 1964). For example, the life cycles of 89.5% of the woody plants in tropical rainforests depend on vertebrate animals for seed dispersal (Jordano 2000). Similarly, it is well accepted that the elimination of pollinators or seed dispersers would have catastrophic consequences within decades, with the elimination of much of terrestrial biodiversity (Thompson 2002; Koh et al. 2004).

Many studies of mutualisms have focused on very specialized interactions between pairs of species. This arises partly because of the attraction toward text-book examples of highly specific interactions such as the ones between orchids, whose flowers perfectly mimic the females of the insect species that pollinate them. This has originated a large body of work on pair-wise coevolution. However, when one looks at nature, mutualistic interactions often involve much larger sets of species. Until recently, the alternative to one-to-one coevolution has been the concept of diffuse coevolution (Janzen 1980). This concept assumes that interactions are not species-to-species, but guild-to-guild. These interactions are very variable from year to year and also through space (Herrera 1988). Since a species is subjected to highly variable and unpredictable selection pressures, diffuse coevolution would thus preclude fine-tuned adaptations. This has been one of the main arguments used in questioning pair-wise coevolution as a prevalent mechanism that explains highly diversified mutualistic interactions.

Some criticism has also been made of the concept of diffuse coevolution because it seems to suggest that these communities are intractable to analysis. To get a real view of how coevolutionary interactions are shaped in species-rich communities, one has first to look at the structure of such complex networks. To do this, one needs appropriate techniques, such as the ones provided by the physics of complex networks and other concepts previously applied to food web studies.

If there is no structure, that is, plants and animals interact randomly, then a concept such as diffuse coevolution may be enough to describe coevolutionary interactions. If there is structure, however, the shape of this structure will certainly lead to an alternative view of coevolution.

In this chapter we will first review the search for structure in food webs. Second, we will turn to mutualistic networks, specifically to recent results showing that they are neither randomly organized nor organized in compartments; instead, they show a very cohesive structure organized in a nested, Chinese-Box fashion. Third, we will discuss the consequences of this nested pattern for the coevolutionary process, community dynamics, and conservation biology. This pattern will also be related to the previous studies on food webs. We will end by listing a series of problems that have to be solved before more progress can be made toward a comprehensive description of structure in ecological networks.

2 FOOD-WEB STRUCTURE

The search for structure has a long tradition in food web research. One of the main research agendas in food webs has been to look for compartments, which has largely influenced our view of community ecology (Paine 1963; May 1972). A structure in compartments or modules implies species interacting strongly within a compartment, with almost no interactions between compartments.

Two approaches to food web research have somehow produced opposite results. First, studies on quantitative food webs (i.e., looking at interaction strength) have emphasized compartments (Paine 1963, 1966). The compartmentalization agenda was also largely imposed by the seminal paper by Robert M. May (1972) on the relationship between stability and complexity in food webs. At the end of the paper, May noticed that model communities tended to be more stable if organized in blocks: "our model multi-species communities, for given average interaction strength and web connectance, will do better if the interactions tend to be arranged in blocks—again a feature observed in many natural ecosystems." In here there are two important messages: first, the result that compartmentalized communities are more stable (which follows from Gardner and Ashby (1970)), and second, the suggestion that real communities are structured in compartments.

The above result was highly influential and spurred a sequence of papers on the subject. Pimm (1979a), for example, looked at theoretical communities assembled under more realistic rules, that is, excluding many of the unreasonable properties found in random communities, such as predators with nothing on which to feed. He concluded that compartmentalization, far from stabilizing a community, destabilizes it. The debate revolved around the consequences of compartmentalization. Alternatively, other papers looked at real data from qualitative food webs (interactions are not weighted by a strength). These papers were aimed at addressing whether communities are compartmentalized, but results were mainly negative (e.g., Pimm and Lawton 1980), although there were exceptions (e.g., Krause et al. 2003). Thus, the concept of compartments in food web research began more as a consequence of Paine's and May's influential papers than as a result of empirical qualitative evidence.

An alternative view of food webs has recently emerged after the analysis of the largest food webs using new techniques such as the connectivity correlation (Melián and Bascompte 2002) and the structure of subwebs (Melián and Bascompte 2004). The former technique, derived from studies on complex networks such as the Internet (Maslov and Sneppen 2002; Pastor-Satorras et al. 2001), looks at the relationship between the number of connections of one species and the average number of connections of the species with which it interacts. For example, protein networks show a negative connectivity correlation so that highly connected proteins tend to interact with weakly connected proteins (Maslov and Sneppen 2002). The dynamic consequence of this structure in compartments is that it enhances the robustness of these networks to the propagation of delete-

FIGURE 1   A plant-pollinator mutualistic network. Plants are represented as bottom nodes, insects as top nodes. This web, compiled by J. M. Olesen and H. Elberling, corresponds to the interactions reported in a single day in an Arctic community in Greenland. Courtesy of J. Olesen, plotted using the FoodWeb3D software by R. Williams.

rious mutations, which tend to be confined within a module. On the contrary, food webs suggest a non-random structure in which generalist species tend to interact among themselves (Melián and Bascompte 2002), and where a collection of small subwebs is connected to a single, highly dense subweb that glues the web together (Melián and Bascompte 2004). The resulting view is very cohesive. This cohesiveness would make the food webs more sensitive to the spread of a contaminant, but more robust to the loss of species.

Thus, the latter results regarding food webs arise the question of whether food webs are more cohesive (therefore, less compartmentalized) than expected. It could be interesting to analyze both qualitative and quantitative data sets to test whether compartments and cohesion are essential features of food webs. A previous study has shown that ignoring weights when aggregating taxa decreases the number of analyzed interactions and can obscure strong relationships that contribute to compartmentalization (e.g., Krause et al. 2003), but more synthetic work is needed here.

We will now turn again to mutualistic networks. The focus of our chapter is to look at the structure of these networks and determine how this structure relates to the one observed in food webs.

3   THE NESTED STRUCTURE OF MUTUALISTIC NETWORKS

There is a marked difference between traditional food webs and mutualistic networks. While traditional food webs are represented as undirected graphs (Cohen 1978; Cohen et al. 1990; Pimm 1982), that is, relationships depicting who-eats-whom through several trophic levels, mutualistic networks are described by means of bipartite graphs (Jordano et al. 2003). Bipartite graphs depict the relationships between (but not within) two distinct sets: plants and animals. Figure 1 represents an example of such a network. Alternatively, mutualistic networks can be described as a matrix, with plants as rows and animals as columns[1] (fig. 2). Each element of the matrix is 1 if that particular plant and animal interact, and 0 otherwise. In this chapter we will only review qualitative information, although each element of this matrix could also describe the strength of the interaction (i.e., the relative frequency of visits, or relative frequency of pollen or seeds dispersed). Plant-animal networks are an adequate way to represent a wide range of situations in which plants and animals coevolve. Some of these examples are more resolved than traditional food webs: one deals with taxonomic species. This and the fact that only two trophic levels are involved, makes these plant-animal networks good candidates for studying organization. Recently, several researchers have focussed on the statistical properties of plant-animal networks (Jordano 1987; Memmott 1999; Memmott and Waser 2002; Vázquez and Aizen 2003, 2004; Jordano et al. 2003; Bascompte et al. 2003). This recent explosion of research has been, in part, spurred by work on the statistical mechanics of complex networks (Amaral et al. 2000; Albert and Barabási 2002). Particularly relevant to the discussion here, mutualistic networks have been found to be highly nested (Bascompte et al. 2003). Nestedness is a concept borrowed from island biogeography to illustrate how a pool of animals is redistributed among a set of islands (Atmar and Paterson 1993). One can use an analogy and imagine that a plant is an "island" that harbors several animal species which feed on it.

A mutualistic matrix is nested if specialists interact with proper subsets of the set of species with which generalists interact. That is, if we move along an axis from the most specialist to the most generalist, we find that the same group of species is repeated within larger groups (see figs. 2(a) and 2(d)). Using an appropriate null model in which the probability of an interaction is proportional to both plant and animal degrees, Bascompte et al. (2003) showed that mutualistic networks are significantly nested. Thus, these networks can be described neither as random collections of interacting species, nor as compartments arising from tight, parallel specialization.

Nestedness implies two properties of these networks related to research on food webs. First, a nested matrix is highly cohesive, because generalist plants and generalist animals interact among themselves. This creates a core in which a small set of species leads the bulk of interactions. Second, these matrices embed asymmetric interactions in the sense that specialist species tend to interact with the most generalist species (this is true for both plants and animals). Vázquez and Aizen (2004) have also detected this asymmetric pattern of specialization.

---

[1]The alternative, transposed arrangement with animals as rows and plants as columns, is also frequently used.

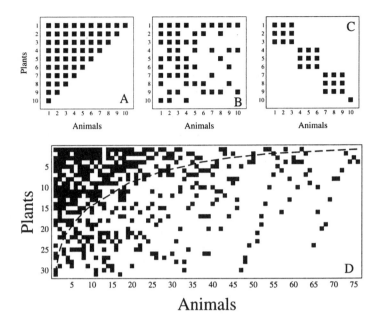

FIGURE 2   A to C represent examples of plant-animal matrices perfectly nested, random, and organized in compartments, respectively. D represents the pollinator community plotted in the previous figure (data now correspond to cumulative interactions over the whole flowering season). This community is significantly nested.

In the following sections we will explore the consequences of this pattern for coevolution, community dynamics, and conservation biology.

Traditional methods of detecting compartments in food webs assume that species within a compartment would share more predators and types of prey than they would with species outside that compartment (Pimm and Lawton 1980; Raffaelli and Hall 1992). Thus, one can calculate the degree of trophic similarity among any two pairs of species (defined as the number of common species they interact with over the total number of species either one or the other interacts with; Pimm and Lawton 1980). Recently, a method for detecting compartments originating from the field of social networking has been used in food webs (Krause et al. 2003). Basically, the method is based on a comparison of the density of interactions within versus between compartments, and uses randomization procedures to test the statistical significance of that comparison (see Krause et al. for details).

In relation to nestedness, the degree of nestedness is calculated using the Nestedness Calculator (AICS Research, University Park, NM; Atmar and Patterson 1993). This software works with a presence-absence matrix and starts by rearranging rows and columns from the most generalist to the most specialist in a combined way that maximizes nestedness. The software calculates an isocline of perfect nestedness, which is the theoretical arrangement of rows, columns, and interactions that forms a perfectly nested matrix. Then, it detects for each row all the unexpected present and absent interactions in the real matrix. That is, it considers the gaps in the left-hand side semimatrix (all interactions should lie in here in the perfectly nested scenario), and the observed interactions in the right-hand side semimatrix (which should be empty in the perfectly nested scenario). For each of these unexpected presences and absences, a normalized distance to the isocline is calculated, and these values are averaged. This global measure is called temperature using an analogy with physical disorder (a completely nested matrix represents order, and a random one represents disorder). Temperature (T) ranges between 0 and 100 degrees. Nestedness (N) can be defined as $N = (100 - T)/100$, so its value ranges between 0 and 1.

The next step after calculating a measure of structure such as nestedness, is to use a null model as a benchmark to assess the significance of the pattern, that is, the chances of getting a similar pattern just by chance. The Nestedness Calculator uses a null model in which each cell of the matrix has the same probability of being occupied, a probability estimated as the fraction of total interactions. However, imagine that we find significant departure. This can be due to multiple reasons. If, for example, we are interested in studying the degree of heterogeneity across species (that is, the variance in the number of interactions per species), the previous null model would be adequate, since we are testing whether we can get, by chance, a certain distribution of links per species given a total number of links, plants, and animals. However, if we are interested in higher levels of structure as nestedness, we can not conclude that a community is significantly nested, because departure from the previous null model could just be due to heterogeneous degree, not to nestedness. When looking for an appropriate null model we have to clearly identify at what level our question is posed. For more details about this subject see Cook and Quinn (1998), Gotelli (2001), Fischer and Lindenmayer (2002), and Vázquez and Aizen (2003) among many others.

The nested pattern in plant-animal mutualisms is opposed to the compartmented view of food webs. This, of course, does not mean that compartments cannot be found in mutualistic networks (see Dicks et al. 2002), but overall nestedness seems a more robust pattern. To assess this, however, one needs a comprehensive approach looking simultaneously for different patterns (see last section). On the other hand, nestedness is compatible with the latest results on food webs reviewed in the previous section which point out their very cohesive structure.

When looking at nestedness, one can also find cases in which a community is significantly less nested than a random one. This occurred in four cases out of 66 studied by Bascompte et al. (2003), all four corresponding to parasitoid food webs: seed feeder miner-parasitoids in Silwood Park (Memmott et al.

2000), and grass-herbivores, insect-parasitoids, and parasitoid-hyperparasitoids in Grass Stems, U.K. (Martinez et al. 1999). Parasitoids tend to be very specialized and organized in well-defined compartments (but see Memmott et al. 1994). Thus a compartmentalized plant-animal matrix would be detected as well by the nestedness analysis. However, we do not intend to say that this is the best way to look at compartments (see last section).

4   IMPLICATIONS OF NESTEDNESS

4.1   IMPLICATIONS FOR COEVOLUTION

The nested structure found in mutualistic networks has clear implications for our understanding of the coevolutionary process. As noted above, nestedness analysis detects two relevant properties of plant-animal networks. First, the nested matrix has a dense core of species. This core is formed by the interactions among generalist plants and animals. It may be formed by a small number of species, but it contains a large fraction of the total number of interactions. This supports John Thompson's view of coevolutionary vortices, that is, a subset of taxa within a given, highly diversified community whose interactions might "drive" the evolution of the whole system. In any community, the identity of the species forming the core is going to determine the selective forces exerted on other, more specialist species which will become attached to such a core. Interestingly enough, this core may change geographically, and this represents a clear link between the two major theories bringing tractability to multispecific coevolutionary studies: network theory and the geographic mosaic theory (Thompson 1994, 2005).

The geographic mosaic of coevolution introduces a landscape approach to coevolution. Populations are distributed in discrete patches, and the sign of an interaction may depend on the presence in that patch of other species. The best known example is *Greya politella* a moth that oviposits inside the ovary of *Litophragma parviflorum*, a widely distributed plant in western North America. Developing *Greya* larvae eat on average from 15–27% of the seeds of an attacked flower. This is an example of antagonism, since it reduces the plant's fitness. However, it is a small reduction, and interestingly enough, since *Greya* oviposits in several flowers, also acts as a mutualism by pollinating *Litophragma* (Thompson 1994). The net sign of the interaction depends on whether there are other co-pollinators in the area. Across the distribution of *Litophragma* there are areas in which the interaction is of a mutualistic type, others where it is antagonistic, and still others in which both effects balance out. The important point here is that the landscape approach brings new ways of understanding such a plant-animal interaction. For example, in some areas there may be mismatches in the adaptation between both species which can only be accounted for if one looks at a broad geographic scale: other patches act as coevolutionary hotspots maintaining such interactions (Thompson 1994).

The geographic mosaic of coevolution is also a theory for plant-animal interactions which can be described neither as pair-wise nor diffuse. While the geographic mosaic puts the emphasis on the landscape component, the network approach emphasizes the highly structured organization of interactions within communities.

It would be interesting to form a bridge between both approaches. This is quite natural if one keeps in mind that nestedness was first introduced into the realm of island biogeography (Atmar and Patterson 1993). This is equivalent to looking at the component of local structure explained by landscape-level properties. For example: are the more widespread species also the more generalist and, therefore, the ones occupying the core of the mutualistic matrix? Or, on the contrary, if there is no relationship between spatial and network structure, is the core of species highly variable among local communities?

A second implication for coevolution is that network structure complements one of the long-standing debates in mutualistic studies: the specialization-generalization debate (e.g., Waser et al. 1996). There have been many discussions about whether specialist interactions are the rule or the exception, or in other words, about the widespread presence of generalism in mutualistic communities. In a highly influential paper, Waser et al. (1996) showed that the bulk of interactions are generalists, and explored the ecological implications of this finding. As Vázquez and Aizen (2003) have correctly stated, even when talking about specialization-generalization, we have to use a null model as a benchmark, because the key question is not whether we observe specialization, but whether the level of specialization we observe is higher than the one we would observe just by chance. Vázquez and Aizen concluded that the number of very specialist and generalist species is significantly higher than the value expected under a simple null model based on a similar probability of interaction across species. However, when controlling for differences across species in the frequency of interactions, the differences disappeared, that is, there is neither more generalization nor more specialization than we would expect under these conditions. An important conclusion of Vázquez and Aizen's (2003) paper is that they identify a close relationship between the degree of generalization of a species and their frequency of interactions, which may bring insight into the mechanisms leading to the observed structure.

The network approach brings a new dimension to the generalization-specialization debate. The focus now is as follows: given a certain distribution of specialists and generalists, how are they related? As noted, nestedness implies that generalist plants and animals interact among themselves more often than expected by chance, and that specialists tend to interact with generalists. In addition, the network approach shows clearly that certain properties of highly diversified mutualisms, such as asymmetry, cannot be understood without using a community-level approach. Understanding specialization-generalization at the species level is insufficient.

4.2   IMPLICATIONS FOR COMMUNITY DYNAMICS

It is well known that structure greatly affects dynamics. This spurred the search for compartments in food webs. There have been two main approaches to the study of stability in food web dynamics: demographic and topologic. Demographic stability is based on the study of local stability in the vicinity of a steady state. It describes whether small fluctuations around this steady state will amplify or die out through time (May 1973). On the other hand, topological stability describes species as nodes of a network, and looks at the effects on network connectivity of removing some nodes. This follows the influential work by Albert et al. (2000) on the Internet, although species deletion stability had already been explored in ecology by Pimm (1979b) using simulations, and by Paine (1980) and others using an experimental approach.

Recently, the complex network approach has produced a series of papers looking at how food webs respond to the loss of species (both randomly chosen or chosen according to their number of links; Albert et al. 2000; Solé and Montoya 2001; Dunne et al. 2002; Jordano et al. 2005). From these studies, it is well known that the pattern of connectivity distribution affects robustness. In this paper we have looked at a deeper level of structure beyond the degree distribution. One can argue that a cohesive pattern such as nestedness is likely to make these networks more robust to the loss, at random, of species or connections. Other things being equal, a cohesive network has alternative states, it is more redundant, and so it will not get fragmented as easily. The core of a nested matrix can be understood from the point of view of percolation theory. In a nested matrix, the density of interactions in the core is largely beyond the percolation threshold, which means that one can find a path from one side to another of the core following observed interactions.

Stability, however, has several sides, and a cohesive pattern, even when it may reduce the risk of network fragmentation, is more susceptible to the spread of a contaminant through the network or to the invasion of a new species into the community (Melián and Bascompte 2002). However, more studies are necessary, particularly studies relating dynamic and topologic approaches to stability.

Also, the nested structure of plant-animal mutualisms implies that rare, specialist species tend to interact with generalists. Other things being equal, generalists tend to be more abundant and less fluctuating since they rely on multiple resources (e.g., Turchin and Hanski 1997). Thus, the nested structure provides pathways for the persistence of rare species.

4.3   IMPLICATIONS FOR CONSERVATION BIOLOGY

The network approach to mutualistic interactions reviewed here has clear implications for management. An expanded concept of biodiversity must address the interactions among species in addition to the species themselves. Interactions are the "glue of biodiversity" (Thomson 2005). One example of how the network approach can be useful in the context of conservation biology is ecosystem restoration, a subject covered in another chapter by Jane Memmott. Conservation biology has often been too narrowly focused on the target species, but thinking at larger scales may be very important. Imagine that we want to preserve an endangered insect species. We have to make sure that the plants it relies on are present in the area. However, because of the asymmetry in the interactions arising from the nested structure, these plants will depend largely on other insects, which are not the target taxa for conservation. Without ensuring the visits of these latter species, the plants may decrease in abundance and push the target insect toward extinction. Knowing that mutualistic networks are nested may help in planning these strategies.

An increasing problem in tropical ecosystems is defaunation (Dirzo and Miranda 1990). Hunting preferentially affects large mammals and birds that are key species in relation to seed dispersal. These species are highly mobile, and so account for the episodes of long-distance dispersal. These episodes, even when not very common, have a huge importance for gene flow (Clarke et al. 1990). The problem is larger than expected, because woody plants may be very long-lived, and are still there for us to see. However, they may be ecological ghosts, because without megafauna, their seeds are wasted beneath the mother tree (Cordeiro and Howe 2003). To assess the community-wide importance of the loss of megafauna, we have to relate the nested structure of mutualistic networks with ecological correlates. For example, are large-bodied frugivores randomly scattered through the matrix of interactions, or do they tend to be the generalists forming the core? If the latter is true, the nested structure implies that losing these few species may lead to a collapse of the network.

Another important issue is related to the propagation of infectious diseases among insects via host plants which are pollinated. This is similar to the role of network topology in sexually transmitted diseases in humans (Liljeros et al. 2001; Pastor-Satorras and Vespigniani 2001). If a disease or a parasite can be transmitted across several insect species, the structure of the network of interactions is largely going to affect the propagation of the disease. For example, theoretical work on computer viruses has shown that the classical eradication thresholds observed in epidemiological models, assuming a random network of interactions, disappear when the topology of the network of interactions follows a power-law distribution (i.e., it is highly skewed so that a small fraction of individuals has a huge number of interactions; Pastor-Satorras and Vespigniani 2001). This line of research should be extended to look at deeper levels of structure such as the ones discussed in here. For example, for a specific connectivity distribution, which network structure reduces the chance of disease propagation? If the network is compartmentalized, the effects will be contained within a compartment. If, on the other hand, it is highly nested, disease will propagate easily throughout the network.

Network structure is also related to the likelihood of biological invasions. The invasion of exotic species is one of the leading factors in the decline of biodiversity following habitat destruction. Increases in commerce and travel have lead to increases of exotic species, with fatal consequences for native plants and animals (Liebhold et al. 1995; Vitousek et al. 1996; Simberloff 2001). In plant communities, the likelihood of an invasion declines as species richness increases (Kinzig et al. 2002). This clearly illustrates how important the characteristics at the level of the target community are. What remains to be studied is whether the way in which a specific number of species is connected also affects the likelihood of an invasion. Memmott and Waser (2002) have looked at the integration of alien plants and insects into a pollination network. They concluded that flowers of alien plants were visited by fewer animal species than flowers of native plants. Also, visitors to alien plants tend to have exceptionally broad diets. This is in agreement with the predictions of a nested community. If alien plants are, on average, more specialist, the prediction is that they will tend to interact with generalists as noted above. This tendency, other things being equal, provides aliens with a more abundant and reliable resource. The question now is whether native communities are more nested than communities containing a large proportion of exotics. If so, dynamic differences between these two communities can be directly inferred. Similarly, when we look at the species level, it would be interesting to compare native and exotic species to see to what degree they contribute to decreasing the level of nestedness of the community. If, for example, exotic species on average tend to decrease the level of nestedness, then their arrival in a community will modify community-wide coevolutionary and stability properties.

Understanding the nested structure of mutualistic networks can explain why the pollination of generalist and specialist plant species is similarly affected by habitat fragmentation (Ashworth et al., 2004). In principle, one would assume that specialist plants are more vulnerable to this destructive process, since any significant decrease in the abundance of its only pollinator would lead to failure in reproduction. However, Aizen et al. (2002) found that similar fractions of generalist and specialist plant species were affected by habitat fragmentation. This can be explained by the asymmetry in the interactions (specialists tend to interact with generalists) arising from a nested structure. Specifically, generalist plants interact with many specialist pollinators that run a high risk of becoming extinct. This would leave a plant with a few generalist pollinators, a situation similar to the one experienced by specialist plants which retain their few generalist pollinators until there are moderate to high levels of habitat destruction (Ashworth et al. 2002). Therefore, when assessing the effects of habitat fragmentation, we should not consider the levels of generalization of plant species in isolation, but in combination with the levels of generalization of its pollinators. In other words, it is a combined property of plants and animals which is relevant to assess the consequences of habitat loss and similar perturbations. Again, from the above one can infer that the nested assembly of plant-animal mutualistic interactions provides mechanisms for the persistence of specialist species.

5 CHALLENGES

The next stage in the search for structure in plant-animal assemblages is to compare both antagonistic and mutualistic networks. Preliminary studies of plant-herbivore interactions seem to suggest that in this case the structure can be better described by compartments (Prado and Lewinsohn 2004; Lewinsohn et al. submitted). However, in order to proceed toward a rigorous comparison across different systems, we need more data on plant-animal antagonistic networks. If a difference exists as these preliminary results seem to suggest, one could explore the different mechanisms acting on these two types of networks to create different structures.

One particular problem is that studies look only for a particular type of structure, either compartments or nestedness. Finding evidence for a particular type of structure does not preclude the existence of other structures (Leibold and Mikkelson 2002; Lewinsohn et al. submitted). Thus, in order to make a serious inference about structure, we need a more comprehensive approach which provides methods of looking simultaneously for different structures and detecting the one that best describes the data. One recently developed approach is based on the study of eigenvalues of properly rearranged matrices (Lewinsohn et al. submitted).

Another serious problem concerns the comparison between plant-animal interaction networks and food webs. As noted above, these are two completely different types of networks from an analytical point of view (Jordano et al. 2005). In order to make them comparable, we have to transform them into a common format. This can be the niche overlap graph (*sensu* Cohen 1978) in which two species in a food web are linked if they share a common prey. Similarly, we would have two niche overlap matrices in plant-animal interactions: one relating two animals if they pollinate or disperse at least a common plant, and the equivalent for the plants (i.e., two plants will be related if they share at least a common pollinator). The task is to see whether measures of structure in original matrices and transformed matrices are related, and to compare measures of structure in transformed matrices for food webs, plant-animal, and other types of networks (Olesen et al. in preparation). Only then we will be able to infer what is common and what is different in the architecture of networks from different types of interactions.

ACKNOWLEDGMENTS

This chapter has been supported by the Spanish Ministry of Education and Science (Grants REN2003-04774 to Jordi Bascompte and REN2003-00273 to Pedro Jordano) and the European Heads of Research Councils and the European Science Foundation through an EURYI award (to Jordi Bascompte). We thank

C. J. Melián, M. A. Fortuna, J. M. Olesen, John Thompson, and T. Lewinsohn for interesting discussions.

REFERENCES

Aizen, M. A., L. Ashworth, and L. Galetto. 2002. "Reproductive Success in Fragmented Habitats: Do Compatibility Systems and Pollination Specialization Matter?" *J. Veget. Sci.* 13:885–892.

Albert, R., H. Jeong, and A.-L. Barabási. 2000. "Error and Attack Tolerance of Complex Networks." *Nature.* 406:378–382.

Albert, R., and A. L. Barabási. 2002. "Statistical Mechanics of Complex Networks." *Rev. Modern Phys.* 74:47–97.

Amaral, L. A., A. Scala, M. Barthelemy, and H. E. Stanley. 2000. "Classes of Small-World Networks." *Proc. Natl. Acad. Sci. USA* 97:11149–11152.

Ashworth, L., R. Aguilar, L. Galetto, and M. A. Aizen. 2004. "Why Do Pollination Generalist and Specialist Plant Species Show Similar Reproductive Susceptibility to Habitat Fragmentation?" *J. Animal Ecol.* 92:717–719.

Atmar, W., and B. D. Patterson. 1993. "The Measure of Order and Disorder in the Distribution of Species in Fragmented Habitat." *Oecologa.* 96:373–382.

Bascompte, J., P. Jordano, C. J. Melián and J. M. Olesen. 2003. "The Nested Assembly of Plant-Animal Mutualistic Networks." *Proc. Natl. Acad. Sci. USA* 100:9383–9387.

Clarke, J. S., M. Silman, R. Kern, E. Macklin, and J. Hilleris-Lambers. 1999. "Seed Dispersal Near and Far: Patterns Across Temperate and Tropical Forests." *Ecology* 80:1475–1494.

Cohen, J. E. 1978. *Food Webs and Niche Space.* Princeton, NJ: Princeton University Press.

Cohen, J. E., F. Briand, and C. M. Newman. 1990. *Community Food Webs: Data and Theory.* Berlin, Germany: Springer-Verlag.

Cook, R. R., and J. F. Quinn. 1998. "An Evaluation of Randomization Models for Nested Species Subsets Analysis." *Oecologia.* 113:584–592.

Cordeiro, N. J., and H. F. Howe. 2003. "Forest Fragmentation Severs Mutualism between Seed Dispersers and an Endemic African Tree." *Proc. Natl. Acad. Sci. USA* 100:14052–14056.

Dicks, L. V., S. A. Corbet, and R. F. Pywell. 2002. "Compartmentalization in Plant-Insect Flower Visitor Webs." *J. Animal Ecol.* 71:32–43.

Dirzo, R., and A. Miranda. 1990. "Contemporary Neotropical Defaunation and Forest Structure, Function and Diversity." *Conserv. Biol.* 4:444–447.

Dunne, J. A., R. J. Williams, and N. D. Martinez. 2002. "Network Structure and Biodiversity Loss in Food Webs." *Ecol. Lett.* 5:558–567.

Ehrlich, P. R., and P. H. Raven. 1964. "Butterflies and Plants: A Study in Co-evolution." *Evolution* 54:1480–1492.

Fisher, J., and B. Lindenmayer. 2002. "Treating the Nestedness Temperature Calculator as a Black Box can Lead to False Conclusions." *Oikos.* 99:193–199.

Gardner, R. M., and W. R. Ashby. 1970. "Connectance of Large Dynamic (Cybernetic) Systems: Critical Value of Stability." *Nature* 228:784.

Gotelli, N. J. 2001. "Research Frontiers in Null Model Analysis." *Global Ecol. Biogeog.* 10:337–343.

Herrera, C. M. 1988. "Variation in Mutualisms: The Spatio-Temporal Mosaic of a Pollinator Assemblage." *Biol. J. Linn. Soc.* 35:92–125.

Janzen, D. H. 1980. "When is it Coevolution?" *Evolution* 34:611–612.

Jordano, P. 1987. "Patterns of Mutualistic Interactions in Pollination and Seed Dispersal: Connectance, Dependence Asymmetries, and Coevolution." *Amer. Natur.* 129:657–677.

Jordano, P. 2000. " Fruits and Frugivory. " In *Seeds. the Ecology of Regeneration in Plant Communities*, ed. M. Fenner, 125–166. Wallingford, UK: CABI Publishing.

Jordano, P., J. Bascompte, and J. M. Olesen. 2003. "Invariant Properties in Coevolutionay Networks of Plant-Animal Interactions." *Ecol. Lett.* 6:69–81.

Jordano, P., J. Bascompte, and J. M. Olesen. 2005. "The Ecological Consequences of Complex Topology and Nested Structure in Pollination Webs." In *Specialization and Generalization in Plant-Pollinator Interactions*, ed. N. Waser and J. Ollerton. Chicago, IL: Chicago University Press.

Kinzig, A. P., S. Pacala, and D. Tilman, eds. 2002. *The Functional Consequences of Biodiversity: Empirical Progress and Theoretical Extensions.* Princeton, NJ: Princeton University Press.

Koh, L. P., R. R. Dunn, N. S. Sodhi, R. K. Colwell, H. C. Proctor, and V. S. Smith. 2004. "Species Coextinctions and the Biodiversity Crisis." *Science* 305:1632–1634.

Krause, A. E., K. A. Frank, D. M. Mason, R. E. Ulanowicz, and W. W. Taylor. 2003. "Compartments Revealed in Food-Web Structure." *Nature* 426:282–285.

Leibold, M. A., and G. M. Mikkelson. 2002. "Coherence, Species Turnover, and Boundary Clumping: Elements of Meta-Community Structure." *Oikos* 97:237–250.

Lewinsohn, T. M., P. Jordano, P. I. Prado, J. M. Olesen, and J. Bascompte. 2005. "Structure in Plant-Animal Interaction Assemblages." Unpublished paper (in preparation).

Liebhold, A., W. L. D. MacDonald, D. Bergdahl, and V. Mastro. 1995. "Invasion by Exotic Forest Pests: A Threat to Forest Ecosystems." *Forest Sci. Monograph* 30:1–49.

Liljeros, F., C. R. Edling, L. A. N. Amaral, H. E. Stanley, and Y. Aberg. 2001. Turchin, P., and I. Hanski. "The Web of Human Sexual Contacts." *Nature* 411:907–908.

Martinez, N. D., B. A. Hawkins, H. A. Dawah, and B. P. Feifarek. 1999. "Effects of Sample Effort on Characterization of Food-Web Structure." *Ecology* 80:1044–1055.

Maslov, S., and K. Sneppen. 2002. "Specificity and Stability in Topology of Protein Networks." *Science* 296:910–913.

May, R. M. 1972. "Will a Large Complex System be Stable?" *Nature* 238:413–414.

May, R. M. 1982. *Stability and Complexity in Model Ecosystems*. Princeton, NJ: Princeton University Press.

Melián, C. M., and J. Bascompte. 2002. "Complex Networks: Two Ways to be Robust?" *Ecol. Lett.* 5:705–708.

Melián, C. M., and J. Bascompte. 2004. "Food Web Cohesion." *Ecology* 85:352–358.

Memmott, J. 1999. "The Structure of a Plant-Pollinator Network." *Ecol. Lett.* 2:276–280.

Memmott, J., and N. M. Waser. 2002. "Integration of Alien Plants into a Native Flower-Pollination Visitation Web." *Proc. Roy. Soc. Lond. B* 269:2395–2399.

Memmott, J., H. C. J. Godfray, and I. D. Gauld. 1994. "The Structure of a Tropical Host Parasitoid Community." *J. Animal Ecol.* 63:521–540.

Memmott, J., N. D. Martinez, and J. E. Cohen. 2000. "Predators, Parasitoids and Pathogens: Species Richness, Trophic Generality and Body Sizes in a Natural Food Web." *J. Animal Ecol.* 69:1–15.

Paine, R. T. 1963. "Trophic Relationships of 8 Sympatric Predatory Gastropods." *Ecology* 44:63–73.

Paine, R. T. 1966. "Food Web Complexity and Species Diversity." *Amer. Natur.* 100:65–75.

Paine, R. T. 1980. "Food Webs: Linkage, Interaction Strength and Community Infrastructure." *J. Animal Ecol.* 49:667–685.

Pastor-Satorras, R., and A. Vespignani. 2001. "Epidemic Spreading in Scale-Free Networks. " *Phys. Rev. Lett.* 86:3200–3203.

Pastor-Satorras, R., A. Vazquez, and A. Vespignani. 2002. "Dynamical and Correlation Properties of the Internet." *Phys. Rev. Lett.* 87:258701.

Pimm, S. L. 1979a. "The Structure of Food Webs." *Theor. Pop. Biol.* 16:144–158.

Pimm, S. L. 1979b. "Complexity and Stability: Another Look at MacArthur's Original Hypothesis." *Oikos* 33:351–357.

Pimm, S. L., and J. H. Lawton. 1980. "Are Food Webs Divided into Compartments?" *J. Animal Ecol.* 49:979–898.

Pimm, S. L. 1982. *Food Webs.* London, UK: Chapman and Hall.

Prado, P. I., and T. M. Lewinsohn. 2004. "Compartments in Insect-Plant Associations and Their Consequences for Community Structure." *J. Animal Ecol.* 73:1168–1178.

Raffaelli, D., and S. J. Hall. 1992. "Compartments and Predation in an Estuarine Food Web." *J. Animal Ecol.* 61:551–560.

Simberloff, D. S. 2001. "Eradication of Island Invasives: Practical Actions and Results Achieved." *Trends Ecol. & Evol.* 16:273–274.

Solé, R. V., and J. M. Montoya. 2001. "Complexity and Fragility in Ecological Networks." *Proc. Roy. Soc. Lond. B.* 268:2039–2045.

Thompson, J. N. 1994. *The Coevolutionary Process.* Chicago, IL: University of Chicago Press.

Thompson, J. N. 2002. " Plant-Animal Interactions: Future Directions." In *Plant-Animal Interactions. An Evolutionary Approach*, ed. C. M. Herrera and O. Pellmyr, 236–247. Oxford, UK: Blackwell.

Thompson, J. N. 2005. *The Geographic Mosaic of Coevolution.* Chicago, IL: University of Chicago Press.

Turchin, P., and I. Hanski. 1997. "An Empirically-Based Model for the Latitudinal Gradient in Vole Population Dynamics." *Amer. Natur.* 149:842–874.

Vázquez, D. P., and M. A. Aizen. 2003. "Null Model Analysis of Specialization in Plant-Pollinator Interactions." *Ecology* 84:2493–2501.

Vázquez, D. P., and M. A. Aizen. 2004. "Asymmetric Specialization: A Pervasive Feature of Plant-Pollinator Interactions." *Ecology* 85:1251–1257.

Vitousek, P. M., C. M. D'Antonio, L. L. Lope, and R. Westbrooks. 1996. "Biological Invasions as Global Environmental Change." *Amer. Sci.* 84:468–478.

Waser, N., M. V. Price, N. M. Williams, and J. Ollerton. 1996. "Generalization in Pollination Systems and Why it Matters." *Ecology* 77:1043–1060.

Integrating Ecological Structure and Dynamics

"In the Neolithic days of animal ecology, that is to say about twenty-five years ago, it seemed reasonable to suppose that every natural food-chain contained within itself the explanation of the control of populations. . . .Each higher consumer layer kept down the numbers of the one below, and each one below limited the numbers of the one above through food supply. That this argument does not go quite in circle was pointed out independently about this time by two mathematicians, Lotka and Volterra, whose equations and suppositions made a deep impression on their contemporaries. Being mathematicians, they did not attempt to contemplate a whole food-chain with all the complications of five stages."

> — Charles Elton,
> *Ecology of Invasions by Animals
> and Plants*, 1958

Diversity, Complexity, and Persistence in Large Model Ecosystems

Neo D. Martinez
Richard J. Williams
Jennifer A. Dunne

Research on how vast numbers of interacting species manage to coexist in nature reveals a deep disparity between the ubiquity of complex ecosystems and their theoretical improbability. Here, we show how integrating models of food-web structure and nonlinear bioenergetic dynamics bridges this disparity and helps elucidate the relationship between ecological complexity and stability. Network structure constraints, including trophic hierarchy, niche contiguity, and looping formalized by the "niche model," greatly increase persistence in complex model ecosystems. Behavioral nonlinearities, particularly competition among consumers and reduced consumption of rare resources, formalized by predator interference and new "Type II.2" functional responses, further encourage persistence of species in complex food webs. Trophic dynamics are also shown to feed back to network structure, resulting in more accurate topologies than those achieved by simple structural models alone. Thus, integrating structure and dynamics of ecological networks yields remarkably comprehensive and ecologically plausible models that highlight the

Ecological Networks: Linking Structure to Dynamics in Food Webs,
edited by Mercedes Pascual and Jennifer A. Dunne, Oxford University Press.
163

importance of network structure, short food chains, and behavioral ecology for ecosystem persistence and stability, and also alters our understanding of the role of omnivory in food webs. This modeling approach provides a potentially powerful framework for exploring the impacts of perturbations on ecosystems, and can be altered to include non-trophic processes, spatial effects, and evolutionary dynamics.

> *Our ultimate goal is to use these consumer-resource models as building blocks for the construction of plausible models of more complicated systems involving many interacting species. In that setting, one needs to be parsimonious with respect to detail, but we hope to do so without too great a sacrifice in realism.*
> —Yodzis and Innes, 1992

1   INTRODUCTION

One of the most important and least settled questions in ecology concerns the roles of diversity and complexity in the functioning and stability of ecosystems (McCann 2000). Scientists still have difficulty explaining why diversity, in terms of vast numbers of species, and complexity, in terms of species' myriad interactions, are ubiquitous in ecological systems (McCann 2000; Kondoh 2003; Brose et al. 2003). Early theoretical considerations suggested that the presence of more feeding links among more species generally reduces the risk of species' dependence on a few resources (MacArthur 1955). By the late 1950s the notion that "complexity begets stability" was considered by many to be a basic ecological theorem (Hutchinson 1959). However, the apparent inevitability of this relationship was severely challenged by simple mathematical models of food-web dynamics which showed that diversity and complexity destabilize idealized ecosystems, either through increasing the chance of positive feedback loops (May 1973) or through additional omnivorous interactions increasing the time needed for perturbed species to return to equilibrium (Pimm and Lawton 1978). Much of the work since those early modeling studies has focused on trying to parse conditions under which ecologists should expect to see (or not see) a positive relationship between diversity/complexity and stability (for review, see Dunne et al. 2005).

Most early work emphasized equilibrium-based modeling (e.g., May 1973) and comparative empiricism (as reviewed in Pimm et al. 1991) with a focus on whole-system analysis (i.e., many species at multiple trophic levels). Later research placed more emphasis on nonlinear modeling and experimental empiricism, with both approaches focusing on parts of ecosystems—small food-web modules in the case of modeling, and single trophic levels in biodiversity/ecosystem function experiments. In general, the nonlinear modeling approach has suggested that increases in complexity, such as the addition of weak or omnivorous interactions (McCann and Hastings 1997; McCann et al. 1998; Fussman and Heber 2002) stabilize ecosystems. Similarly, experimental work suggests that increases in diversity, in terms of numbers of species and functional groups (Naeem et al. 1994; Tilman et al. 2001), also stabilizes ecosystems. However, one of the few experimental tests of complexity/stability in speciose, multi-trophic level communities showed that complexity, defined in terms of species richness and number of interactions, destabilized microcosm assemblages (Fox and McGrady-Steed 2002). This and other studies suggest that there is still an important disparity to be addressed between the improbability of diverse, complex, stable ecosystems in theory and their pervasiveness in nature. In particular, it is unclear whether the stabilizing effects of omnivory (McCann and Hastings 1997), weak links (McCann et al. 1998; Berlow 1999), and diversity (Naeem et al. 1994; Tilman et al. 2001) found in small modules or single trophic levels also apply to large networks with many species at multiple trophic levels.

Here, we address these issues by examining species persistence in nonlinear dynamical models of large complex ecological networks. Our model (Williams and Martinez 2004b) builds on research that replaces unrealistic modeling assumptions prevalent in early studies (e.g., food webs are random networks, populations are at equilibrium—May 1973), with more empirically supported and mechanistically based assumptions (Yodzis and Innes 1992; McCann et al. 1998). This recent approach to modeling explicitly incorporates the nonlinearities, non-equilibrium behavior, and non-random topologies that many ecologists now believe characterize natural ecosystems. However, few analyses have examined the nonlinear dynamics of model systems with more than ten species (but see Drossel et al. 2001; Kondoh 2003).

We present results from an integrated model of ecosystem structure and dynamics, which is used to examine food-web networks with up to fifty species. The structural "niche model" component successfully predicts the network structure of the largest and most complex food webs in the primary literature (Williams and Martinez 2000; Camacho et al. 2002; Dunne et al. 2004). The dynamical bioenergetic model component (based on Yodzis and Innes 1992) successfully simulates persistent and non-persistent stable, cyclic, and chaotic dynamics (Williams and Martinez 2004b) that are often found in nature (Kendall et al. 1998). We explore the interplay of structure and nonlinear dynamics by systematically varying diversity, complexity, and function to "elucidate the devious strategies which make for stability in enduring natural systems" as suggested by May (1973). Diversity refers to the number of species in a food web, and complexity is quantified as connectance, the proportion of potential links in a food web which are actually realized (links/species^2). Function refers to processes associated with species' interactions including rates of consumption and preferences for different prey. The relatively high dimensionality of the model makes it impossible to fully explore the parameter space here. However, by focusing on key aspects of the model that speak most closely to ongoing theory and experimentation, we arrive at several intriguing, if provisional, insights. In general, the model suggests that recently discovered network structure properties, as well as longer-standing functional properties of ecological interactions, appear to promote stability and persistence in large complex ecosystems.

2 METHODS

Our bioenergetic network model constructs food webs in two steps. The first step specifies the structure of a food-web network using one of three different stochastic models, which are described briefly below and in more detail in chapter 2 (Dunne this volume; Williams and Martinez 2000; Dunne et al. 2004). The second step uses a nonlinear bioenergetic model to compute the dynamics of the network (Williams and Martinez 2004b). This integrated approach allows us to explore the impact of structure on dynamics as well as the impact of dynamics on structure.

2.1 STRUCTURAL MODELS AND FOOD-WEB TOPOLOGY

All three structural models require the number of species in the system (S) and the number of trophic links (L) in terms of directed connectance ($C = L/S^2$) as input parameters, but vary in the degree to which they constrain network organization. In the random model (Cohen et al. 1990; Solow and Beet 1998), any link among S species occurs with the same probability equal to C of the empirical web. This creates webs as free as possible from biological structuring while maintaining observed S and C. The modified (Williams and Martinez 2000) cascade model (Cohen et al. 1990) creates a hierarchical structure by assigning each species a random value drawn uniformly from the interval [0,1] and giving each species a probability $p = 2CS/(S-1)$ of consuming only species with values less than its own. The niche model (Williams and Martinez 2000) similarly assigns each species a randomly drawn "niche value." The species consume all species with niche values within one contiguous range. The size of the range is chosen from a beta distribution with a mean $= C$. The range is located by uniformly and randomly assigning its center to be less than the consumer's niche value. Because the center can be close to the consumer's niche value, the strict hierarchy of the cascade model is relaxed, and cannibalism and looping can occur. Niche model networks that contain energetically unsustainable closed loops such as pairs of mutual predators with no other prey items (which sometimes occurs in low diversity, low-connectance webs) are eliminated from analysis.

When describing food webs, we employ several conventions. Top species have resources but no consumers. Intermediate species have resources and consumers. Basal species have consumers but no resources. Omnivores feed from more than one trophic level and herbivores eat only basal species. To remove the confounding variability of the number of basal species, omnivory and herbivory are the fraction of consumers that are omnivores and herbivores respectively. Similarly, to better measure the trophic height of the consumers independent of the fraction of basal species, mean trophic level is the mean of all consumer species' trophic levels. Among a variety of definitions of trophic level, we use a modification of previous trophic-level definitions (Levine 1980; Adams et al. 1983) that weights each consumer's prey equally (Williams and Martinez 2004a). A species' connectivity is its total number of links (both incoming and outgoing) divided by the mean connectivity $(2L/S)$ of the network.

2.2   BIOENERGETIC MODEL OF NONLINEAR FOOD-WEB DYNAMICS

The dynamic model closely follows previous work (Yodzis and Innes 1992; McCann and Yodzis 1995; McCann and Hastings 1997; McCann et al. 1998) but is generalized to n species and arbitrary functional responses. Extending earlier notation (Yodzis and Innes 1992) to n-species systems, variation of B_i, the biomass of species i, over time t, is given by

$$B_i'(t) = G_i(B) - x_i B_i(t) + \sum_{j=1}^{n} \left(\frac{x_i y_{ij} \alpha_{ij} F_{ij}(B) B_i(t) - x_j y_{ji} \alpha_{ji} F_{ji}(B) B_j(t)}{e_{ji}} \right).$$
(1)

The first term $G_i(B) = r_i B_i(t)(1 - B_i(t)/K_i)$ is the gross primary production rate of species i where r_i is the intrinsic growth rate that is non-zero only for basal species, and K_i is the carrying capacity. The second term is metabolic loss where x_i is the mass-specific metabolic rate. The third and fourth terms are gains from resources and losses to consumers respectively. The rate y_{ij} is the maximum at which species i assimilates species j per unit metabolic rate of species i. The term α_{ij} is the relative preference of species i for species j compared to the other prey of species i and is normalized so that the sum of $\alpha_{ij}(0 \leq \alpha_{ij} \leq 1)$ across all j is 1 for consumer species and 0 for basal species. Non-zero α_{ij}'s are assigned according to the topology specified by the structural models. A non-dimensional functional response, $F_{ij}(B)$, that may depend on resource and consumer species' biomasses (Box 1), gives the fraction of the maximum ingestion rate of predator species i consuming prey species j. The term e_{ij} is the conversion efficiency with which the biomass of species j lost due to consumption by species i is converted into the biomass of species i. Dividing the last term by e_{ij} converts the biomass assimilated by consumer j into biomass lost by resource i. Parameter values in these equations have been estimated from empirical measurements (Yodzis and Innes 1992) and there are wide ranges of biologically plausible values.

The form of the functional response $F_{ij}(B)$ can have a large impact on predator-prey dynamics. While a variety of functional responses have been proposed in the literature, our model uses two basic families of functional responses (F_H and F_{BD}, Box 1; Martinez and Williams 2004b) that have both mechanistic and empirical justifications (Skalski and Gilliam 2001). The F_H functional response (Box 1, eq. (2)) is based on a parametrized form (Real 1977, 1978; Yodzis and Innes 1992) of Holling's type II and III responses (Holling 1959a,b). F_H generalizes earlier multispecies type II responses (McCann et al. 1998; Fussman and

Heber 2002). Type II responses have been used in many studies of the dynamics of small food-web modules (Yodzis and Innes 1992; McCann and Yodzis 1995; McCann and Hastings 1997; McCann et al. 1998; Post et al. 2000; Fussman and Heber 2002). The F_{BD} response (Box 1, eq. (3)) models predator interference (Skalski and Gilliam 2001) by extending earlier models (Beddington 1975; DeAngelis et al. 1975) to consumers of multiple species. Predator interference and type III responses are known to stabilize small food-web modules (DeAngelis et al. 1975; Murdoch and Oaten 1975; Hassell 1978; Yodzis and Innes 1992) but have not previously been used to study the dynamics of relatively species-rich systems. In addition, small deviations from the type II response such as our "type II.2 response" ($q = 0.2$), intermediate between type II and III responses, are a recent innovation, but have only been applied to food-web models with 10 or fewer species (Williams and Martinez 2004b).

We simplify the dynamical model through our choice of parameter values. First, we set a single value for the parameters $K_i = 1, r_i = 1, x_i = 0.5, y_{ij} = 6, e_{ij} = 1$, and $B_{0ij} = 0.5$. Simulations that draw these parameters from normal distributions with specified means and standard deviations ($e_{ij} > 1$ not allowed) gave similar results to fixed parameter simulations (results not shown). Second, even though functional responses can differ for each link in the network (Williams and Martinez 2004b), we specify a single value of q_{ij} or c_{ij}, so each link within a network is of the same type.

Unless stated otherwise, we assume that predator species have equal preference (α_{ij}) for all their prey. If n_i is the number of prey that species i consumes, $\alpha_{ij} = 1/n_i$ for each species j in the diet of species i. We also systematically vary the α_{ij} of omnivores to examine the effects of skewing diets to higher or lower trophic level prey. The range of α_{ij} is defined by a preference skewness $k = \alpha_{i\,max}/\alpha_{i\,min}$, where $\alpha_{i\,max}$ and $\alpha_{i\,min}$ are the preferences for the prey items of species i with the maximum and minimum trophic levels TL_{max} and TL_{min}, respectively. For each prey species j of species i, we define $b_{ij} = 1 + (k-1)(TL_j - TL_{min})/(TL_{max} - TL_{min})$, where TL_j is the trophic level of prey item j. The preference of species i for prey item j is then $\alpha_{ij} = b_{ij}/\sum_l b_{il}$, where the sum is across all prey items of species i. When $k = 1$, all prey preferences of an omnivore are equal, when $k < 1$, low trophic level prey are preferred, and when $k > 1$, high trophic level prey are preferred.

Each simulation begins by building an initial random, cascade, or niche model web of a certain size (S_0) and connectance (C_0). The integrated structure/dynamics model then computes which species persist with positive biomass greater than a local extinction or "exclusion" threshold of 10^{-15} after 4000 time steps. Following any exclusions, a "persistent web" with S_P species and connectance C_P remains. The initial biomasses of species are stochastic (uniformly random between 1 and 10^{-15}), as are elements of the structural models. Therefore, we repeated this procedure a large number of times so that statistical properties of the integrated structure-dynamic model resulting from systematically varied parameters can be ascertained. In particular, we systematically varied di-

Box 1. Functional response modeling

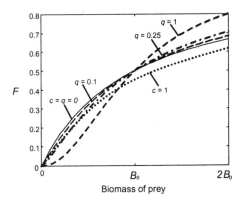

The effects of control parameters q (F_H, eq. (2)) and c (F_{BD}, eq. (3)) on fractions of maximal consumption rates (F) are shown. Where $q_{ij} = c_{ij} = 0$, the functional response is a standard type II response, and where $q_{ij} = c_{ij} = 1$, the functional response is a standard type III or predator interference response, respectively. F_H of predator i consuming prey j is

$$F_{Hij}(B) = \frac{B_j(t)^{1+q_{ij}}}{\sum_{k=1}^{n} \alpha_{ik} B_k(t)^{1+q_{ij}} + B_{0ji}^{1+q_{ij}}} \tag{2}$$

where B_{0ji} is the half saturation density of species j when consumed by species i and q_{ij} controls the form of F_H. The functional response decelerates and accelerates feeding on relatively rare and abundant resources as q increases and decreases, respectively, as shown in the figure above. The range $0 < q_{ij} \leq 1$ generalizes F_H so that it can smoothly vary from standard type II ($q_{ij} = 0$) to standard type III responses ($q_{ij} = 1$).

F_{BD} of predator i consuming prey j is

$$F_{BDij}(B) = \frac{B_j(t)}{\sum_{k=1}^{n} \alpha_{ik} B_k(t) + (1 + c_{ij} B_i(t)) B_{0ji}}. \tag{3}$$

Similar to F_H, F_{BD} has a control parameter $c_{ij} \geq 0$ that quantifies the intensity of predator interference. Empirical studies suggest $c \approx 1$ (Skalski and Gilliam 2001). Note that F_{BD} depends on the density of consumers that pushes the half saturation density (B_0) of the dotted $c = 1$ line left or right as the consumer density decreases or increases, respectively.

versity (S_0), complexity (C_0), the functional response control parameters (q_{ij} and c_{ij}), and a predator's preferences among prey (α_{ij}) to study effects of food-web structure on dynamics and persistence, as well as effects of dynamics on food-web structure. For each model iteration, absolute persistence $P_A = S_P$ and relative persistence $P_R = S_P/S_0$ are calculated. Overall persistence P is the mean value of P_R across a set of iterations. Topological properties of the persistent webs are then compared to different versions of niche model webs. Here, we focus on the distribution of trophic levels and connectivity among species by examining the fractions of top, intermediate, basal, omnivorous, and herbivorous species, mean trophic level, and the standard deviation of the connectivity of each species.

3 RESULTS

We analyzed the behavior of the dynamic network model with respect to the combined variation of several key parameters. The model's high dimensionality, resulting from the model's many parameters, prevents full examination of all the combinations of parameter values. Instead, we present a sequence of results that describes the effects of varying a few parameters and then fix these parameters and analyze effects of varying other parameters. Fixing the parameters at different values changes the results quantitatively. Therefore, we report overall behaviors that resist qualitative changes due to alternative choices.

Perhaps most importantly, varying network structure and the functional response control parameters profoundly affects persistence. Figures 1(a) and 1(b) show the effect of varying q and c on 30-species webs with an intermediate level of $C_0 = 0.15$ for food webs with initial topologies built using the random, cascade, and niche models. All other input parameters are constant across all trials of the stochastic models unless otherwise indicated. Most or all species go extinct in every trial of random webs and q and c have little if any effect on their relative persistence ($P < 0.05$). The structural constraints provided by the cascade model and especially the niche model increase P by more than an order of magnitude. In addition to this enormous effect of network structure, a large change in persistence occurs when q is increased from 0 to 0.1 (fig. 1(a)). In this range, cascade-web P increases 32% from 0.34 to 0.44 and niche-web P increases 44% from 0.43 to 0.62. Compared to cascade webs, niche webs are 27% to 50% more robust for any fixed q from 0 to 0.3 and more strongly increase in persistence for $q > 0$. Figure 1(b) shows that predator interference causes a similar change in the persistence of 30-species webs when c varies across a biologically reasonable range (Skalski and Gilliam 2001). The effect of c on persistence is similar to the effects of q but, unlike q's asymptotic effects, increasing c continually increases persistence across the whole range of values examined. Due to the similar effects of q and c, we present further results only for intermediately robust responses with $q = 0.2$ or $c = 1.0$, a choice that highlights the effects of altering other model parameters in a representative manner.

Relative persistence ($P_R = P_A/S_0$) of niche-model webs decreases linearly both with increasing initial network size (S_0) and with increasing initial connectance (C_0) (fig. 1(c) and 1(d)) as shown by linear regressions of P_R as a

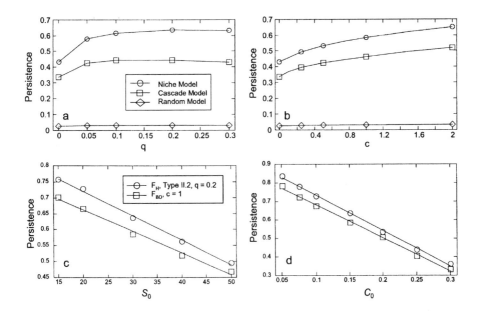

FIGURE 1   1(a) and 1(b): Mean overall persistence of model food webs vs. functional response control parameter for networks built using the niche, cascade, and random models. In (a) q controls the parametrized Holling functional response (F_H, Box 1, eq. (2)); in (b) c controls the Beddington-DeAngelis predator interference functional response (F_{BD}, Box 1, eq. (3)). All networks initially have $S_0 = 30$ and $C_0 = 0.15$. 1(c) and 1(d): Mean overall persistence of niche model food webs versus (c) initial network size S_0 for networks with $C_0 = 0.15$, and (d) initial network connectance C_0 for networks with $S_0 = 30$. Responses shown for a single value of the two types of functional responses, Type II.2, F_H, $q = 0.2$; and Type BD, F_{BD}, $c = 1$. The regression lines are (c) Type II.2: $P = 0.874 - 0.00770S_0$, $r^2 = 0.996$; BD: $P = 0.799 - 0.00682S_0$, $r^2 = 0.992$ and (d) Type II.2: $P = 0.927 - 1.923C_0$, $r^2 = 0.998$; BD: $P = 0.862 - 1.799C_0$, $r^2 = 0.997$. Values shown and used for regression analyses are averages of 500 trials.

function of the product S_0C_0, the network's initial value of L/S. For the type II.2 response ($q = 0.2$) with constant $C_0 = 0.15$, $P_R = 0.87 - 0.05S_0C_0$ ($R^2 = 0.48$, $n = 2500$); with constant $S_0 = 30$, $P_R = 0.93 - 0.06S_0C_0$ ($R^2 = 0.23$, $n = 3500$). Despite the negative effect of S_0 on P_R, absolute persistence (P_A) increases with S_0 from roughly 11 when $S_0 = 15$ to approximately 25 when $S_0 = 50$.

We compared variation in C_P with S_P among persistent webs that were initially constructed with the niche model to two other sets of model webs (fig. 2). These sets were created by starting with a set of niche webs using fixed parameters $S_0 = 30$ and $C_0 = 0.15$ and then randomly deleting species (Solé and Montoya 2001; Dunne et al. 2002a) to create networks with the same S as the persistent webs. Two deletion algorithms were used. One deletes species entirely at random and the other randomly deletes only non-basal "consumer" species (Dunne et al. 2002a). The value C of niche webs increases with the number of entirely random deletions but varies little when basal species are protected (fig. 2). Despite the strong negative effects of C_0 on P, C_P of the most robust webs ($S_P > 21$, $P_R > 0.7$) is typically greater than the C of niche webs subjected to random deletions (fig. 2). This suggests that structurally peculiar subsets of niche webs with relatively high C yield remarkably persistent networks (Dunne et al. 2002a).

Both S and C affect many topological properties of empirical and niche-model webs (Williams and Martinez 2000; Camacho et al. 2002; Dunne et al. 2002b; Williams et al. 2002). We examined how dynamic extinctions affect network topology by controlling for these effects and comparing the persistent webs with two sets of 1000 niche webs (fig. 2). One set had the initial values of $S_0 = 30$ and $C_0 = 0.15$ as inputs and non-basal species were randomly deleted until $S = S_p$. This compares persistent webs of a certain size to similarly sized niche webs subjected to randomized extinctions that leave C relatively unchanged ($C \approx C_0 \approx C_P$, fig. 2). The second set was created using the values $S = S_P$ and $C = C_P$ as inputs into the niche model, allowing comparison between persistent webs of a certain size and similarly sized niche webs not subject to extinctions.

Compared to either set of niche webs, persistent webs consistently have consumers with lower mean trophic levels, and higher fractions of basal species, especially in the largest, most persistent webs ($S_P > 25$, fig. 3(a) and 3(b)). These properties of persistent webs vary with S_P in the same direction but less strongly as the properties vary with S in niche webs. The fractions of consumer species that are herbivores or omnivores are higher in the persistent webs than in the niche webs (fig. 3(c) and 3(d)). This helps explain the lower mean trophic levels of persistent webs. The differences in herbivore and basal species richness tend to lose their statistical significance as webs get smaller, while the differences in mean trophic level also get smaller but remain significant. The fraction of omnivorous consumers was often slightly ($5 - 10\%$) though not significantly higher in the highly robust persistent webs ($S_P > 25$), whereas there was a slight deficit of omnivores in less robust persistent networks ($S_P < 15$). The standard deviations of node connectivity were similar between persistent and niche webs but random deletions increased standard deviations above those in persistent webs (fig. 3(e)). This similarity also applies to the standard deviation of the number of incoming and outgoing links taken separately, properties previously referred to as generality and vulnerability, respectively (Williams and Martinez 2000). Overall, these results indicate that more persistent webs are shorter and fatter than niche webs, since persistent webs have more basal and herbivore species as well as consumers with lower trophic levels.

We examined omnivory more finely by altering the skewness of omnivores' preference for prey at different trophic levels. Such skewness has profound effects on overall persistence, P (fig. 4), similar to the effects of varying the functional

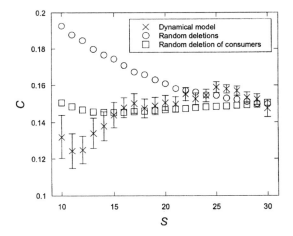

FIGURE 2 Mean connectance C of model food webs versus dynamically persistent network size S, with error bars showing plus and minus two standard errors of the estimated mean. The points without error bars show the mean connectance of 1000 niche model networks that have species deleted at random or have consumer species deleted at random. All initial networks are built using the niche model with $S_0 = 30$, $C_0 = 0.15$, and the dynamical model uses our Holling Type II.2 functional response where $q = 0.2$ (Box 1, eq. (2)).

response parameter q. Niche webs are most persistent ($P \approx 0.42$ when $q = 0$ and $P \approx 0.64$ when $q = 0.2$) when omnivores prefer lower trophic-level resources but avoid near exclusive consumption of the lowest trophic-level resources ($0.2 <$ skewness < 0.8). Persistence drastically falls to as low as $P \approx 0.25$ when $q = 0$ and $P \approx 0.34$ when $q = 0.2$ when omnivores more strongly prefer upper trophic-level resources (skewness $= 10$).

4 DISCUSSION

4.1 EFFECTS OF STRUCTURE ON DYNAMICS

Our results illustrate how the structure of ecological networks may influence their function by showing the effects of diversity and complexity on *in silico* ecosystem dynamics. May's early and remarkably durable theory based on linear stability analyses of random networks proposed that S and C have hyperbolically negative effects on stability (May 1973). Qualitatively similar effects occur in our nonlinear analyses of more ecologically realistic networks, but the effects are linear rather than hyperbolic, perhaps due to the differences between linear

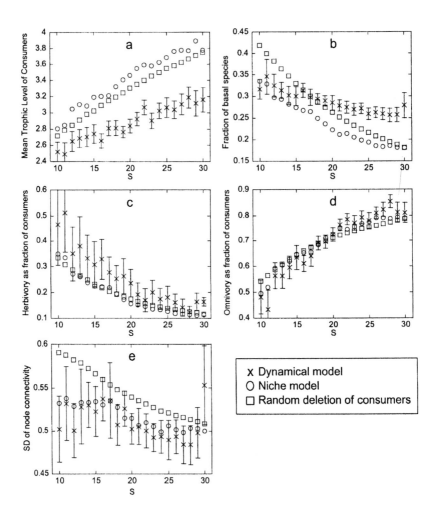

FIGURE 3   Mean and variation of model food-web properties versus persistent network size S. Error bars show plus and minus two standard errors of the estimated mean. Points without error bars show the mean property value in 1000 niche model networks with the same size and connectance as the dynamical model networks and in 1000 niche model networks with the same initial size and connectance as the dynamically constrained networks that then had consumer species deleted at random. Properties shown are (a) mean trophic level of consumers, (b) fraction of basal species, (c) fraction of consumers that are herbivores, (d) fraction of consumers that are omnivores, and (e) standard deviation of node connectivity. Initial networks are built using the niche model with $S_0 = 30$, $C_0 = 0.15$, and the dynamical model uses our Holling "type II.2" functional response with $q = 0.2$ (Box 1, eq. (2)).

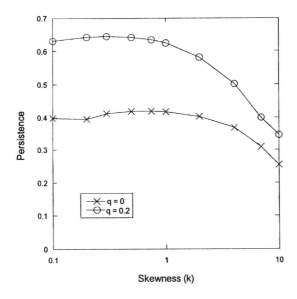

FIGURE 4   Mean ($n = 500$) overall persistence of model food webs vs. skewness k of
the prey preference of omnivores. When $k = 1$, all prey preferences of an omnivore are
equal; when $k < 1$, low trophic level prey are preferred and when $k > 1$, high trophic
level prey are preferred (see methods). All networks initially have $S_0 = 30$, $C_0 = 0.15$,
and the dynamical model uses parametrized Holling Type II ($q = 0$) and II.2 ($q = 0.2$)
functional responses (Box 1, eq. (2)).

stability and nonlinear persistence. Connectance (C) affects persistence much
more strongly than does diversity (S). This is illustrated by the regressions in
which variance in C explains over twice as much variance of P_R as does variance
in S. This greater importance of C than S to persistence had been previously
noted but the negative effects of C observed here are opposite the previously
noted positive effects (Dunne et al. 2002a; Fussman and Heber 2002; Kondoh
2003). Analyzing the effects of deleting species or otherwise challenging persistent
webs to study their robustness may clarify this discrepancy.

Beyond the classic effects of S and C on dynamics, our study illustrates
the overriding importance of the arrangement of links among species (fig. 5).
Random webs have almost no persistence, and the hierarchical ordering of the
cascade model vastly increases persistence. The contiguous niches, cannibalism,
and looping in the niche model allow even more persistence in food-web networks.
The hierarchical ordering of the cascade and niche models is easily interpreted as
a mechanistic formalization of energy flowing from plants to upper trophic levels.

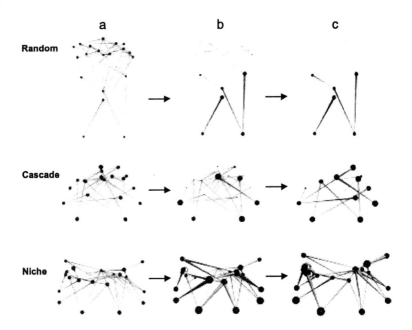

FIGURE 5 (a) Screen shots showing images of examples of random, cascade, and niche model food webs, each with $S_0 = 20$ and $C_0 = 0.15$. The dynamical model is then run on each structure. (b) Shows a moment while dynamics are running, and (c) shows a moment of the final persistent dynamical structure for each food web. The relative persistence ($P_r = S_p/S_o$) for each web is $P_\text{random} = 0.30$, $P_\text{cascade} = 0.60$, $P_\text{niche} = 0.90$. Images were produced with FoodWeb3D, written by R.J. Williams and provided by the Pacific Ecoinformatics and Computational Ecology Lab (www.foodwebs.org).

Models that ignore such distinctions between plants and animals by making all species capable of growing without consuming other species (Kondoh 2003) fail to detect the significance of nonrandom and hierarchical network structure (Brose et al. 2003). Niche space as formalized by the niche model is much less easily interpreted and deserves more study to understand which evolutionary, ecological, and mathematical factors underlie the model's improved empirical fit (Williams and Martinez 2000; Dunne et al. 2004) and increased persistence (figs. 1 and 5).

4.2 EFFECTS OF DYNAMICS ON STRUCTURE

This research also illuminates how the functioning of ecological networks influences their structure by examining the effects of nonlinear dynamics on the topology of complex food webs. We show for the first time that the stabilizing effects of both predator interference and respective decelerated and accelerated feeding on rare and abundant resources found in small modules of two species also apply to much larger networks with 30 or more species. This enables large complex food webs to sustain many more species than networks governed by standard type II responses. This remarkable persistence greatly increases the potential to add other ecological processes such as facilitation, age-structured populations, migration, and environmental stochasticity to models of large ecological networks, which should further facilitate exploration of their effects on ecological structure and dynamics. We also show that small and perhaps empirically undetectable changes in functional responses (e.g., changes from $q = 0.0$ to $q = 0.2$, Box 1) foster greatly increased persistence in model ecosystems (Williams and Martinez 2004b). This suggests that tiny amounts of prey switching behavior by consumers (Post et al. 2000; Kondoh 2003) or refuge-seeking behavior by resource taxa (Holling 1959a; Sarnelle 2003) can have large effects on the structure and dynamics of complex ecological networks, and may act as some of nature's more prevalent and important stabilizing strategies.

More effects of network function on network structure are seen in comparisons between persistent webs and webs generated by structural models free from biomass dynamics. Persistent webs typically have similar C to that in niche webs whose consumers are randomly deleted, but have lower C than that in niche webs subjected to random deletions of any species. More strikingly, persistent webs have higher fractions of basal species and consumers with lower mean trophic levels than do niche webs. This is consistent with the niche model's overestimation of empirically observed food-chain lengths (Williams and Martinez 2000), assuming that empirical webs have more persistent topologies than do niche webs. While the standard deviation of node connectivity shows few differences between niche webs subjected to dynamic loss of species and random loss of consumers, more detailed investigation of degree distributions (Dunne et al. 2002b) could illuminate differences hidden by our relatively coarse analysis.

Given the niche model's overestimation of the mean trophic level of consumers in large persistent webs by almost a whole level (fig. 3(a)) and its underestimation of the fraction of herbivores by ~0.07 (fig. 3(c)), we tested the niche model against these properties of the seven empirical webs originally compared to the niche model (Williams and Martinez 2000). Table 1 shows that the niche model consistently overestimates the mean trophic level by 0.2–2.4 levels and underestimates the fraction of herbivores by 0.01–0.32. Apparently, dynamics alters these properties of niche webs to produce network structures even more similar to empirically observed properties. The empirically observed fraction of basal species is well explained by the niche model (Williams and Martinez 2000), so the higher fraction of basal species observed in the dynamically constrained networks (fig. 3(b)) appears to conflict with empirical findings. This discrepancy may be due to highly aggregated and poorly described basal species in the empirical data. For example, basal species in the St. Martin Island food web (Goldwasser and Roughgarden 1993) are categories of plant material such

TABLE 1   Errors of niche model predictions of the fraction of herbivores (Herbivory) and mean trophic level (TL) of consumers in empirical food webs. S is the number of trophic species. C is directed connectance. Error is measured both as the difference between the model's mean property and the empirically observed property (in parentheses) and in more rigorously comparable terms of the number of model standard deviations that the empirically observed property differs from the model's mean (Williams and Martinez 2000).

Food Web	S	C	Herbivory	TL Consumers
St. Martin Island	42	0.12	-2.7 (-0.15)	1.4 (0.79)
Bridge Brook Lake	25	0.17	-3.9 (-0.19)	1.5 (1.23)
Coachella Valley	29	0.31	-1.3 (-0.04)	0.6 (1.24)
Chesapeake Bay	31	0.072	-0.2 (-0.01)	0.6 (0.21)
Skipwith Pond	25	0.32	-7.8 (-0.29)	0.1 (2.39)
Ythan Estuary	78	0.061	-4.1 (-0.20)	1.6 (0.60)
Little Rock Lake	92	0.12	-12.7 (-0.32)	2.5 (1.52)
		Mean	-4.62 (-0.17)	1.17 (1.14)
		Std error	1.65 (0.04)	0.30 (0.27)

as seeds and leaves. Many basal taxa in the Bridge Brook Lake (Havens 1992) food web are trophically identical in terms of having the exact same set of consumers, suggesting that the trophic links are poorly resolved (Martinez et al. 1999). Therefore, the fraction of basal species in the observed trophic-species networks and the niche model's fit to these fractions could be methodological artifacts of taxonomic and trophic resolution. The importance of basal species for persistence emphasizes the need for high quality data resolved evenly at all trophic levels (Cohen et al. 1993). Alternatively, artifacts of the dynamical model might cause the discrepancy (Brose et al. 2003). Our models assume that basal species do not compete for shared resources. Adding competition among basal species might lower the fraction of basal species in the persistent webs.

4.3   THE ROLE OF OMNIVORY

One of the more confusing interdependencies between food-web structure and dynamics concerns the issue of omnivory. There is a close positive and confounding relationship between omnivory and C in earlier studies (McCann and Hastings 1997; Fussman and Heber 2002), since increasing C typically makes consumers more omnivorous and increasing omnivory typically increases C. We help clarify this issue by controlling for the strong effects of C on persistence (fig. 2) and showing that the prevalence of omnivorous consumers in persistent webs is usually similar to that in niche webs (fig. 3(d)), which is much less than in cascade webs (Williams and Martinez 2000). If structural omnivory has an unusually strong positive effect on persistence, one would expect higher omnivory in the most persistent niche webs and more persistence in cascade webs. This is not generally supported by our results.

Contemporary modeling studies also tend to confound increasing omnivory with lowering consumers' trophic levels by increasing omnivory in a restricted fashion. That is, omnivory that lowers a consumer's trophic level is typically created by adding short paths that enable carnivores to consume primary production (McCann and Hastings 1997; Fussman and Heber 2002). Omnivory that increases a consumer's trophic level, for example, by adding carnivorous links to an herbivore's diet, is typically avoided. Omnivores that prefer higher trophic level prey strongly decrease persistence compared to omnivores lacking such preference, while variable preference for low levels has much less effect (fig. 4). These findings, combined with consumers' lower trophic levels and higher prevalence of basal species and herbivores in the most persistent niche webs, suggest that shortening food chains and reducing trophic levels account for the stabilizing effects previously attributed to omnivory. In contrast, omnivory strongly decreases persistence in food webs when omnivores engage in the empirically unusual (Williams and Martinez 2004a) destabilizing behavior of preferring prey at higher trophic levels.

5 CONCLUSIONS AND FUTURE DIRECTIONS

Our analyses address several historically perplexing aspects of the remarkable complexity and persistence of natural ecosystems and show how more empirically prevalent aspects of trophic interactions (Williams and Martinez 2000; Skalski and Gilliam 2001; Sarnelle 2003; Williams and Martinez 2004a) may confer persistence on large complex ecosystems. Both food-web structure, characterized by the empirically successful the niche model, and food-web function, characterized by decelerated consumption of rare resources (Sarnelle 2003), predator interference (Skalski and Gilliam 2001), and omnivores' preferences for lower trophic-level prey (Williams and Martinez 2004a), greatly increase the diversity and complexity that persists in ecological networks. Some of the increased persistence resulting from including these factors appears to have been mistakenly attributed to unqualified omnivory. The strong effects of predator interference and decelerated and accelerated feeding on relatively rare and abundant resources, respectively, suggests that other behaviors that reduce consumption of rare resources, for example, prey switching (Post et al. 2000; Kondoh 2003), will also stabilize large complex networks. In contrast, responses that increase consumption of rarer and higher trophic level resources, such as economic exploitation of relatively rare carnivorous fishes (Pauly et al. 2002), can be expected to decrease persistence of species within ecosystems.

Perhaps even more important than these results is that the models described here provide new, more sophisticated, and flexible tools for exploring crucial issues such as the impacts of various types of perturbations on ecosystem structure and nonlinear dynamics, as well as the influence of structure and dynamics on mitigating ecosystem responses to perturbations. For example, integrated structure/dynamics models can be used to explore which properties of species (e.g., trophic level, generality, vulnerability) make them more effective invaders and which properties of complex networks (e.g., connectance, distribution of species among trophic levels) make them more resistant to invasions. Similarly, this modeling approach can be used to explore which properties of species, interactions, or networks are likely to make ecosystems more or less robust to biodiversity loss (Solé and Montoya 2001; Dunne et al. 2002a). Climate change impacts can be investigated by simulating the effects of temperature change on metabolic rates. The dynamic consequences of consumer-resource (predator-prey) body-size ratios can be explored by examining the effects of metabolic rates that reflect body-size ratios and metabolic types (i.e., ectotherm, endotherm, vertebrate, and invertebrate) found in natural systems. These and other research questions represent important future directions for structure/dynamics modeling that could be explored with relatively minor modifications of the methods described here.

With more significant modifications, our modeling approach can be used to explore other processes such as behavior modification, detrital loops, adaptation, coevolution, mutualism, and competition. Addressing these types of issues would involve moving from food webs containing only trophic interactions to broader ecological networks that include non-trophic interactions. For example, the consequences of resource sharing and competition among basal species can be explored by replacing the independent logistic growth of basal species with basal growth that is dependent on explicit dynamics of flows into and out of limiting nutrient pools, including differential uptake by plants according to their growth and relative consumption rates (Brose et al. in press). Another major modification would be to add explicit detrital dynamics to account for biomass shed and excreted by organisms. That organic matter becomes available to microbes and other detritivores, which are consumed by higher trophic level organisms, and whose activity helps determine nutrient availability for photosynthetic species. Such modifications of the model presented here will likely involve new functional responses characterizing consumption as a function of nonliving resources. Other changes in functional responses can allow exploration of nontrophic influences (e.g., ecological engineering, Jones et al. 1997; indirect effects, Peacor and Werner 2001) and evolution rates (Yoshida et al. 2003) on consumption rates and population dynamics. The addition of nutrient and detrital dynamics should also provide a powerful framework for exploring network evolution (McKane and Drossel Chapter 9). By modeling the emergence of biological innovation within an interaction network, feedbacks through short and long chains of direct and indirect effects help determine the success or failure of new traits and species, and alter the structure and dynamics of the network.

This basic model could also be altered to account for spatially explicit processes. The simplest approach is to add functions such as density-dependent migration. More ambitiously, complex networks could be made spatially explicit by placing the networks within cells in a landscape. Migration could occur between adjacent cells and network structure in a cell could depend on the spatial ranges of species and other network nodes. This *trophic circulation model* approach would be analogous to better known global circulation models of weather and climate. Similar *NPZ* models (Franks and Walstad 1997) are already used by oceanographers to model the nonlinear dynamics of nutrients, phytoplankton, and zooplankton in a spatially explicit manner (Franks 2002). However, such models, focused on relatively simple modules, face the classic problem of dynamical instability (Denman 2003). Our results suggest that scaling up beyond simple spatially explicit modules may be achieved by incorporating realistic network structure, non-type II responses, and omnivory skewed towards lower trophic levels. While the vast number of parameters and computational intensity required may hinder scaling up within a spatially explicit framework, rapid advances in informatics and computing may facilitate advances in the near future (Green et al. 2005).

Regardless of how they are modified or augmented, models of complex systems are still simplifications of nature. In order to create a plausible and useful simplification of natural systems, we base our integrated structure/dynamics model on simple empirical regularities and processes that are well documented in the literature. This strategy has produced novel insights into the complexity and stability of diverse, multi-trophic level ecosystems and should continue to facilitate research that includes other well-documented regularities and processes. We encourage the continued exploration of high diversity model systems that go beyond traditional module or single trophic-level approaches. Such models of complex systems have greater fidelity to the diversity in natural ecosystems that field ecologists study every day. More research based on these types of models as well as empirical and experimental tests of their findings could significantly extend and refine our understanding of the persistence and stability of complex networks of species. Such integrated studies can facilitate exciting new insights regarding trophic and non-trophic processes in the complex ecosystems that sustain the stunning, yet tragically diminishing, levels of diversity in nature.

Models that incorporate more detail than ours quickly require so much information about any given real population that very substantial empirical programs are needed to provide it. That is not to say that such models or such programs are to be avoided: quite to the contrary, often they are necessary. However, when constraints of time or research resources call for maximum realism from minimum data, plausible models such as we have discussed here may be a valid recourse.
—Yodzis and Innes, 1992

6 ACKNOWLEDGMENTS

This work was supported by NSF grants ITR-0326460, DBI-0234980, DEB-0083929, and DEB/DBI-0074521. The authors thank the Santa Fe Institute for support and hospitality. N. D. Martinez thanks the NSF-funded IGERT Program in Nonlinear Systems and the Telluride House, both at Cornell University, for support and hospitality.

REFERENCES

Adams, S. M., B. L. Kimmel, and G. R. Plosky. 1983. "Sources of Organic Matter for Reservoir Fish Production: A Trophic-Dynamics Analysis." *Can. J. Fisheries and Aquatic Sci.* 40:1480–1495.

Beddington, J. R. 1975. "Mutual Interference between Parasites or Predators and Its Effects on Searching Efficiency." *J. Animal Ecol.* 51:331–340.

Berlow, E. L. 1999. "Strong Effects of Weak Interactions in Ecological Communities." *Nature* 398:330–334.

Brose, U., E. L. Berlow, and N. D. Martinez. In press. "From Food Webs to Ecological Networks: Linking Nonlinear Trophic Interactions with Nutrient Competition." In *Dynamic Food Webs: Multispecies Assemblages, Ecosystem Development and Environmental Change*, ed. J. Moore, P. C. deRuiter, V. Walters, and J. C. Moore. Theoretical Ecology Series. Elsevier/Academic Press.

Brose, U., R. J. Williams, and N. D. Martinez. 2003. "Comment on 'Foraging Adaptation and the Relationship between Food-Web Complexity and Stability.'" *Science* 301:918b–918c.

Camacho, J., R. Guimerà, and L. A. N. Amaral. 2002. "Robust Patterns in Food Web Structure." *Phys. Rev. Lett.* 88:228102.

Cohen, J. E., F. Briand, and C. M. Newman. 1990. *Community Food Webs: Data and Theory.* Berlin: Springer-Verlag.

Cohen, J. E., R. A. Beaver, S. H. Cousins, D. L. DeAngelis, L. Goldwasser, K. L. Heong, R. D. Holt, A. J. Kohn, J. H. Lawton, N. Martinez, R. O'Malley, L. M. Page, B. C. Patten, S. L. Pimm, G. A. Polis, M. Rejmánek, T. W. Schoener, K. Schoenly, W. G. Sprules, J. M. Teal, R. E. Ulanowicz, P. H. Warren, H. M. Wilbur, and P. Yodzis. 1993. "Improving Food Webs." *Ecology* 74:252–258.

DeAngelis, D. L., R. A. Goldstein, and R. V. O'Neill. 1975. "A Model for Trophic Interaction." *Ecology* 56:881–892.

Denman, K. L. 2003. "Modelling Planktonic Ecosystems: Parametrizing Complexity." *Prog. in Oceanography* 57:429–452.

Drossel, B., P. G. Higgs, and A. J. McKane. 2001. "The Influence of Predator-Prey Population Dynamics on the Long-Term Evolution of Food Web Structure." *J. Theor. Biol.* 208:91–107.

Dunne, J. A. "The Network Structure of Food Webs." This volume.

Dunne, J. A., R. J. Williams, and N. D. Martinez. 2002a. "Network Structure and Biodiversity Loss in Food Webs: Robustness Increases with Connectance." *Ecol. Lett.* 5:558–567.

Dunne, J. A., R. J. Williams, and N. D. Martinez. 2002b. "Food-Web Structure and Network Theory: The Role of Connectance and Size." *Proc. Natl. Acad. Sci.* 99:12917–12922.

Dunne, J. A., R. J. Williams, and N. D. Martinez. 2004. "Network Structure and Robustness of Marine Food Webs." *Marine Ecol. Prog. Ser.* 273:291–302.

Dunne, J. A., U. Brose, R. J. Williams, and N. D. Martinez. 2005. "Modeling Food-Web Structure and Dynamics: Implications for Complexity-Stability." In *Aquatic Food Webs: An Ecosystem Approach*, ed. A. Belgrano, U. Scharler, J. A. Dunne, and R. E. Ulanowicz, 117–129. Oxford University Press.

Franks, P. 2002. "NPZ Models of Plankton Dynamics: Their Construction, Coupling to Physics, and Application." *J. Oceanography* 58:379–387.

Franks, P., and L. Walstad. 1997. "Phytoplankton Patches at Fronts: A Model of Formation and Response to Transient Wind Events." *J. Marine Resh.* 55:1–29.

Fox, J. W., and J. McGrady-Steed. 2002. "Stability and Complexity in Microcosm Communities." *J. Animal Ecol.* 7:749–756.

Fussman, G. F., and G. Heber. 2002. "Food Web Complexity and Chaotic Population Dynamics." *Ecol. Lett.* 5:394–401.

Goldwasser, L., and J. Roughgarden. 1993. "Construction and Analysis of a Large Caribbean Food Web." *Ecology* 74:1216–1233.

Green, J. L., A. Hastings, P. Arzberger, F. Ayala, K. L. Cottingham, K. Cuddington, F. Davis, J. A. Dunne, M.-J. Fortin, L. Gerber, and M. Neubert. 2005. "Complexity in Ecology and Conservation: Mathematical, Statistical, and Computational Challenges." *Bioscience* 55:501–510.

Hassell, M. P. 1978. *The Dynamics of Arthropod Predator-Prey Systems.* Princeton, NJ: Princeton University Press.

Hassell, M. P., and G. C. Varley. 1969. "New Inductive Population Model for Insect Parasites and Its Bearing on Biological Control." *Nature* 223:1133–1136.

Havens, K. 1992. "Scale and Structure in Natural Food Webs." *Science* 257:1107–1109.

Holling, C. S. 1959a. "The Components of Predation as Revealed by a Study of Small-Mammal Predation of the European Pine Sawfly." *Can. Entom.* 41:293–320.

Holling, C. S. 1959b. "Some Characteristics of Simple Types of Predation and Parasitism." *Can. Entom.* 91:385–399.

Jones, C. G., H. J. Lawton, and M. Shachak. 1997. "Positive and Negative Effects of Organisms as Physical Ecosystem Engineers." *Ecology* 78:1946–1957.

Kendall, B. E., J. Prendergast, and O. N. Bjornstad. 1998. "The Macroecology of Population Dynamics: Taxonomic and Biogeographic Patterns in Population Cycles." *Ecol. Lett.* 1:160–164.

Kondoh, M. 2003. "Foraging Adaptation and the Relationship between Food-Web Complexity and Stability." *Science* 299:1388–1391.

Levine, S. 1980. "Several Measures of Trophic Structure Applicable to Complex Food Webs." *J. Theor. Biol.* 83:195–207.

MacArthur, R. H. 1955. "Fluctuation of Animal Populations and a Measure of Community Stability." *Ecology* 36:533–536.

Martinez, N. D., B. A. Hawkins, H. A. Dawah, and B. Feifarek. 1999. "Effects of Sampling Effort on Characterization of Food-Web Structure." *Ecology* 80:1044–1055.

May, R. M. 1973. *Stability and Complexity in Model Ecosystems*. Princeton, NJ: Princeton University Press.

McCann, K. 2000. "The Diversity-Stability Debate." *Nature* 405:228–233.

McCann, K., and A. Hastings. 1997. "Re-evaluating the Omnivory-Stability Relationship in Food Webs." *Proc. Roy. Soc. Lond. B* 264:1249–1254.

McCann, K., and P. Yodzis. 1995. "Biological Conditions for Chaos in a Three-Species Food Chain." *Ecology* 75:561–564.

McCann, K., A. Hastings, and G. R. Huxel. 1998. "Weak Trophic Interactions and the Balance of Nature." *Nature* 395:794–798.

McKane, A. J., and B. Drossel. "Models of Food Web Evolution." This volume.

Murdoch, W. W., and A. Oaten. 1975. "Predation and Population Stability." *Adv. Ecol. Resh.* 9:1–131.

Naeem, S., L. J. Thompson, S. P. Lawler, J. H. Lawton, and R. M. Woodfin. 1994. "Declining Biodiversity Can Affect the Functioning of Ecosystems." *Nature* 368:734–737.

Neutel, A.-M., J. A. P. Heesterbeek, and P. C. de Reuter. 2002. "Stability in Real Food Webs: Weak Links in Long Loops." *Science* 296:1120–1123.

Paine, R. T. 1966. "Food Web Complexity and Species Diversity." *Amer. Natur.* 100:65–75.

Pauly, D., V. Christensen, S. Guénette, T. J. Pitcher, U. R. Sumaila, C. J. Walters, R. Watson, and D. Zeller. 2002. "Toward Sustainability in World Fisheries." *Nature* 418:689–695.

Peacor, S. D., and E. E. Werner. 2001. "The Contribution of Trait-Mediated Indirect Effects to the Net Effects of a Predator." *Proc. Natl. Acad. Sci. USA* 98:3904–3908.

Pimm, S. L., and J. H. Lawton. 1978. "On Feeding on More than One Trophic Level." *Nature* 275:542–544.

Post, D. M., M. E. Conners, and D. S. Goldberg. 2000. "Prey Preference by a Top Predator and the Stability of Linked Food Chains." *Ecology* 81:8–14.

Real, L. A. 1977. "The Kinetics of Functional Response." *Amer. Natur.* 111:289–300.

Real, L. A. 1978. "Ecological Determinants of Functional Response." *Ecology* 60:481–485.

Sarnelle, O. 2003. "Nonlinear Effects of an Aquatic Consumer: Causes and Consequences." *Amer. Natur.* 161:478–496.

Skalski, G. T., and J. F. Gilliam. 2001. "Functional Responses with Predator Interference: Viable Alternatives to the Holling Type II Model." *Ecology* 82:3083–3092.

Solé, R. V., and J. M. Montoya. 2001. "Complexity and Fragility in Ecological Networks." *Proc. Roy. Soc. Lond. B* 268:2039–2045.

Solow, A. R., and A. R. Beet. 1998. "On Lumping Species in Food Webs." *Ecology* 79:2013–2018.

Tilman, D., P. B. Reich, J. Knops, D. Wedin, T. Mielke, and C. Lehman. 2001. "Diversity and Productivity in a Long-Term Grassland Experiment." *Science* 294:843–845.

Williams, R. J., and N. D. Martinez. 2000. "Simple Rules Yield Complex Food Webs." *Nature* 404:180–183.

Williams, R. J., and N. D. Martinez. 2004a. "Limits to Trophic Levels and Omnivory in Complex Food Webs: Theory and Data." *Amer. Natur.* 163:458–468.

Williams, R. J., and N. D. Martinez. 2004b. "Stabilization of Chaotic and Non-Permanent Food Web Dynamics." *Eur. Phys. J. B* 38:297–303.

Williams, R. J., E. L. Berlow, J. A. Dunne, A.-L. Barabási, and N. D. Martinez. 2002. "Two Degrees of Separation in Complex Food Webs." *Proc. Natl. Acad. Sci.* 99:12913–12916.

Yodzis, P. 2000. "Diffuse Effects in Food Webs." *Ecology* 81:261–266.

Yodzis, P., and S. Innes. 1992. "Body-Size and Consumer-Resource Dynamics." *Amer. Natur.* 139:1151–1173.

Yoshida, T., L. E. Jones, S. P. Ellner, G. F. Fussmann, and N. G. Hairston. 2003. "Rapid Evolution Drives Ecological Dynamics in a Predator-Prey System." *Nature* 424:303–306.

Exploring Network Space with Genetic Algorithms: Modularity, Resilience, and Reactivity

Diego Ruiz-Moreno
Mercedes Pascual
Rick Riolo

The relationship between stability and complexity in food webs was originally addressed with random networks of links. Non-random structures are evident in data, and in webs resulting from a variety of modeling approaches, including assembly and coevolution. As a basis for interpreting the population dynamics of such networks, a better understanding is needed of the large space of possible structures and their associated dynamical properties. We illustrate here the use of genetic algorithms to explore this large space, by focusing on two dynamical properties related to the stability of equilibria: resilience and reactivity, for long- and short-term responses of the system, respectively. These properties define the "fitness" criteria used for the search of the structural space. We analyze the resulting patterns of clustering and interaction strength distributions in the most resilient and less reactive networks, relative to random ones. Historically, the effect of network modularity on long-term stability has been of interest but it still remains theoretically unclear and empirically controversial. Our main finding is that modularity in community matrices relates to short-term responses in the transients. We discuss related recent work on clustering and complexity measures (from information theory) obtained for large networks in the brain. Similar approaches to the one presented here, albeit more computationally intensive, should be applicable to the nonlinear dynamics of ecological networks.

1 INTRODUCTION

The complexity-stability debate has a very long history in the study of food webs, and addresses the fundamental question of whether ecosystems that are species rich and highly connected by interactions among these species are also better able to withstand perturbations. Both species richness and food web connectance were considered early on as measures of complexity. The early intuition of ecologists that more complex networks would also be more stable (McArthur 1955; Elton 1958; Hutchinson 1959), was contradicted by the well-known results of May (1971, 1972) showing that this is not necessarily the case in random networks. Much has been said about the different definitions of stability that could explain this mismatch between theory and biological intuition, and about the limitations of considering random networks and small perturbations close to equilibrium (see Pimm 1984 and McCann 2000 for review). Improvements in stability have been obtained by modifying the basic hypotheses of May (Pimm and Lawton 1980), by considering, for example, low levels of connectance, the absence of unrealistic links among species, the low frequency and observed patterns of omnivory, or particular constraints on biomass transfer (De Angelis 1975). More recently, theoretical studies have demonstrated that stable complex networks are possible (Kokkoris et al. 1999; Rozdilsky and Stone 2001; Kokkoris et al. 2002; Brose et al. 2003; Jansen and Kokkoris 2003; Kondoh 2003; Martinez et al. this volume). One key question is, what structural properties make these large networks both feasible and more stable.

Different approaches may be used to address this question. One approach considers the progressive assembly of a community of interacting species from a larger pool of potential members by successive invasions (e.g., Law and Blackford 1992). With an assembly approach and Lotka-Volterra dynamics, Kokkoris et al. (1999) support the previously proposed idea that weak interactions promote stability (McCann et al. 1998). When the pool of component species is not prescribed *a priori*, the explicit treatment of evolution is necessary to generate diversity (see Chapter by McKane and Drossel for an example). Assembly and evolutionary approaches restrict the size of the enormous parameter space of large networks because the path of the system itself, as it evolves or assembles, "selects" a subset of possible parameters. Alternatively, particular choices of parameters can be selected on the basis of observations in real systems, for example, on allometry (e.g., Emmerson and Raffaeli 2004) or interaction strength (Neutel et al. 2002). De Ruiter et al. (1995) and Neutel et al. (2002) demonstrate that a specific pattern observed in the distribution of interaction strength in soil food webs, namely that weak links tend to occur in long loops, contributes to the stability of food webs. However, particular structural patterns are not always apparent without an *a priori* hypothesis and can require the observation of regularities in many systems. Theory can contribute insights into structural features that might matter, and should be probed for, in real data. To investigate the full nonlinear dynamics of food webs not restricted to the proximity of equilibria, small systems with only a few species can be simulated. For example, McCann et al. (1998) found that weak interactions decrease both population variability and the likelihood of chaos in the nonlinear dynamics of low-dimensional food webs. Extrapolation to large networks remains computationally prohibitive given the enormous parameter space of the full nonlinear model. Parallel computation was used by Fussman and Heber (2002) to address the likelihood of chaotic dynamics in food webs with several trophic levels and more species, even though species numbers were still small by comparison with those of real communities. Martinez et al. (this volume) consider the nonlinear dynamics of larger communities by specifying the initial binary network of links with a static model of structure, the niche model, and then restricting the actual values of the parameters in this network with simplifying assumptions. Mathematical analyses of such large networks are rare (but see Cohen et al. 1990a and Chen and Cohen 2001a for a combination of simulations and analytical stability criteria).

Here, we use a complementary theoretical approach and tackle the structure-dynamics connection by asking, not whether a particular property promotes stability, but what are the structural properties of the most stable systems (see also Haydon 2000). We illustrate how genetic algorithms can be used to explore the huge parameter space of possible parameters by selecting networks with specific dynamical properties of interest. Two such properties are considered, resilience and reactivity (Chapter 1, Box 1) that characterize the response of the system to small perturbations close to equilibrium, where the dynamics are essentially linear and governed by the so-called community matrix. Resilience is a property of the long-term behavior of the system and measures the rate at which perturbations eventually decay or grow. By contrast, reactivity is a property of the transient behavior or initial response, measuring whether a perturbation will initially grow and move away from equilibrium. We describe how reactivity and resilience behave as a function of connectance and species richness in systems selected for high stability (low reactivity and high resilience). Although resilience has already been considered in a large number of studies, including the classic one by May on stability and complexity, our approach allows us to relax the typical assumption of a random structure and to specifically compare the *most* stable systems to inquire about properties of their dynamics. We also examine the resulting structural properties of the selected systems, in particular, modularity and the distribution of interaction strength.

In a modular structure, a *cluster*, block or compartment consists of a set of species with strong interactions among themselves, and weak or no interactions outside this group. So far, theory provides no clear guidance as to whether modularity enhances stability. May (1973) first suggested that the arrangement of interactions in perfect blocks could increase local stability, but Pimm (1982) demonstrated that this is not necessarily the case, particularly if connectance is kept constant, and that the opposite effect was likely. Pimm and Lawton (1980) argued that "compartmented models are no more likely to be stable than randomly organized models, if care is taken to exclude biologically unreasonable phenomena from the analyses." Empirical evidence is also divided. Pimm and Lawton (1980) found some evidence for modularity in real food webs but by averaging a probability of modularity over 12 empirical binary webs, they concluded that evidence for interactions being structured into compartments was weak. From further analysis of the same webs, Raffaelli and Hall (1992) concluded that "there was evidence for significant compartmentation in several real webs, all of which are reasonably well documented, aquatic and benthic." In the same study, however, another well documented estuarine web, the Ythan web, showed no such structure, in agreement with the lack of no strong functional interactions and the argument by Paine (1980) on the role of keystone species (1980). In this argument, compartments of resources and their consumers would "behave as a functional unit" and be maintained by the strong interaction between a keystone predator and a competitively dominant prey, both of which do not belong to the group itself. Compartments in nature have also been related to the existence of spatial habitat structure (Pimm and Lawton 1980; Pimm 1982; Polis et al. 1997; Holyak 2000; Holt 2002). More recently, Krause et al. (2003) applied methods from social networks and successfully identified compartments in a number of real food webs. They argued that this structure plays a central role in stability by retaining the impact of disturbances within compartments, although no evidence is given in support of this intuitive and appealing hypothesis.

Here, we ask whether compartments are a typical feature of the selected systems, and therefore, whether they are required for high resilience and low reactivity. We show that reactivity but not resilience generates a modular structure. Thus, the transient response of the network to perturbations, and not its long-term response, appears to be closely related to modularity. We discuss our results in light of the recent application of genetic algorithms to the structure-dynamics relationship in other large networks in physics (Variano et al. 2004) and neurobiology (Tononi et al. 1994; Tononi et al. 1998a; Tononi et al. 1998b.)

2 METHODS: RESILIENCE AND REACTIVITY

Consider the case of a multispecies community with n populations whose respective abundances are denoted by $N_i(t)$. The elements of "*Community Matrix*" A describe the effects of species j upon species i near equilibrium (May 1971, 1973).

These elements are given by:

$$a_{ij} = \left(\frac{\partial F_i}{\partial N_j} \right)^* \tag{1}$$

where $F_i = f(N_1, N_2, \ldots, N_n)$ is the function specifying the rate of change $(dN_i)/(dt)$ in the original nonlinear system. The entries of A are generally normalized so that,

$$-1 \le a_{ij} \le 1 \quad \forall i, j \tag{2}$$

$$a_{ii} = -1 \quad \forall i \tag{3}$$

By imposing the condition that

$$\text{sign}(a_{ij}) = -\text{sign}(a_{ji}) . \tag{4}$$

the interactions within the community are restricted to those between predators and their prey. Properties of the community matrix quantify the transient and long-term response of the community to perturbations near equilibrium. In particular, we use here the following well-known definition of resilience

$$\text{Resilience} = -\text{Re}(\lambda_1(A)) \tag{5}$$

which measures the (asymptotic) decay rate of perturbations as given by the real part of the dominant eigenvalue of A. The larger the resilience, the faster the perturbations decay. Thus, systems with a positive resilience value are denoted as (locally) *stable*. Note however that resilience characterizes the long-term behavior of the system, and when positive, the long-term return of the system to equilibrium. It does not characterize the short-term transient behavior, immediately after a disturbance, and in particular it does not tell us whether a perturbation will initially grow, even if it eventually decays.

In order to study the short-term transient behavior of ecological systems Neubert and Caswell (1997) present several measures. One of them is known as *Reactivity* and defined as the maximum amplification rate, over all initial perturbations x_0 (to the asymptotically stable equilibrium of a linear system), immediately following the perturbation.

$$\text{Reactivity} = \max_{\|x_0\| \ne 0} \left(\frac{1}{\|x\|} \cdot \frac{d \|x\|}{dt} \right) \Big|_{t=0} . \tag{6}$$

With some algebraic manipulation,

$$\text{Reactivity} = \lambda_1(H(A)) \tag{7}$$

where $H(A)$ is the *Hermitian Part of* A,

$$H(A) = \frac{A + A^T}{2} . \tag{8}$$

Given a system with a stable equilibrium point, it is *reactive* when reactivity is positive. In this case, perturbations can initially grow in magnitude, independently of their initial size.

3 FOOD-WEB SPACE

Even when the nonlinear system is simplified by considering only its linear approximation close to equilibrium, there are $O(n^2)$ different parameters to consider. To explore this large state space, a *genetic algorithm* was developed (see Box 1 for definitions). For predefined values of species richness and connectance, this algorithm searches for matrices with high resilience and low reactivity, starting from a randomly generated set of food webs. Given the objectives of the search, a simple (generational) genetic algorithm was chosen, retaining an elite subpopulation from generation to generation. The elite was restricted to contain a single individual, the one returning the best value for the objective function at each generation. Each genome corresponds to a community matrix, and therefore, to a food web. The population size of matrices was fixed at a constant value P. Mutation was implemented in two different ways. The first one modifies the values of the community matrix entries, and is, therefore, called here a *strength mutation*. It produces small variability in the value of a_{ij} at a constant rate by sampling from a normal distribution. The new value is chosen according to a set of restrictions (see eqs. (2) and (4)) that guarantee that the new genome is still a valid food web. The second one modifies the presence and position of the links themselves, and is, therefore, called a *structure mutation*. New links between randomly selected species (a_{ij} and a_{ji}) can be generated (i.e., set to nonzero values) and old links removed (set to 0) while still maintaining a valid food web (see eqs. (2) and (4)). To avoid changes in connectance every time a link is removed, a new link is added, and vice versa.

Several objective functions were developed to rank the matrices and choose the elite one. We describe here results for three criteria:

- *The resilience criterion* selects for a large numerical value of resilience. A binary tournament selects the best genome (see Box 2).
- *The reactivity criterion* requires first that both genomes involved in the binary tournament have a positive value for resilience. If the difference between these values is less than 10%, then the genome having the smaller reactivity value is considered the winner, otherwise the one with the larger resilience is selected. If only one of the genomes has a positive resilience, it is automatically considered the winner. If both have negative resilience, one is selected at random.
- *The random criterion* makes no difference between genomes. The results of the binary tournament are completely random.

Each of these objective functions, or criteria, was the core of an experiment. The first experiment, Experiment 0, used the random criterion to establish a base-line for comparisons. Experiments 1 and 2 applied the resilience and reactivity objective functions respectively. For each experiment, 50 cases were considered for each combination of *species richness* (or *community size*, $n = 32, 60, 100, 200$) and proportion of links among species (or *connectance*, $C = 0.15, 0.3, 0.5, 0.7, 0.9$). In all cases, the population size for the genetic algorithm had the same fixed value and the number of generations was at least 1500 and no larger than 5000. Population measures, as well as the elite matrix and the top subset of the highly ranked matrices, were saved every 100 generations. For each experiment the evolution of such measures was calculated for both the best (elite) individual and the whole population.

For the selected matrices, we consider properties of the distribution of interaction strength, including its mean and variance, relevant to the proposed stabilizing role of weak links. We also examine the occurrence of weak links in cycles, to address their proposed stabilizing role in long loops. Finally, a clustering algorithm (see Box 3) is used to determine whether a modular structure develops.

3.1 EXPERIMENT 1: RESILIENCE

The influence of both species richness and connectance on resilience was studied by using the resilience objective function. The results confirm the previously described negative effect of both these parameters (fig. 1). The negative influence of connectance is weak in small communities and strong in large ones. Together, increments in connectance and species richness have a very dramatic impact on the resilience of the resulting community. For each run of this experiment, the final whole population's average resilience values were higher (65% on average) and different (ANOVA, $p = 0.05$) from those obtained in the random experiment.

With regard to the distribution of interaction strength, the genetic algorithm is able to find stable communities whose means and standard deviations are not significantly different from those of the random case. Furthermore, differences regarding the mean of the final distribution of interaction strength from that of the random experiment are not significant (ANOVA, $p = 0.05$). Therefore, we ask, where in the food web do the weak links occur with particular reference to the loops or cycles? A characteristic of the food webs is, indeed, the presence of loops or cycles. A loop describes a pathway of interactions from a given species through the network back to itself without visiting any other species more than once (Neutel et al. 2002). While cycles are rare in earlier studies (Cohen et al. 1990b), probably due to the lack of data, more recent studies describe their presence (Martinez 1991; Neutel et al. 2002). Here, the quantity of loops is positively related to both connectance and species richness (fig. 2), and differs significantly (ANOVA, $p = 0.05$) from the random case. The lengths of such cycles usually involve 3 or 4 species, irrespective of the values of connectance and species rich-

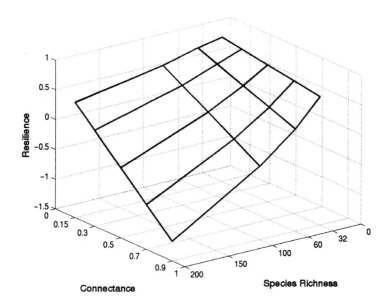

FIGURE 1   Average values for resilience reached by the whole population of genomes for several values of connectance ($C = 0.15, 0.3, 0.5, 0.7, 0.9$) and species richness ($S = 32, 60, 100, 200$). It is clear that increments in connectance for a given number of species, or increments in species richness for a given connectance, lead to a decrease in resilience. Moreover, a joint increment has the strongest effect.

ness. A detailed analysis of the position of weak links reveals that, as in Neutel et al. (2002), weak links occur in the cycles, although the weakest links are not necessarily in the cycles.

Regarding modularity (fig. 3), the arrangement of the links does not generate a degree of compartmentalization statistically different from that of random webs (ANOVA, $p = 0.05$). There is no evidence that improvements in resilience alone lead to the formation of clusters (for constant connectance).

Notice that in this experiment for the cases with species richness = 200 and connectance = 0.9 (as in fig. 1), negative resilience values indicate that the selected food webs are not stable. This means that the genetic algorithm was not able to find stable webs within the maximum searching time, which suggests that such food webs may have a very low probability of occurrence. An analogous situation occurs for the reactivity objective function.

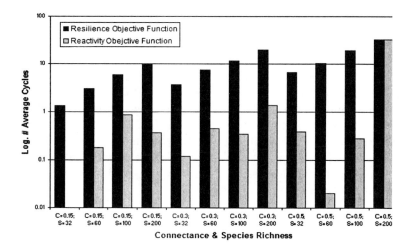

FIGURE 2 Logarithm of the average number of cycles (found in the genome population at the final time step) for the different values of connectance (C), species richness (S), and objective function. Results for the resilience objective function are shown with black bars, while gray bars are used for the reactivity objective function. X axes represent connectance and species richness. In the case where resilience is used as the objective function, the number of cycles exhibits a high positive correlation with both species richness and connectance. In the case of the reactivity objective function, there is no correlation and the number of cycles is much smaller, except for the more intricate situation ($C = 0.5$ and $S = 200$). (The random case is not included because the number of cycles is zero.)

3.2 EXPERIMENT 2: REACTIVITY

Reactivity concerns the largest initial "reaction" or deviation from equilibrium of the food web in response to a small perturbation away from equilibrium. Although it is therefore likely that connectance and species richness matter, it is not evident *a priori*, whether a larger or more interconnected food web should be more reactive. Complexity could dampen the impact of the disturbance leading to low reactivity. These possibilities were explored by applying the reactivity objective function for the same values of connectance and species richness as those of the previous experiment.

The results of that search show that both species richness and connectance are positively correlated with reactivity. More complex food webs have the strongest transient responses to disturbance. This response appears more sensitive to species richness (for a given connectance) than the other way around. Moreover, a joint increase in species richness and connectance shows a dramatic impact on reactivity (fig. 4).

As expected for this experiment, the genetic algorithm found webs with smaller reactivity than random webs, but an unexpected consequence of the search was the slightly lower values of resilience relative to the ones in experiment 1 (a small difference of approximately 2%). In addition, the dispersion of the genome population was also lower. This increase in resilience did not modify the patterns described in the previous section (fig. 1).

Another unexpected result relative to experiment 1 was the lower occurrence of cycles (fig. 2), and the absence of a relationship between cycles and connectance or species richness. Nevertheless, whenever cycles were present, the weakest links occurred in the cycles.

More importantly, matrices selected for low reactivity produce a distinct pattern of modularity, with a larger number of clusters (fig. 3). There are two to six clusters more than in the resilience case, depending on the specific values of connectance and species richness. When connectance is higher, this difference is lower, because in general, a higher connectance implies stronger links and, therefore, fewer clusters. These results indicate that matrices with high resilience and lower reactivity should exhibit an internal structure similar to a block matrix. However, it is important to recognize that while strong interactions determine the formation of clusters, strong interactions are not the majority. From the distribution of interaction strength for the three experiments (fig. 5), it is clear that most interactions have mean strength, and that there is no trend in the proportion of weak and strong interactions. There is no significant difference for the distribution of interaction strength between the random and reactivity cases, although the latter exhibits a slightly sharper distribution ($\Delta SD = 0.05$). Thus, dynamical differences among the networks result from the relative positions of the links and the distribution of interaction strengths on these links.

3.3 DISCUSSION

Resilience and *reactivity* refer to properties of local stability that concern different times of the response to small perturbations, namely long-term and short-term responses, respectively. Not surprisingly, these two dynamical properties generate very different structural patterns. We found that resilience increases the presence of cycles and slightly decreases the modularity of the system, with typically fewer clusters than for the random case. In contrast, selection for low reactivity generates the inverse pattern, with matrices that exhibit almost no cycles but more modularity than expected at random. The implication is that modular networks confer stability in the transients of linear systems, and therefore, dampen deviations from equilibrium in the transients of nonlinear ones. Our main finding that modularity relates to stability in the transients is consistent with the earlier proposal that clusters tend to dampen disturbances, preventing their spread through the food web, even though this idea was typically based

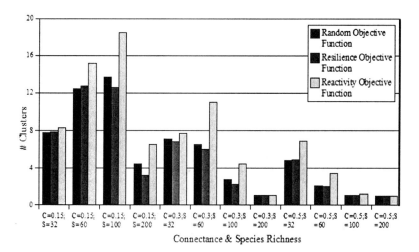

FIGURE 3   Number of clusters (average over the genome population at the final time step) for the different values of connectance (C), species richness (S), and objective function. Black bars represent the random case. The results for the resilience objective function are shown with dark gray bars, while light gray bars are used for the reactivity objective function. X axes represent connectance and species richness. Basically, there is no difference between the resilience and random case, while the reactivity case always exhibits a higher number of clusters. The number of clusters presents maximum values for mean community size, and as connectance is increased the maximum is shifted to smaller communities.

on long-term stability (May 1971, 1973; Pimm and Lawton 1980; Krause et al. 2003).

The proportions of strong and weak interactions do not show significant departures from the random case. The mean and variance of interaction strengths do not differ either, contrary to recent findings (Jansen and Kokkoris 2003). Such a difference could arise because Jansen and Kokkoris (2003) consider matrices with feasible equilibria (within a set of random matrices) whereas our approach involves an evolutionary search and considers only those matrices with the highest stability. Jansen and Kokkoris (2003) also recognize that stable matrices have a "special internal structure" that could not be fully mapped to the mean and variance. This implies that, in agreement with De Ruiter et al. (1995), the arrangement of interaction strength on the network is likely to be more important to stability than the values of the strengths per se. We found that weak links tend to occur in cycles for the low reactivity matrices, even though the number of cycles is small. More importantly, this arrangement is responsible for effectively structuring the network into clusters. Finally, we have

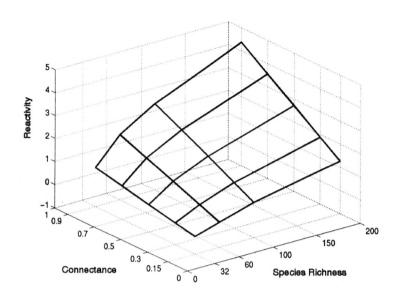

FIGURE 4 Average values for reactivity reached by the final population of genomes for several values of connectance $(C = 0.15, 0.3, 0.5, 0.7, 0.9)$ and species richness $(S = 32, 60, 100, 200)$. It is clear that increments in connectance for a given number of species, or increments in species richness for a given connectance, lead to an increment in the values of reactivity. Moreover a joint increment has the strongest effect.

revisited the stability-complexity relationship, with species diversity and food web connectance as measures of complexity (May 1971, 1973). Again, our approach examines this relationship not in random matrices but in those whose structure confers the highest stability. Our findings confirm that resilience decreases for increasing species richness or connectance. But importantly, the effect of size on resilience is weak among the "top" networks, which is consistent with their lack of modularity. Similarly, reactivity increases with species richness and connectance, a result concerning the transients. Thus, species richness and connectance are destabilizing factors even when comparisons are made in the region of food web space conferring the highest stability. However, species richness is only weakly destabilizing.

We are aware of only one other study of reactivity and complexity in food webs (Chen and Cohen 2001b). This study, which generates the network of links with the cascade model, a stochastic static model of structure, also finds that reactivity typically increases with species richness. Beyond reactivity itself, Chen and Cohen (2001b) consider other measures related to transient responses and

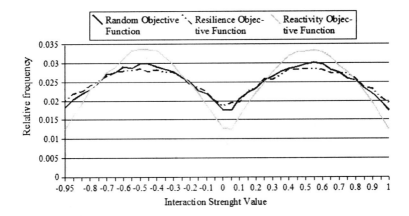

FIGURE 5   Distribution of Interaction Strength Values. The black continuous line corresponds to the results obtained with the random objective function; the dashed black line corresponds to the resilience case and the light gray line, to the reactivity case. Each "case" corresponds to an average of the distribution of interaction strength of the elite corresponding to the last time step, independently of connectance and species richness. Low frequencies at both extremes and close to zero are amplified because a strength mutation can neither generate an interaction strength with an absolute value bigger than one, nor *change* the sign of an interaction (generate an interaction bigger than zero). There is no difference between the random case and the resilience one, but reactivity generates a sharper distribution. Also, it is clear that the most frequent interaction strengths are around 0.5 (in absolute value).

to the amplification of perturbations, namely the size and timing of maximum amplification (Neubert and Caswell 1997). They further allow for different types of possible links, including consumer-victim interactions as in our webs, but also donor-controlled and recipient-controlled interactions. They find that for a given number of species, the maximum size of the transient response can increase or decrease with connectance, depending on the types of links present in the web. We have examined a fitness function that, for reactive matrices, incorporates the size of the response and, for consumer-victim interactions, the results are similar to those reported for network selection based solely on reactivity. Thus, a critical issue appears to be, as emphasized by Chen and Cohen (2001b), what types of interactions predominate.

Another recent study that applies a genetic algorithm to examine the linear dynamics of large networks raises interesting questions on the role of modularity when community interactions go beyond traditional predator-prey links (Variano et al. 2004). Networks in that study were generic in the sense that many types of interactions were allowed, without consideration of any particular type of system.

Thus, links between two nodes could be bi-directional or uni-directional, and interaction strengths (i.e., the corresponding entries in the matrix) could take any value between −1 and 1, including 0. Matrices were selected in that study with an objective function that rewarded negative eigenvalues, with only the signs of these eigenvalues taken into account. The selection process progressively eliminates the unstable manifold of the equilibrium (that is, directions in the phase-space of the system along which perturbations move away from equilibrium). Thus, long-term stability was addressed, although without consideration of the return times to equilibrium. Interestingly, the selection process led to modularity including two main types of structures: components and strongly connected components (SCCs). The former are simply disconnected blocks of non-interacting nodes in the network. The latter are more interesting and consist of subsets of nodes whose interactions involve cycles, with reciprocal connections required between nodes of a single SCC. It is unclear why their results differ from ours on the relationship between modularity and long-term stability. One possible explanation is the lack of control for variation in connectance in their study, while we do not allow for such a change and compartments that are fully disconnected; another, the restriction to one type of interaction (predator-prey) in our study, with only consideration of compartments and not of SCCs. But notice that of the two types of structures they see in their "evolving" networks, SCCs are more prevalent than compartments (Variano, personal communication). SCCs have no parallel in our analyses because we restrict the network to predator-prey interactions and the clustering algorithm considers a single type of link (those leading from prey to predators or the other way around). Furthermore, the predator-prey interactions are of one single type, always with a negative effect of the predator on the prey's population growth rate and inversely, a positive effect of the prey on the predator's population growth rate. In other words, we do not allow for unidirectional links. It would be interesting to extend the type of interactions considered here to include, for example, donor-controlled predation as well as mutualisms and competition. In addition, networks that explicitly reflect flows and recycling should be especially interesting with respect to the existence of SCCs. A recent paper by Allesina et al. (in press) describes the occurrence of SCCs in this type of ecological network. Their dynamical consequences remain to be explored.

The study by Variano et al. (2004) considers another form of perturbation they describe as robustness, in which a link is either selected at random or specifically targeted. The weight of this link (or interaction strength) is then mutated by assignment of a random value between −1 and 1. They find that as the selection process proceeds and modularity develops, this perturbation has a weaker and weaker effect on stability (resilience). Similarly, it would be interesting to examine with our matrices how food webs respond to additions and deletions of species along trajectories of the genetic algorithm.

Evolutionary algorithms have also been used to explore the large space of network configurations for another biological system of tremendous complexity, the network of neurons in the brain (Tononi et al. 1998b). These studies explore the structures associated with high complexity and high integration of such networks, and introduce definitions of these quantities based on information theory. Thus, networks are selected for their ability to process information. The premise is that the proper functioning of the brain requires both an ability to process information and a capacity for high integration. But these are opposite requirements in the sense that "the former requires the firing of specialized groups of neurons, the latter requires that their joint activity be highly coherent" (Tononi et al. 1998b). The proposed complexity measure is high when both these needs are achieved. Interestingly, the neural networks selected for high complexity exhibit clear modularity (Sporns and Tononi 2002). Whether these notions are relevant to ecosystems remains an open question, but ecological networks must be able to isolate the spread of perturbations, while maintaiing sufficient connectivity for biomass and energy fluxes.

We have illustrated the application of genetic algorithms for exploring structures related to specific dynamical properties in linear systems. The next step is to consider properties of the full nonlinear dynamics, such as population variability and species persistence. Clearly, the computational difficulties are nontrivial, as simulations of large dimensional systems would need to be performed to evaluate these selection criteria. In the above-cited studies for networks of neurons, the evolutionary algorithm was applied to the nonlinear model with results similar to those found for the linear approximation (Sporns and Tononi 2002). Similar questions for the nonlinear dynamics of food webs remain an interesting challenge for the future.

ACKNOWLEDGMENTS

We thank Ed Baskerville for help with the initial code to implement the evolutionary search algorithm. This work was supported by a Centennial Fellowship of the James S. McDonnell Foundation to Mercedes Pascual. The Center for the Study of Complex Systems provided computing resources and support for Rick L. Riolo.

Box 1: Genetic Algorithms—Basic Definitions

Genetic Algorithms belong to the class of stochastic search methods. Whereas most stochastic search methods operate on a single solution to the problem at hand, genetic algorithms operate on a population of solutions.

To use a genetic algorithm, it is necessary to encode solutions to the problem in question in a structure that can be stored in the computer. This object is denoted as a *genome*.

The genetic algorithm creates an entire initial population of genomes, and then generates new individuals (offspring genomes) through reproduction with variation, i.e., mutation and recombination of parental traits.

The genetic algorithm uses a selection criterion to select the best individuals that serve as the parents for the subsequent generation. An *objective function* evaluates each individual and assigns to each a score, which serves to rank the entire population from best to worst. The *mutation* operator introduces a certain amount of randomness into the search to make possible the exploration of genome space. The *crossover* operator promotes the recombination of parental traits.

With regard to temporal dynamics, the *Simple Genetic Algorithm (SGA)* is a *generational* algorithm, in which the entire population is replaced each generation. Between generations, a *selection method* selects individuals for mating. In some SGAs, the best individuals, the *elite*, are carried unchanged from one generation to the next. But if the selection method picks only the best individual, the population would quickly converge to that single individual. To avoid this dead end, selection should be biased toward better individuals, but also should choose some that are not quite as good (incorporating in this way a degree of genetic diversity in the population). Some of the more common selection methods include roulette wheel selection (the likelihood of picking an individual is proportional to the individual's score), tournament selection (a number of individuals are chosen randomly, then the best of these is/are chosen for mating), and rank selection (the best individual(s) is/are selected every time).

Box 2: Genetic Algorithms—Temporal Dynamics

The genetic algorithm starts with the random creation of a whole population of genomes, and proceeds to "evolve" the population in discrete generations. The passage from one generation to the next consists of the following set of stages:

Stage 1: the population at generation t is ranked based on the objective function (from white to black in the figure).

Stage 2: the elite is selected and is copied from *generation t* to *generation t + 1*. The rest of the population for *generation t + 1* remains undefined.

Stage 3: the rest of the population at time $t+1$ is produced in our particular case by a Binary Tournament (two individuals from generation t are chosen and the best is selected to be copied to the next generation). In general, the resulting population for *generation t + 1* will be unordered. Note that the elite will participate (and win) several tournaments because selection is proportional to the individual's score as defined by the objective function.

Stage 4: all individuals in the population, except for the elite, suffer mutations.

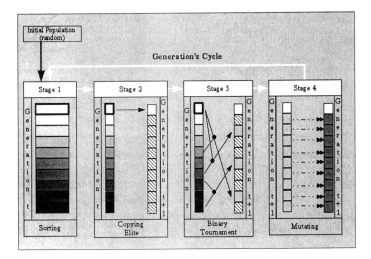

These four stages or the generation cycle are repeated until a maximum time or a particular objective function value is reached. In the final step, the last population is ranked again.

Box 3: Food-Web Clustering

A version of the non-hierarchical clustering method, denoted as k-means clustering algorithm (Wong and Lane 1983), was used to examine the interactions among species for each food web generated by the genetic algorithm. The interactions under consideration were either predator-prey or prey-predator.

This version of the algorithm determines clusters according to a *threshold* value that defines a *strong interaction*. The interactions of the food web under analysis are ordered and stored in a stack from the weakest to the strongest. Then the following steps are applied:

1. If the stack is not empty, then the strongest interaction is removed from the stack.

2. If none of the species linked by this interaction are in a cluster, create a new cluster with both species.

3. If one of the species is in a cluster and the other one is not in any cluster, then add the latter to the cluster of the first species.

4. If both are in different clusters, join the clusters only if there is a strong interaction between them.

5. Return to step 1.

In the figure, the left side shows a schematic representation of a simple (and unrealistic) food web, in which circles represent species, black arrows represent strong interactions and white arrows are non-strong interactions (smaller than the threshold). In order to maintain simplicity, this representation shows only interactions from the prey to the predator. Clusters are formed by species "connected" mainly through strong interactions and are show as gray rectangles. However, non-strong interactions can be included within the clusters in two ways: First, it can be included when a species is incorporated into a cluster due to a particular strong link, but it also has non-strong links with species that are already in that cluster (an example of this situation is shown for the species with the double white circle). Second, because the algorithm does not allow clusters of only one species, when a species has only non-strong interactions with others species, it will be incorporated into a cluster based on its stronger link, and, hence, weak links could end up incorporated into such a cluster (although in this figure all links were simplified to only two values or colors, an example of this situation is the species with the white circle).

The right side of the figure shows the matrix representation of the same schematic food web. As for the network, dark gray squares represent no interaction, black squares represent strong interactions, white squares non-strong interactions, and light gray regions indicate clusters.

Box 3 continued on next page

Box 3 Continued

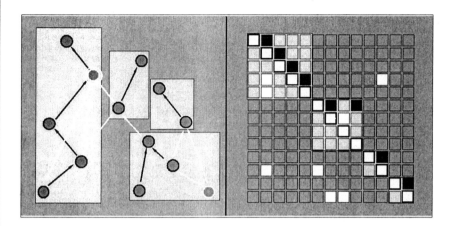

Besides the threshold value, the number of clusters obtained with this method depends on the distribution of interaction strength in the food web, and on the way species are connected. It also varies with connectance and species richness. A very low connectance can create a high number of clusters, just because there are not enough links connecting the species. For uniformly distributed interaction strengths, a higher connectance or higher species richness implies a lower number of clusters.

REFERENCES

Allesina, S., A. Bodini, and C. Bondavalli. In press. "Ecological Subsystems via Graph Theory: The Role of Strongly Connected Components." *Oikos.*

Brose, U., R. J. Williams, and N. D. Martinez. 2003. "Comment on 'Foraging Adaptation and the Relationship between Food-Web Complexity and Stability.'" *Science* 301:918b.

Chen, X., and J. E. Cohen. 2001a. "Transient Dynamics and Food-Web Complexity in the Lotka-Volterra Cascade Model." *Proc. Roy. Soc. Lond. Ser. B* 268:869–877.

Chen, X., and J. E. Cohen. 2001b. "Global Stability, Local Stability, and Permanence in Model Food Webs." *J. Theor. Biol.* 212(2):223–235.

Cohen, J. E., F. Briand, and C. M. Newman. 1990a. "Community Food Webs." In *Biomathematics*, vol. 20. Berlin: Springer-Verlag.

Cohen, J. E., T. Luczak, C. M. Newman, and Z.-M. Zhou. 1990b. "Stochastic Structure and Nonlinear Dynamics of Food Webs: Qualitative Stability in a Lotka-Volterra Cascade Model." *Proc. Roy. Soc. Lond. B.* 240:607–627.

De Angelis, D. L. 1975. "Stability and Connectance in Food Web Models." *Ecology* 56:238–243.

de Ruiter, P. C., A. Neutel, and J. C. Moore. 1995. "Energetics, Patterns of Interaction Strengths, and Stability in Real Ecosystems." *Science* 269:1257–1260.

Elton, C. S. 1958. *The Ecology of Invasions by Animals and Plants.* New York: Macmillan.

Emmerson, M. C., and D. Raffaelli. 2004. "Predator-Prey Body Size, Interaction Strength and the Stability of A Real Food Web." *J. Animal Ecol.* 73:399–409.

Fussman, G. F., and G. Heber. 2002. "Food Web Complexity and Chaotic Population Dynamics." *Ecol. Lett.* 5:394–401.

Haydon, D. T. 2000. "Maximally Stable Model Ecosystems can be Highly Connected." *Ecology* 81:2631–2636.

Holt, R. 2002. "Food Webs in Space: On the Interplay of Dynamic Instability and Spatial Processes." *Ecol. Resh.* 17:261–273.

Holyoak, M. 2000. "Habitat Subdivision Causes Changes in Food Web Structure." *Ecol. Lett.* 3:509–515.

Hutchinson, G. E. 1959 "Homage to Santa Rosalia or Why are There so Many Kinds of Animals?" *Amer. Natur.* 870:145–159.

Jansen, V. A. A., and G. D. Kokkoris. 2003. "Complexity and Stability Revisited." *Ecol. Lett.* 6:498–502.

Kokkoris, G. D., A. Y. Troumbis, and J. H. Lawton. 1999. "Patterns of Species Interaction Strength in Assembled Theoretical Competition Communities." *Ecol. Lett.* 2:70–74.

Kokkoris, G. D., V. A. A. Jansen, M. Loreau, and A. Y. Troumbis. 2002. "Variability in Interaction Strength and Implications for Biodiversity." *J. Animal Ecol.* 71:362–371.

Kondoh, M. 2003. "Foraging Adaptation and the Relationship between Food-Web Complexity and Stability." *Science* 299:1388–1391.

Krause, A. E., K. A. Frank, D. M. Mason, R. E. Ulanowicz, and W. W. Taylor. 2003. "Compartments Revealed in Food-Web Structure." *Nature* 426:282–285.

Law, R., and J. C. Blackford. 1992. "Self-Assembling Food Webs: A Global Viewpoint of Coexistence of Species in Lotka-Volterra Communities." *Ecology* 73:567–579.

MacArthur, R. H. 1955. "Fluctuations of Animal Populations and a Measure of Community Stability." *Ecology* 36:533–536.

Martinez, N. D. 1991. "Artifact or Attributes? Effects of Resolution on the Little Rock Lake Food Web." *Ecol. Monogr.* 61(4):367–392.

May, R. M. 1971. "Stability in Multi-Species Community Models." *Math. Biosci.* 12:59–79.

May, R. M. 1972. "Will a Large Complex System be Stable?" *Nature* 238:413–414.

May, R. M. 1973. *Stability and Complexity in Model Ecosystems.* Princeton, NJ: Princeton University Press, Princeton.

McCann, K. S. 2000. "The Diversity-Stability Debate." *Nature* 405:228–233.

Neubert, M. G., and H. Caswell. 1997. "Alternatives to Resilience for Measuring the Responses of Ecological Systems to Perturbations." *Ecology* 78(3):653–665.

Neutel, A., J. A. P. Heesterbeek, and P. C. de Ruiter. 2002. "Stability in Real Food Webs: Weak Links in Long Loops." *Science* 296:1120–1123.

Paine, R. T. 1980. "Food Webs, Linkage Interaction Strength, and Community Infrastructure." *J. Animal Ecol.* 49:667–685.

Pimm, S. L. 1982. *Food Webs.* London: Chapman and Hall.

Pimm, S. L. 1984. "The Complexity and Stability of Ecosystems." *Nature* 307:321–326.

Pimm, S. L., and J. H. Lawton. 1980. "Are Food Webs Divided into Compartments?" *J. Animal Ecol.* 49:879–898.

Polis, G. A., W. B. Anderson, and R. D. Holt. 1997. "Toward an Integration of Landscape and Food Web Ecology: The Dynamics of Spatially Subsidized Food Webs." *Ann. Rev. Ecol. & System.* 28:289–316.

Raffaelli, D. G., and S. J. Hall. 1992. "Compartments and Predation in an Estuarine Food Web." *J. Animal Ecol.* 61:551–560.

Rozdilsky, I. D., and L. Stone. 2001. "Complexity Can Enhance Stability in Competitive Systems." *Ecol. Lett.* 4:397–400.

Sporns, O., and G. Tononi. 2002. "Classes of Network Connectivity and Dynamics." *Complexity* 7(1):28–38.

Tononi, G., O. Sporns, and G. M. Edelman. 1994. "A Measure for Brain Complexity: Relating Functional Segregation and Integration in the Nervous System." *Proc. Natl. Acad. Sci.* 91:5033–5037.

Tononi, G., G. M. Edelman, and O. Sporns. 1998a. "Complexity and Coherency: Integrating Information in the Brain." *Trends in Cog. Sci.* 2(12):474–484.

Tononi, G., A. R. McIntosh, D. P. Russell, and G. M. Edelman. 1998b. "Functional Clustering: Identifying Strongly Interactive Brain Regions in Neuroimagin Data." *Neuroimage* 7:133–149.

Variano, E. A., J. H. McCoy, and H. Lipson. "Networks, Dynamics, and Modularity." *Phys. Rev. Lett.* 92(18):188701-1–188701-4.

Watts, D. J., and S. H. Strogatz. 1998. "Collective Dynamics of Small-World Networks." *Nature* 393:440–442.

Williams, R. J., and N. D. Martinez. 2000. "Simple Rules Yield Complex Food Webs." *Nature* 404:180–183.

Williams, R. J., E. L. Berlow, J. A. Dunne, A.-L. Barabási, and N. D. Martinez. 2002. "Two Degrees of Separation in Complex Food Webs." *PNAS* 99:12913–12916.

Wong, M. A., and T. Lane. 1983. "A *k*th Nearest Neighbor Clustering Procedure." *J. Roy. Stat. Soc. Lond. B.* 45(3):362–368.

Food-Web Structure and Dynamics: Reconciling Alternative Ecological Currencies

James F. Gillooly
Andrew P. Allen
James H. Brown

1   INTRODUCTION

There is a longstanding history of research on trophic relationships in ecology. Historically, most researchers have adopted one of two distinct ecological currencies to develop theoretical foundations and test theories against data.

Community ecologists have traditionally used the currency of abundance, N, to characterize the structure and dynamics of food webs. They have typically taken single species, or small groups of ecologically similar "trophic species," as the fundamental units of analysis. They have based their conceptual approaches on the theories of population dynamics, species interactions, and evolution by natural selection. Together these theories comprise the concept of Malthusian-Darwinian dynamics, which represents the outcome of three kinds of fundamental individual- and population-level processes:

1. Facultative adaptive changes in performance of individuals. These are of two basic types: (a) shifts in diet and other patterns of food acquisition by con-

Ecological Networks: Linking Structure to Dynamics in Food Webs, edited by Mercedes Pascual and Jennifer A. Dunne, Oxford University Press.   209

sumers, including changes in the kinds of prey items in the diet e.g., as predicted by models of optimal foraging and search images (see Krebs and Davies 1978), and changes in rates of food intake, e.g., as predicted by models of functional responses and satiation (see Holling 1959); and (b) shifts in behavior of prey to avoid predators, including such things as aggregation, microhabitat selection, and alteration in the timing of activity.

2. Numerical responses of populations. These include the wide range of population dynamics for coupled resource-consumer systems that have been described empirically and modeled theoretically (e.g., MacArthur 1955; Yodzis and Innes 1992; McCann et al. 1998; Turchin 2003).

3. Evolutionary adaptive shifts in resource-consumer interactions. These include such things as behavioral and evolutionary adaptations of consumers to utilizing new resources and increasing intake and assimilation of particular food types, and adaptations of prey to altering exposure, apparency, palatability, and digestability by predators.

Community ecologists have traditionally used these Malthusian-Darwinian dynamic concepts and more explicit N-currency models to address questions about the lengths, species diversities, and trophic structures of food chains, and the topology and dynamical stability of species interaction networks.

Ecosystem ecologists, on the other hand, have traditionally used the currency of energy, E, to characterize trophic relationships. They have typically considered trophic levels, dietary guilds, or functional groups—and not species populations—as the fundamental units of analysis. They have developed theories and models based on the physical principles of thermodynamics and mass, energy, and stoichiometric balance. They have typically addressed questions about fates of energy and materials in ecosystems, such as how differences in abiotic conditions and biotic composition affect the fluxes and pools of energy, water, and nutrients and the trophic structure and dynamics of particular terrestrial, freshwater, and marine ecosystems.

These two contrasting N- and E-currency approaches reflect longstanding historical specializations and divisions within ecology. The theoretical foundations of the N-currency approach were developed by Alfred Lotka (1925), Vito Volterra (1926), Charles Elton (1933), and Robert MacArthur (1955), among others. This seminal work laid the foundation for a flurry of activity in the 1970s and 1980s in which publications by Cohen (1977; Cohen and Briand 1984), May (1986), and Pimm and Lawton (1980; Pimm 1982; Lawton 1989) figured prominently. Theoretical foundations of the E-currency approach were most clearly articulated in Raymond Lindeman's classic paper on the trophic-dynamic aspect of ecology (Lindeman 1942). This approach was subsequently adopted and elaborated in a major research effort led by Eugene Odum (1955), Howard Odum (1957), Frank Golley (1960), Charles Kendeigh (1961), and their students. Despite the efforts of Lotka (1925), Elton (1933), and Hutchinson (1959), among others, a synthesis of these two divergent perspectives has yet to materialize.

Brown (1981) has attributed this longstanding division to the contrasting perspectives of Eugene Odum and Robert MacArthur, the two most influential
ecologists throughout the last half of the twentieth century. Fundamental differences in the approaches of these two schools of ecology led to spirited, sometimes
heated debate. Each school pointed to cases where the other's theories seemed
to conflict with empirical data and the intuition of field naturalists.

Within the last decade or so, there has been renewed interest in achieving a
more synthetic framework for trophic ecology (Yodzis and Innes 1992; Hall and
Raffaelli 1993; Polis and Winemiller 1996; Kerr and Dickie 2001; Drossel and
McKane 2003; Berlow et al. 2004; Williams and Martinez 2004). In particular,
there has been progress made in linking the N and E currencies (Yodzis and
Innes 1992; McCann et al. 1998; Kerr and Dickie 2001).

We believe recent advances toward a metabolic theory of ecology (e.g., Allen
et al. 2002; Gillooly et al. 2002; Brown et al. 2004; Savage et al. 2004; Allen et al.
2005) can build on this progress and thus provide a means of achieving greater
synthesis. In particular, metabolic theory has developed models that predict the
combined effects of three variables on metabolic rate—body size, temperature,
and stoichiometry—based on first principles of biology, physics, and chemistry
(West et al. 1997; Gillooly et al. 2001). They yield explicit, quantitative expressions that directly link the ecological currencies of abundance and energy. As we
demonstrate below, extensions of these models also predict rates of population
growth and interspecific interactions because metabolic rate ultimately governs
these processes.

2 CONCEPTUAL CHALLENGES IN TROPHIC ECOLOGY

Trophic interactions are often represented as simplified abstractions referred to
as food webs. A food web depicts trophic interactions among species as a network
of connected, interacting nodes. However, there are many different definitions for
nodes or their interactions. A node can represent a single species population or
an aggregation of multiple species that together comprise a guild, functional
group, or trophic level.

Aggregation avoids the monumental tasks of identifying all of the species
in the food web, determining their abundances, and characterizing their diets. Trophic interactions among nodes may be represented either qualitatively,
based on the presence or absence of interactions between nodes, or quantitatively.
Quantitative representations depict the magnitudes of interactions among nodes
as flows of energy or biomass, E, or as changes in numbers of individuals, N.

More standardized definitions and measurements in describing food webs
would be helpful in achieving a more unified conceptual approach to trophic ecology. Qualitative models achieve notational and computational simplicity, but at
the expense of sacrificing details necessary to ensure that Malthusian-Darwinian
dynamics and mass and energy balance are upheld. N-currency models are con-

sistent with Malthusian-Darwinian dynamics, but may violate mass and energy balance. For example, N-currency models do not account for the differential contributions to predators of prey that vary in body size and elemental composition. E-currency models, on the other hand, enforce mass and energy balance, but do not account for adaptive shifts in diet with changes in the abundances of alternative prey, or for varying impacts of consumption of alternative prey on population sizes and growth rates of predators. These examples illustrate both the strengths and weaknesses of N- and E-currency approaches. Neither is sufficient to capture all of the principles that govern the dynamics of food webs in nature. We therefore suggest that trophic ecology would benefit from integrating N-currency and E-currency approaches.

So, how are these currencies integrated? Is it possible to move toward an approach to food webs that explicitly incorporates both approaches? We believe metabolic theory offers some potential for doing so. It could be helpful in developing models of trophic ecology that characterize the nodes, and their interactions, in terms of both abundance and energy. Using metabolic theory, interactions can be measured in terms of the birth, growth, and death of individual organisms, or in terms of the production and consumption of energy and materials.

Metabolic theory has two main components. The first is a characterization of how three variables, body size, temperature, and stoichiometry, affect the metabolic rates of individual organisms, and hence rates of resource acquisition and allocation (West et al. 1997; Gillooly et al. 2001). The second is a characterization of how energy and materials combine to affect the pools and fluxes of resources in organisms and their environments (Brown et al. 2004; Savage et al. 2004; Allen et al. 2005). Thus, metabolic theory provides a means of linking Malthusian-Darwinian dynamics to the flow of energy and materials through food webs.

3   THEORETICAL FRAMEWORK: A SIMPLE MODEL

We illustrate a potential application of metabolic theory to food webs by presenting simple formulations that describe the storage and flux of energy and materials within and between two nodes in a food web (fig. 1). We develop a model that builds on the pioneering work of Yodzis and Innes (1992; see also Peters 1983; Kerr and Dickie 2001). Yodzis and Innes developed a two-species consumer-resource model, which incorporated the quarter-power allometries in food requirements, energy expenditure, and population dynamics in population interaction models. We build on some of these ideas by explicitly incorporating the effects of temperature and stoichiometry on interactions between nodes, and then making explicit the relationships between N and E currencies in these trophic interactions.

In developing this model, we make two simplifying assumptions. First, we assume that the food web is at steady state, which implies that the abundances

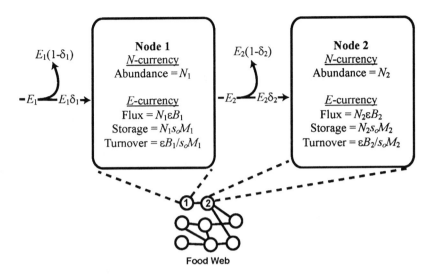

FIGURE 1   Depiction of just two nodes in a food web, predator and prey, and interactions between these nodes. The focus on just two nodes is intended to show how
metabolic theory can be used to model the storage and flux of energy and materials
within and between nodes of a food web in terms of both the number of individuals
and energy. The paired symbols N_1 and N_2 represent abundances, E_1 and E_2 represent
energy supply rates, B_1 and B_2 represent metabolic rates, δ_1 and δ_2 represent energy
assimilation efficiencies for nodes 1 and 2 (i.e., prey and predators, respectively). The
parameter ε denotes the fraction of metabolic energy allocated to growth, and the parameter s_o denotes the energy content of biomass. Both parameters are assumed to be
the same for predators and prey in our simplified formulation.

of prey and predators remain constant through time. Second, we assume that
the composition of biomass at the two nodes is identical, thus avoiding the ne-
cessity to write explicit expressions for all the substances fluxing between nodes.
These two assumptions allow us to impose mass and energy balance, and to thus
explicitly link individuals to the structure and function of the community.

Given these assumptions, the model derivation proceeds as follows. First,
we use the size and temperature dependence of individual metabolic rate, B,
to obtain expressions for the flux, turnover, and storage of energy and materials in individual organisms. Second, we derive N-currency expressions for the
abundances of prey and predators, assuming that all of the biomass produced
by prey at node 1 is used to fuel the metabolism of predators at node 2. Third,
by imposing mass and energy balance on this system using the two simplifying
assumptions discussed above, we derive expressions for energy flux that inte-

grate both ecological currencies, N and E. And finally, fourth, we quantify the relationships between these two currencies based on the size and temperature dependence of individual metabolic rate.

3.1 LINKING METABOLIC RATE TO INDIVIDUAL-LEVEL ENERGY STORAGE AND FLUX

Metabolic rate is the fundamental biological rate. It is the rate of transformation of energy and materials within an organism. As a consequence of mass and energy balance, metabolic rate determines the rate of uptake of energy and materials from the environment and the allocation of these resources to survival, growth, and reproduction. The combined effects of body size and temperature on mass-specific metabolic rate, B/M (watts/g), can be expressed as

$$\frac{B}{M} = b_0 M^{-1/4} e^{-e_a/kT} \qquad (1)$$

where b_0 is a normalization constant independent of body size and temperature (watts/g$^{3/4}$), and M is individual body mass (g) (Gillooly et al. 2001). The Boltzmann or Arrhenius factor, $e^{-E_a/kT}$, characterizes the exponential effects of temperature on metabolic rate, where E_a is the activation energy of heterotrophic metabolism ($\sim$0.65 eV), k is Boltzmann's constant (8.62×10^{-5} eV/K), and T is absolute temperature in kelvin degrees.

The amount of energy stored in an organism, S (joules), can be expressed as

$$S = s_0 M \qquad (2)$$

where s_0 is a normalization constant independent of body size and temperature that characterizes the number joules of energy contained in a gram of biomass. Energy flux in the form of biomass, F (watts), can be expressed as

$$F = \varepsilon B = \varepsilon b_0 M^{3/4} e^{-E_a/kT} \qquad (3)$$

where ε is the fraction of metabolic energy allocated to growth. Empirical data indicate that the parameter ε is a constant independent of body size and temperature (Ernest et al. 2003). For what follows, we will assume that b_0, s_0, ε, T, and E_a are identical for prey and predators. We will also assume that the body sizes of prey and predator, M_1 and M_2, are identical for all organisms comprising each of the two nodes. These assumptions can be relaxed without substantively changing our conclusions, but yield more complicated expressions.

3.2 EXPRESSIONS FOR ABUNDANCE USING THE N-CURRENCY

For illustrative purposes, we model changes in the abundances of prey and predators, N_1' and N_2' (individuals/m^2), through time, t (sec), using a pair of coupled differential equations:

$$\frac{dN_1'}{dt} = r_1 N_1' \left(1 - \left(\frac{N_1'}{N_1^*}\right)\right) - f_1(N_1', N_2') \tag{4}$$

and

$$\frac{dN_2'}{dt} = f_2(N_1', N_2') \tag{5}$$

where r_1 is the intrinsic rate of population increase for prey (1/sec), and N_1^* is the carrying capacity for prey, at steady state, when predators are absent ($N_2' = 0$). The functions $f_1(N_1', N_2')$ and $f_2(N_1', N_2')$ can take a variety of forms depending on the numerical and functional responses of predators to prey. Regardless of the functional forms of $f_1(N_1', N_2')$ and $f_2(N_1', N_2')$, the only biologically plausible solutions to these equations are those where the steady-state abundance of prey in the presence of predators, N_1, is less than the steady -tate abundance of prey in the absence of predators, N_1^* (i.e., $N_1 \leq N_1^*$). This is because $f_1(N_1, N_2) \geq 0$ due to consumption of prey by predators.

3.3   EXPRESSIONS FOR ENERGY FLUX THAT INTEGRATE THE N- AND E-CURRENCIES

We express fluxes between nodes in terms of energy by assuming that the prey and predator populations maintain steady-state abundances of N_1 and N_2, respectively, and by assuming that predators consume all of the biomass produced by prey at node 1 (fig. 1). At node 1, total energy flux, B_1^{Tot} (watts/m^2), is described by the following expression:

$$B_1^{\text{Tot}} = N_1 B_1 = N_1(1 - \varepsilon)B_1 + N_1 \varepsilon B_1 = \delta E_1 \tag{6}$$

where B_1 is the metabolic rate of an individual at node 1, $N_1(1 - \varepsilon)B_1$ is the total metabolic energy allocated to maintenance, $N_1 \varepsilon B_1$ is the total metabolic energy allocated to growth, E_1 is the energy supply rate to node 1 (watts/m^2), and δ_1 is the dimensionless efficiency of energy assimilation by individuals at node 1. Assuming that predator abundance is controlled by energy supply from node 1, total energy flux at node 2, B_2^{Tot} (watts/m^2), is equal to

$$B_2^{\text{Tot}} = N_2 B_2 = \delta_2 E_2 = \delta_2(N_1 \varepsilon B_1) = \delta_2 \varepsilon \delta_1 E_1 \tag{7}$$

where B_2 is the metabolic rate of an individual at trophic level 2, $E_2 = N_1 \varepsilon B_1$ is the energy supply rate to node 2, and δ_2 is the dimensionless efficiency of energy assimilation by individuals at node 2. Equation (6) shows that total energy flux by prey at node 1 is equal to the sum of the individual fluxes (i.e., metabolic rates) at that node. This sum is in turn constrained by the energy supply rate, E_1. Equation (7) shows that the total energy flux by predators at node 2 is constrained by the rate of biomass production by prey at node 1, $N_1 \varepsilon B_1$.

3.4 LINKING THE N- AND E-CURRENCY MODELS

Having defined expressions that link the flux and storage of energy in individual organisms to body size and temperature (eqs. (1)–(3)), that link interactions between predators and prey to their abundances (eqs. (4) and (5)), and that link abundance to energy availability (eqs. (6) and (7)), we can now explicitly characterize how body size and temperature constrain relationships between abundance and energy availability at a node through their effects on individual metabolic rate. Specifically, the abundances of prey and predator nodes can respectively be quantified by the following expressions:

$$N_1 = \frac{\delta_1 E_1}{B_1} = \frac{\delta_1 E_1}{b_0} M_1^{-3/4} e^{E_a/kT} \tag{8}$$

and

$$N_2 = \frac{\delta_2 E_2}{B_2} = \frac{\delta_2 \varepsilon \delta_1 E_1}{B_2} = \frac{\delta_2 \varepsilon \delta_1 E_1}{b_0} M_2^{-3/4} e^{E_a/kT} . \tag{9}$$

Equations (8) and (9) impose constraints on steady-state solutions to eqs. (4) and (5) based on the size and temperature dependence of individual metabolic rate (eq. (1)), and resource availability in the environment, E_1. Holding resource availability constant, these equations also predict how increases in both temperature and body size reduce carrying capacity at a node through their effects on the energetic demands of individuals (Savage et al. 2004; Allen et al. 2005).

Using the equations above, we can now express biomass storage in terms of the size and temperature dependence of individual metabolic rate. Following eqs. (2), (8), and (9), energy storage in the form of biomass (J/m^2) is equal to

$$S_1 = N_1 S_0 M_1 = \delta_1 E_1 s_0 \left(\frac{M_1}{B_1} \right) = \delta_1 E_1 \frac{s_0}{b_0} M_1^{1/4} e^{E_a/kT} \tag{10}$$

and

$$S_2 = N_2 S_0 M_2 = \delta_2 \varepsilon \delta_1 E_1 s_0 \left(\frac{M_2}{B_2} \right) = \delta_2 \varepsilon \delta_1 E_1 \frac{s_0}{b_0} M_2^{1/4} e^{E_a/kT} . \tag{11}$$

Equations (10) and (11) demonstrate the fundamental importance of mass-specific metabolic rate (B_1/M_1) and (B_2/M_2) in controlling the storage of biomass at nodes. More specifically, they show how storage increases allometrically with increasing body size, and declines exponentially with increasing temperature, due to the effects of these variables on metabolic rate (Allen et al. 2005).

Together, eqs. (1)–(11) link the N and E currencies within and between the nodes of food webs through their relationships to individual level metabolic rate, and the primary factors controlling metabolic rate, body size and temperature. In doing so, they provide a means of capturing both Malthusian-Darwinian dynamics, characterized by the currency of N, and the laws of thermodynamics and mass and energy balance, characterized by the currency of E.

4   REFLECTIONS AND CONCLUSIONS

The model presented above makes a number of simplifying assumptions so as to reduce the complexity to a bare minimum. In particular, the model derivation assumes that the nodes are at dynamic equilibrium, in other words, steady state. The model predicts how steady-state values of N and E will change in response to spatial or temporal changes in abiotic environmental variables, such as resource supply and temperature, and to shifts in biotic composition, such as changes in body size distributions or in replacements of ectotherms by endotherms at a node. So, for example, if the nodes are comprised of ectotherms, an increase in environmental temperature due to global warming will exponentially increase all rates of population change and energy flux as described by eq. (1).

However, it is important to recognize that metabolic rate, and thus body size and temperature, also constrain parameters governing nonequilibrium dynamics. Specifically, the intrinsic rate of population increase (r_1 in eq. (4)), the rate of individual mortality, and the rate of ontogenetic growth are all similarly constrained by metabolic rate and, therefore, show the same size and temperature dependence as in eq. (1) (West et al. 2001; Gillooly et al. 2002; Savage et al. 2004). So, for example, a shift in body size at a node, such as may occur by the invasion and replacement by a smaller species, should alter both N- and E-currency rates as given by eq. (1). It is conceptually straightforward, but computationally more complicated, to increase the model's realism by incorporating these more complex dynamics.

More explicitly linking the two fundamental currencies, N and E, may yield new insights into some longstanding questions in trophic ecology. For example, interaction strengths between nodes are often measured in terms of N. However, understanding the consequences of these interactions for energy flow through food webs requires expressing interaction strength in terms of E. The currency chosen to evaluate interaction strength can have profoundly different implications. So, for example, consumption of a rare prey species may be an unimportant N-currency interaction for an abundant predator species if this prey simply supplies energy (biomass) and there are alternative prey species available. On the other hand, consumption of this rare prey may represent an important E-currency interaction if it provides the sole or primary supply of an essential nutrient for the predator species.

Energy and material flow impose important structural (topological) and dynamical constraints on attributes of entire food webs, such as the number of trophic levels (Lindeman 1942) and size-abundance relationships within and across trophic levels (Kerr and Dickie 2001; Jennings and Mackinson 2003; Brown et al. 2004). So, for example, there has been an ongoing debate as to whether food chain length is limited by attenuation of energy supply or dynamical population instability at higher trophic levels (Post et al. 2000). Arguments have traditionally been made by efforts to show that one or the other phenomenon could limit the persistence of the highest trophic level, rather than by examining the joint effect of N- and E-currencies and ensuring that the fundamental principles of population persistence and mass and energy balance are both obeyed. Additionally, the body sizes of the organisms occupying different nodes or trophic levels have enormous implications for the rates of flows and the sizes of pools for both N- and E-currencies. For example, plant communities subject to similar abiotic conditions but comprised of different-sized individuals, such as an adjacent forest and early successional old field, are predicted by metabolic theory and observed empirically to have similar rates of energy flux (productivity) but dramatically different energy storage pools and, hence, correspondingly different energy turnover rates (Allen et al. 2005). For another example, the biomass invariance observed empirically across pelagic marine food chains (e.g., Sheldon et al. 1977; Kerr and Dickie 2001) can potentially be explained as a straightforward consequence of the effect on energy flux of the approximately four orders of magnitude increase in body size between successive trophic levels (Brown and Gillooly 2003; Brown et al. 2004).

We believe that the approach taken here offers potential for a more synthetic approach to trophic ecology for two reasons. First, by explicitly linking Malthusian-Darwinian dynamics to mass and energy balance, the structure and dynamics of food webs can be better conceptualized theoretically and better studied empirically. Second and more generally, this approach serves to break down the specialization that has historically separated community and ecosystem ecology. This separation was due largely to treating the N- and E-currencies as separate and distinct entities. The emerging metabolic theory of ecology makes explicit the linkages between N and E by quantifying the material and energetic demands of survival, growth, and reproduction.

REFERENCES

Allen, A. P., J. F. Gillooly, and J. H. Brown. 2005. "Linking the Global Carbon Cycle to Individual Metabolism." *Functional Ecology* 19:202–213.

Allen, A. P., J. H. Brown, and J. F. Gillooly. 2002. "Global Biodiversity, Biochemical Kinetics, and the Energetic-Equivalence Rule." *Science* 297:1545–1548.

Allen, A. P., J. F. Gillooly, and J. H. Brown. 2005. "Linking the Global Carbon Cycle to Individual Metabolism." *Functional Ecology* 19(2):202–213.

Berlow, E., A-M. Neutel, J. E. Cohen, P. deRuiter, B. Ebenman, M. Emmerson, J. W. Fox, V. A. A. Jansen, J. I. Jones, G. D. Kokkoris, D. O. Logofet, A. J. McKane, J. M. Montoya, and O. Petchey. 2004. "Interaction Strengths in Food Webs: Issues and Opportunities." *J. Animal Ecol.* 73:585–598.

Brown, J. H. 1981. "Two Decades of Homage to Santa-Rosalia: Toward a General Theory of Diversity." *Amer. Zool.* 21:877–888.

Brown, J. H. and J. F. Gillooly. 2003. "Ecological Food Webs: High-Quality Data Facilitate Theoretical Unification." *PNAS* 100:1467–1468.

Brown, J. H., J. F. Gillooly, A. P. Allen, V. M. Savage, and G. B. West. 2004. "Toward a Metabolic Theory of Ecology." *Ecology* 85:1771–1789.

Cohen, J. 1977. "Food Webs and the Dimensionality of Trophic Niche Space." *Proc. Natl. Acad. Sci. USA* 74:4533–4536.

Cohen, J., and F. Briand. 1984. "Trophic Links of Community Food Webs." *Proc. Natl. Acad. Sci. USA* 81:4105–4109.

Drossel, B., and A. J. McKane. 2003. "Modelling Food Webs." In *Handbook of Graphs and Networks: From the Genome to the Internet*, ed. S. Bornholdt and H. G. Schuster, 1–34. Berlin: Wiley.

Elton, C. 1933. *The Ecology of Animals.* London: Methuen.

Ernest, S. K. M., B. J. Enquist, J. H. Brown, E. L. Charnov, J. F. Gillooly, V. M. Savage, E. P. White, F. A. Smith, E. A. Hadley, J. P. Haskel, S. K. Lyons, B. A. Maurer, K. J. Niklas, and B. Tiffney. 2003. "Thermodynamic and Metabolic Effects on the Scaling of Production and Population Energy Use." *Ecol. Lett.* 6:990–995.

Gillooly, J. F., J. H. Brown, G. B. West, V. M. Savage, and E. L. Charnov. 2001. "Effects of Size and Temperature on Metabolic Rate." *Science* 293:2248–2251.

Gillooly, J. F., E. L. Charnov, G. B. West, V. M. Savage, and J. H. Brown. 2002. "Effects of Size and Temperature on Developmental Time." *Nature* 417:70–73.

Golley, F. B. 1960. "Energy Dynamics of a Food Chain of an Old-Field Community." *Ecol. Monogr.* 30:187–206.

Hall, S. J., and D. G. Raffaelli. 1993. "Food Webs: Theory and Reality." *Adv. Ecol. Res.* 24:187–237.

Holling, C. S. 1959. "Some Characteristics of Simple Types of Predation and Parasitism." *Canadian Entomologist* 91:385–398.

Hutchinson, G. E. 1959. "Homage to Santa Rosalia or Why are There so Many Kinds of Animals." *Amer. Natur.* 93:145–159.

Jennings, S., and S. Mackinson. 2003. "Abundance-Body Mass Relationships in Size-Structured Food Webs." *Ecol. Lett.* 6:971–974.

Kendeigh, S. C. 1961. *Animal Ecology.* Englewood Cliffs, NJ: Prentice-Hall.

Kerr, J. T., and L. M. Dickie. 2001. *Biomass Spectrum.* New York: Columbia University Press.

Krebs, J. R., and N. B. Davies 1978. *Behavioural Ecology: An Evolutionary Approach.* Oxford: Blackwell Scientific.

Lawton, J. H. 1989. "Food Webs." In *Ecological Concepts*, ed. J. M. Cherrett, 43–78. Oxford: Blackwell.

Lindeman, R. L. 1942. "The Trophic-Dynamic Aspect of Ecology." *Ecology* 23:399–417.

Lotka, A. J. 1925. *Elements of Physical Biology.* Baltimore, MD: Williams & Wilkins.

MacArthur, R. H. 1955. "Fluctuations of Animal Populations, and a Measure of Community Stability." *Ecology* 36:533–536.

May, R. M. 1986. "The Search for Patterns in the Balance of Nature: Advances and Retreats." *Ecology* 67:1115–1126.

McCann, K., A. Hastings, and G. R. Huxel. 1998. "Weak Trophic Interactions and the Balance of Nature." *Nature* 395:794–798.

Odum, E. P. 1955. "Trophic Structure and Productivity of a Windward Coral Reef Community on Eniwetok Atoll." *Ecol. Monogr.* 25:291–320.

Odum, H. T. 1957. "Trophic Structure and Productivity of Silver Springs, Florida." *Ecol. Monogr.* 27:55–112.

Peters, R. H. 1983. *The Ecological Implications of Body Size.* Cambridge, MA. Cambridge University Press.

Pimm, S. L. 1982. *Food Webs.* London: Chapman and Hall.

Pimm, S. L., and J. H. Lawton. 1980. "Are Food Webs Divided into Compartments?" *J. Animal Ecol.* 49:879–898.

Polis, G., and K. Winemiller, eds. 1996. *Food Webs: Integration of Patterns and Dynamics.* New York: Chapman and Hall.

Post, D. M., M. L. Pace, and N. G. Hairston, Jr. 2000. "Ecosystem Size Determines Food-Chain Length in Lakes." *Nature* 405:1047–1049.

Savage, V. M., J. F. Gillooly, J. H. Brown, G. B. West, and E. L. Charnov. 2004. "Effects of Body Size and Temperature on Population Growth." *Amer. Natur.* 163:430–441.

Sheldon, R. W., W. H. Sutcliffe, and M. A. Paranjape. 1977. "Structure of Pelagic Food-Chain and Relationship between Plankton and Fish Production." *J. Fish. Res. Board of Can.* 34:2344–2353.

Turchin, P. 2003. *Complex Population Dynamics: A Theoretical/Empirical Synthesis.* Princeton, NJ: Princeton University Press.

Volterra, V. 1926. "Fluctutations in the Abundance of a Species Considered Mathematically." *Nature* 118:558–560.

West, G. B., J. H. Brown, and B. J. Enquist. 1997. "A General Model for the Origin of Allometric Scaling Laws in Biology." *Science* 276:122–126.

Williams, R., and N. Martinez. 2004. "Limits to Trophic Levels and Omnivory in Complex Food Webs: Theory and Data." *Amer. Natur.* 163:E458–E468.

Yodzis, P., and S. Innes. 1992. "Body Size and Consumer-Resource Dynamics." *Amer. Natur.* 139:1151–1175.

Ecological Networks as Evolving, Adaptive Systems

"A second and much more metaphysical general point is perhaps worth a moments discussion. The evolution of biological communities, though each species appears to fend for itself alone, produces integrated aggregates which increase in stability. There is nothing mysterious about this; it follows from mathematical theory and appears to be confirmed to some extent empirically."

— G. Evelyn Hutchinson,
Homage to Santa Rosalia,
or Why are There so Many Animals?, 1959

Models of Food-Web Evolution

Alan J. McKane
Barbara Drossel

While it is often possible to model the nature of the dynamics with which constituents of a fixed network interact with each other, it is frequently much more difficult to specify the range and nature of the topologies and interaction strengths of possible networks. Here we argue, in the context of food webs, that it is necessary to go beyond simply studying dynamics on a network with a fixed number of constituents and fixed interaction strengths, and to couple this dynamics to processes connected with the construction of the network. This could include dynamics for the introduction and deletion of constituents and for the changing of interaction strengths, depending on the states of the local constituents. In the case of food webs, where the constituents are species and the interactions are predator-prey relationships, this involves going beyond conventional population dynamics to include mechanisms which allow for the introduction and extinction of species and changing of feeding preferences. We review a model of food-web evolution that we have been involved in developing over the last few years, which encapsulates this

Ecological Networks: Linking Structure to Dynamics in Food Webs,
edited by Mercedes Pascual and Jennifer A. Dunne, Oxford University Press.   223

philosophy. We also discuss a variety of other models that have been put
forward.

1 INTRODUCTION

The complexity and seemingly unlimited variation found in ecological communities prove endlessly fascinating. It is a daunting task to try to establish any general principles to govern the structure of such systems. One aspect of multispecies
communities that may have universal attributes is the network of predator-prey
links: the community food web. Empirical work carried out over the last decade
or two indicates that a range of measured quantities that characterize the web
display measurable trends (Drossel and McKane 2003). Food-web data are no-
toriously difficult to collect, but it is improving all the time, as is our ability
to identify important characteristics which are common to some food webs and
those which are specific to a particular web.

If there are similarities between food webs in such diverse communities (for
instance, marine, desert or lake communities), then these must be due to rather
fundamental aspects of the community structure; perhaps general principles
which govern predator-prey relationships, flexibility in the choice of prey, body
size of predators versus prey, and so on. If this is so, then it follows that we
should be able to reproduce the universal aspects of the food-web data by building models which incorporate these attributes, but neglect less essential details.
Of course, the difficulty lies in determining what these attributes are. However,
the number of possibilities is not so large as to prohibit a systematic investigation. In this way it should be possible to identify the important ingredients that
dictate basic food-web structure.

To pursue this program requires us to "construct" food webs based on various
general principles, and compare the results with empirical webs. One way forward
is to construct typical networks which represent food webs by assigning links
between species according to a rule which captures some structural attributes
of real webs. A large number of model webs are generated using this algorithm,
and average properties of these webs compared to empirical webs. The best
known example that exemplifies this approach is the niche model (Williams and
Martinez 2000). We will not be discussing models of this type here: we believe
that a more realistic web structure will be obtained by allowing the web to evolve.
In this approach the web will again be constructed according to reasonable rules,
but now these rules will govern the dynamics of web evolution. This has an
advantage, since there will be properties of the webs which are important, but
which are not known *a priori*, and which only emerge when the webs are allowed
to evolve. An example of such a property is the existence of many weak links
in evolved webs (Drossel et al. 2004; Quince et al. 2005b)—a feature that also
seems to be present in real webs (McCann et al. 1998; Berlow et al. 2004). This
property of webs was not suspected until relatively recently. There may be many

other attributes which are currently unknown, and so cannot be included in the construction of static webs, but which are instead emergent properties and so will be present in dynamically constructed webs.

We will also not be concerned with *assembly models*, in which new species are introduced into the system from a *species pool* simulating the process of immigration (Drake 1988, 1990; Law and Morton 1996; Morton and Law 1997; Lockwood et al. 1997; Law 1999). However, the species pool does not consist of species which have co-evolved. Instead they typically consist of species which are labeled *plants*, *herbivores*, *carnivores*, and so on, and so have already been assigned positions in the web to some extent. The interaction coefficients are assigned according to some plausible rule, perhaps involving body size or allowing for a mix of specialists and generalists. The same objections as were raised in the discussion of static models above, can be raised here: much of the structure and properties of the system are put in by hand instead of being allowed to emerge dynamically.

This article is, therefore, devoted to models of food webs in which the network is created purely dynamically. This involves both the introduction and elimination of species and/or individuals and the determination of the existence and strengths of predator-prey links between species. The introduction of new species/individuals can be by immigration from another geographical region or by speciation of existing species in the community. Extinction would naturally occur when the number of individuals fell to zero. A model incorporating all of these aspects of web dynamics has not yet been constructed. It would presumably be an individual-based model (IBM) with speciation due to changes in characteristics of an individual and immigration due to the arrival of one or more individuals from another location. The model would have to operate over a large range of time scales from the very short time intervals on which link strengths might change through time scales on which immigration might occur and to the very long evolutionary time scales. Such a model would take up a formidable amount of computing power to gain results.

To add to the computational problem of analyzing such a complicated model, there is also the problem of identifying the origin of any novel behavior that would be observed. Instead, if successive models of increasing complexity are built, it is easier to understand the predictions of the models, since one can keep track of them as they appear in successive versions. Two obvious simplifications that we can make to the full model described above are first, to study the non-spatial version of the model and second, to look at a population-level model, rather than an IBM. This would mean that new species would not be introduced by immigration, and that since there would not be discrete individuals, an artificial cutoff below which species are assumed extinct $N^{\min}$ (of order one) would have to be introduced. For a similar reason, speciation would consist of the reduction of the number of individuals of one species by N^{child} (also of order one), and the creation of a new species containing N^{child} individuals.

A model of this kind would still be quite complicated, because of the requirement that it span such a wide range of time scales. One of the simplest ways to deal with this is to try to separate out the most important mechanisms into processes occurring on different time scales. On short time scales we expect that the number of species will be fixed and so the dynamics will essentially be that of conventional population dynamics. Of course, there will be some crucial differences to the majority of population dynamics found in the literature. For instance, there will be a large number of species (not just two or three), that will have among them a complex set of predator-prey relationships. The reason why the population dynamics of a large number of species have been little studied in the past is because there would be literally hundreds of parameters in the model (the interaction strengths), which would make it impossible to draw any conclusions. In the approach we are advocating here, the interaction strengths between the species are determined by dynamics at different scales: the existence of links (that is, if two species are linked in a predator-prey relationship) by the dynamics of web formation, which occur at a longer time scale, and the strength of the link by this, and also by adaptive dynamics in which individuals vary the amount of effort that they put into preying on species depending on the amount of resources they obtain. In the absence of dynamics for the existence and strengths of links, there is no way to predict interaction strengths other than through some kind of plausible rules, which is the approach that we have already argued against in the case of static models.

From these comments it is clear that the choice of dynamics on time scales different than the ecological time scale—the one on which population dynamics operate—will have a significant impact on the result of calculations using population dynamics. The converse is also true, so there will be a considerable amount of feedback between these various types of dynamics. While population dynamics is a relatively well-studied subject (albeit with only a small number of species present) (Maynard Smith 1974; Pielou 1977; Roughgarden 1979), the other types of dynamics we have mentioned have hardly been studied at all. This is largely because population dynamics can be simply formulated in terms of coupled ordinary differential equations for the population sizes of the species in the system, the other types of dynamics to which we have alluded to will involve a variable number of species (due to the speciation and extinction dynamics) and a link strength which will vary adaptively to take into account other changes in the system (adaptive foraging). Both of these are difficult, if not impossible, to deal with analytically, and, moreover, there is no obvious choice for the form of the dynamics. Therefore, they will have to be introduced in the context of numerical simulations and modeling choices will have to be made.

Although we have tried to provide a discussion of the issues surrounding the modeling of food webs in this introductory section, we have naturally concentrated on issues which are relevant to the approach to modeling food webs with which we are familiar. In section 2 we will describe this approach in more detail and, in particular, describe the model we have been investigating for the last few years (Drossel et al. 2001; Quince et al. 2002; McKane 2004; Drossel et al. 2004; Quince et al 2005a; Quince et al. 2005b). In section 3, we explore the predictions of our model. Using this model as a basis, we will discuss other evolutionary models and their predictions in section 4. We will conclude in section 5 with a summary and a look to the future.

2   A MODEL OF EVOLUTIONARY DYNAMICS

In section 1, we have highlighted some of the desirable properties that a model of the evolution of food webs should have. Here we will make some choices that lead to a particular model and attempt to justify them.

One of the novel aspects of modeling this system is that the space of variables which describe the system (the number of species in the community, for example) keeps changing due to speciations and extinctions. The space of "possible species" is essentially infinite dimensional, but at any given time we are only exploring a small part of it and the small part we are exploring keeps changing with time. In addition we need to characterize species in some way that allows us to produce child species which are "mutations" of parent species. Therefore, the child species should be "close" to the parent species in some way. One way to deal with both these problems is to characterize species by traits or *features*, which are behavioral or phenotypical characteristics of that particular species (Caldarelli et al. 1998; Drossel et al. 2001). This is a macroevolutionary description—there is no attempt at a genetic description, even though ultimately this is supposed to be the origin of the change. We suppose that every species is characterized by L features out of a possible set of K features. We make no attempt to identify these features with any real attributes of individuals belonging to any species (being warm blooded or nocturnal, for instance); they are just taken to be integers $1, 2, \ldots, K$. Note that they could equally well be taken to be sets of intervals on the real axis of a multidimensional space, or some other set of distinct objects. One requirement is that there should be (effectively) an infinite number of possible species. If we take $L = 10$ and $K = 500$, as we often do in our simulations, then since $\binom{500}{10} \sim 10^{20}$, this is the case. The representation of species in terms of features allows us to define two important quantities which describe the relationship between any two species i and j. The first one of these is the *overlap* q_{ij}, which is a measure of the similarity of the two species and is defined as the number of features that are shared by species i and species j, divided by the number of features, L. The second quantity is the score of species i against species j, S_{ij}, which indicates how well species i is adapted to prey on j. We would expect that the score should depend on the quality of the features of one of the species against the other, and, therefore, we first need to introduce a more primitive notion of the score between constitutive features. These scores are more fundamental and will not change with time; they will be fixed at the start of a simulation run. The simplest assumption is to define a score matrix for feature α against feature β, denoted by $m_{\alpha\beta}$, as a $K \times K$ antisymmetric matrix of random numbers. The numbers are random, since we have no other information with which to choose them. We usually make them follow a Gaussian distribution with zero mean and unit variance, but clearly other choices are possible. A preliminary investigation of the consequences of making other choices for the structure of the matrix $m_{\alpha\beta}$ suggests that the exact form is not crucial for the eventual nature of the resulting food web. In Box 1, we illustrate how the scores between species are obtained as the sum of the scores of each feature of i against each feature of j. Note that if i is adapted to be a prey of j, the definition (1) returns a value zero; the strength of the interaction between i and j can be determined from S_{ji}, which in this case is guaranteed to be positive. Below, we will see how the scores affect the population dynamics and the link strength.

A speciation event consists of changing one of the features at random to another random feature. For example, suppose that $K = 9$ and $L = 3$ (not realistic choices, but conveniently small for illustrative purposes), then species i having a population of $N_i(t)$ individuals at time t might be characterized by the features $\{4, 7, 8\}$ and species j having a population $N_j(t)$ might be characterized by the features $\{2, 5, 6\}$. A speciation might then take place and result in an individual belonging to species i creating a new species, i', which would have the features $\{4, 6, 8\}$. The child species i' differs by a single feature from the parent species i. Clearly the order of features is not important, there should not be repeated features in a particular species, and if a speciation results in a species already present in the system, it should not be designated as a new species. After this change, the population of species i will be $N_i - 1$, of species j it will still be N_j, and of species i' it will be 1. The overlap between parent and child species is $q_{ii'} = (L - 1)/L$. The speciation dynamics is illustrated in Box 2.

The extinction dynamics is very simple: whenever the population level falls below a certain number $N^{\min}$, then eliminate the species from the system. We have frequently taken $N^{\min} = 1$ in simulations. In addition, the environment (from which basal species extract resources) is denoted as species 0. It has the same structure as other species, except that it is not subject to speciation: once it is chosen at the beginning of a simulation it remains unchanged. These rules specify the web dynamics.

The other kind of dynamics which is novel occurs on much shorter time scales: the change in the link strengths between a predator and its prey which takes place because the predator is trying to optimize its intake of resources against a background of changes in, for instance, population sizes. An example would be a predator that had access to two prey species which were very similar, but one of which had just suffered a decrease in population size. All else being equal, the predator would presumably now switch to putting more effort into preying on the prey that had increased in population. This is adaptive foraging: the continuous adaptation of foraging strategy to try to optimize the gain in resources for a given amount of effort. To formulate this mathematically, let us

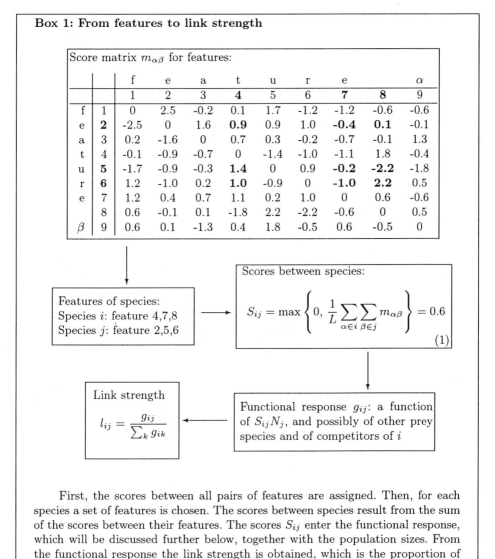

Box 1: From features to link strength

Score matrix $m_{\alpha\beta}$ for features:

		f	e	a	t	u	r	e		α
		1	2	3	4	5	6	7	8	9
f	1	0	2.5	-0.2	0.1	1.7	-1.2	-1.2	-0.6	-0.6
e	2	-2.5	0	1.6	**0.9**	0.9	1.0	**-0.4**	**0.1**	-0.1
a	3	0.2	-1.6	0	0.7	0.3	-0.2	-0.7	-0.1	1.3
t	4	-0.1	-0.9	-0.7	0	-1.4	-1.0	-1.1	1.8	-0.4
u	5	-1.7	-0.9	-0.3	**1.4**	0	0.9	**-0.2**	**-2.2**	-1.8
r	6	1.2	-1.0	0.2	**1.0**	-0.9	0	**-1.0**	**2.2**	0.5
e	7	1.2	0.4	0.7	1.1	0.2	1.0	0	0.6	-0.6
	8	0.6	-0.1	0.1	-1.8	2.2	-2.2	-0.6	0	0.5
β	9	0.6	0.1	-1.3	0.4	1.8	-0.5	0.6	-0.5	0

Features of species:
Species i: feature 4,7,8
Species j: feature 2,5,6

Scores between species:

$$S_{ij} = \max\left\{0, \frac{1}{L}\sum_{\alpha \in i}\sum_{\beta \in j} m_{\alpha\beta}\right\} = 0.6 \tag{1}$$

Link strength

$$l_{ij} = \frac{g_{ij}}{\sum_k g_{ik}}$$

Functional response g_{ij}: a function of $S_{ij}N_j$, and possibly of other prey species and of competitors of i

First, the scores between all pairs of features are assigned. Then, for each species a set of features is chosen. The scores between species result from the sum of the scores between their features. The scores S_{ij} enter the functional response, which will be discussed further below, together with the population sizes. From the functional response the link strength is obtained, which is the proportion of prey j in predator i's diet.

introduce two quantities which are central to the model. The first is the fraction of effort (or available searching time) that species i puts into preying on species j. We denote this by $f_{ij}(t)$, and obviously these efforts must satisfy $\sum_j f_{ij}(t) = 1$ for all i. The second quantity is the functional response, which plays a central role in theories of population dynamics. It is the rate at which one individual of species i consumes individuals of species j and is denoted by $g_{ij}(t)$. The choice

---

Box 2: The speciation process

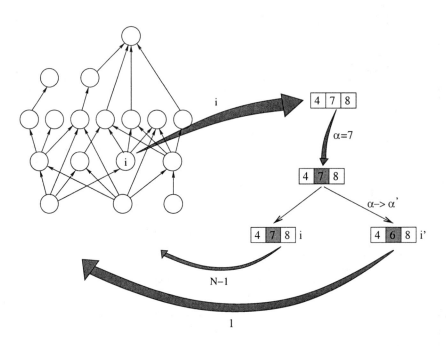

Species i is picked at random to speciate. A further random choice picks out feature α belonging to species i to change into feature α' to create species i'. If there were N_i individuals of species i before speciation, $N_i - 1$ are put back into the system after speciation, along with 1 individual of species i'.

---

of efforts of any species i such that the gain per unit effort g_{ij}/f_{ij} is equal for all prey j can be shown to be an evolutionarily stable strategy (ESS) (Drossel et al. 2001); that is, if a population has efforts chosen in this way, there is no other choice of efforts that can do better. Using the normalization condition for the f_{ij} this condition may be written as

$$f_{ij}(t) = \frac{g_{ij}(t)}{\sum_k g_{ik}(t)}. \tag{2}$$

Since in any theory of population dynamics the functional response would be specified, this equation determines the efforts, and so the dynamics of adaptive foraging. As the efforts are identical to the proportion of prey j in predator i's diet, they are also a useful measure of link strength.

Having specified these novel types of dynamics, all that remains is to define what we have called "conventional" population dynamics. The problem, however, is that while the population dynamics of one or two species have been studied for over a century, there have been few studies of multispecies communities, and even fewer where there is a large number of species, each of which will typically have several predators and prey, the identity of which will vary with time. Therefore, we had also to develop this part of the dynamics. We chose to do this by beginning with the general balance equation

$$\frac{dN_i(t)}{dt} = [\text{Rate of increase of new individuals of species } i \text{ due to predation}]$$

$$- [\text{Rate of decrease of individuals of species } i \text{ due to predation}]$$

$$- [\text{Death rate for species } i \text{ in the absence of other species}] . \qquad (3)$$

We will assume that the resources invested in a particular species is proportional to the number of individuals of that species, and will choose the constant of proportionality to be 1. Since $g_{ij}(t)$ is the rate at which one individual of species i consumes individuals of species j, $N_i(t)g_{ij}(t)$ is the rate at which species i as a whole consumes individuals of species j. Summing over all the prey j of species i gives the total rate of gain of resources for species i. However, only a fraction, λ, will be used to create new individuals of species i. Thus the first term on the right-hand side of eq. (3) is $\lambda \sum_j N_i(t)g_{ij}(t)$. By a similar argument, the second term on the right-hand side of eq. (3) is $\sum_j N_j(t)g_{ji}(t)$, where the sum on j is now over all *predators* of i. Finally, the third term is $d_i N_i(t)$, where d_i is a (constant) death rate. For simplicity we will take all the death rates equal to 1. Clearly, it is possible to be more sophisticated about the choice of death rates, as well as about the ecological efficiency, λ, but for the present at least we will content ourselves with these simple choices.

The population dynamics is now completely specified if the functional response $g_{ij}(t)$ is given. If there is no predator-prey relationship between species i and j, g_{ij} is zero, so it also defines the network at a given time. The form that $g_{ij}(t)$ takes for a given network is quite complicated, since it needs to reflect several biological mechanisms. The functional response depends on the scores S_{ij}. A simple linear functional response would take the form, $g_{ij}(t) = S_{ij}N_j(t)$; however, many choices are possible, most of which build in a greater degree of biological realism. An analysis of some of these choices (Drossel et al. 2004) reveals that the ratio-dependent functional response has many advantages ranging from purely theoretical ones, such as satisfying consistency requirements (Drossel et al. 2001), to empirical ones, such as giving rise to evolved webs which do not collapse to networks with a single level (Drossel et al. 2004). In Box 3 we build the functional response up gradually by constructing a form for $g_{ij}(t)$ in three simple situations. These are: a predator-prey pair and a triplet of species consisting of one predator with two prey and two predators with the same prey.

This describes the essentials of the model. In practice, a simulation run begins with a randomly generated species 0 (the environment) and species 1 which evolves according to the rules described in the section. There are, inevitably, a

Box 3: The functional response

In the case of a system with a single predator with a single prey species, the ratio-dependent functional response is taken to be (Drossel et al. 2001)

$$g_{ij}(t) = \frac{S_{ij} N_j(t)}{b N_j(t) + S_{ij} N_i(t)}. \tag{4}$$

The constant b controls the effectiveness of predation, and is one of the few parameters of the model (another is λ, the ecological efficiency).

The generalization of eq. (4) to a system of three species—two prey species consumed by one predator—might be thought straightforward: if i is the predator and j and j' are the prey, $g_{ij'}$ could be simply obtained by replacing j by j' in eq. (4). However, this does not satisfy the consistency condition that if the prey are equivalent from the predator's point of view (i.e., $S_{ij} = S_{ij'}$), then the dynamics of the predator population should be identical to the case where there is just one population of size $N_j + N_{j'}$. This condition can be satisfied if the efforts $f_{ij}(t)$ and $f_{ij'}(t)$ are used in constructing the functional response. It is straightforward to check that the choices

$$g_{ij}(t) = \frac{S_{ij} f_{ij}(t) N_j(t)}{b N_j(t) + S_{ij} f_{ij}(t) N_i(t)}, \ g_{ij'}(t) = \frac{S_{ij'} f_{ij'}(t) N_{j'}(t)}{b N_{j'}(t) + S_{ij'} f_{ij'}(t) N_i(t)}, \tag{5}$$

do satisfy this consistency requirement. Another system that brings out an important aspect of the structure of the functional response, is the case of two predators, i and k, having a single shared prey species, j. We would then expect the functional responses to be reduced by the competition from the other predator, that is,

$$g_{ij}(t) = \frac{S_{ij} N_j(t)}{b N_j(t) + S_{ij} N_i(t) + \alpha S_{kj} N_k(t)}, \ g_{kj}(t) = \frac{S_{kj} N_j(t)}{b N_j(t) + S_{kj} N_i(t) + \alpha S_{ij} N_i(t)}. \tag{6}$$

Here α is a constant which reflects the degree of competition between species i and species j. Typically, this will increase with the number of features i and k have in common.

Such considerations, applied to very simple web structures, suggest the following form for the functional response between any two species i and j belonging to an arbitrary web:

$$g_{ij}(t) = \frac{S_{ij} f_{ij}(t) N_j(t)}{b N_j(t) + \sum_k \alpha_{ki} S_{kj} f_{kj}(t) N_k(t)}, \tag{7}$$

where

$$\alpha_{ki} = c + (1 - c) q_{ki} \quad (0 \le c \le 1). \tag{8}$$

Here k labels all predators of j, including i. Notice that if the species are identical, $k = i$, $\alpha_{ii} = c + (1 - c) = 1$ and if they have no features in common $\alpha_{ki} = c$. The competition c is another one of the parameters of the model.

few technical points which are concerned with the running of the simulation, but which are not important to the overall understanding of the model. We have glossed over these here; anyone interested in them should seek out the original paper on the model (Drossel et al. 2001). The result of a typical run, with the model parameters in an appropriate range, is shown in figure 1. It seems to have many of the broad characteristics of real webs, and as we will discuss in section 3, the average values of quantities of interest over an ensemble of these webs is in good agreement with food-web data.

In section 4, we will review other evolutionary approaches to food-web modeling and compare these with the model we have just described.

3   COMPARISON WITH DATA

Although we have been advocating a program of modeling food-web evolution, it is far from clear that a first attempt at formulating a model designed for this purpose would be successful. Even if a stable web were formed, it might be clearly incorrect: having only basal species or far too many levels, for instance. So the first test for any model webs is not strictly a comparison with data, but simply that the webs constructed using the model are *reasonable*, in the sense that they look like real food webs. Once this has been achieved, more stringent tests can be applied. These are carried out by performing a large number of runs with the same parameter values, and averaging measured characteristic web properties over all the webs which have been generated. Finally, if it appears that the model is producing realistic webs, it can begin to be used to study other aspects of food webs. These may act as other tests for the model, or they may be "numerical experiments," the analogs of which would be difficult or even impossible to carry out in the field.

We now present some results obtained with the model discussed in section 2. As we have mentioned already, it turns out that a web can be grown only for certain choices of the functional response $g_{ij}(t)$. So, for example, if the simplest type of population dynamics (Lotka-Volterra dynamics) was chosen, a typical food web which results (after enough time has been allowed to elapse so that a stationary structure can form) is shown in figure 2 (Drossel et al. 2004). All species feed on the internal resources and on the other species as well. For short, transients periods the web may have species on a second level, but it soon collapses back down to a web with a single level. Moreover, simulations that started from a complex web with several trophic levels which was stable under the population dynamics, soon collapse under the evolutionary dynamics. Such a collapse also occurred with other types of functional responses, such as the Holling (1965) (see Hassell 1978) or Beddington (1975) forms. Only with ratio-dependent functional response (Arditi and Ginzburg 1989; Hanski 1991; Arditi et al 1992; Akcakaya et al. 1995; Abrams and Ginzburg 2000) or with the introduction of artificial mechanisms which allow only the best predators to feed on a given prey species (Caldarelli et al. 1998), were realistic food-web structures found. It appears that the ability of predators to concentrate on the prey that

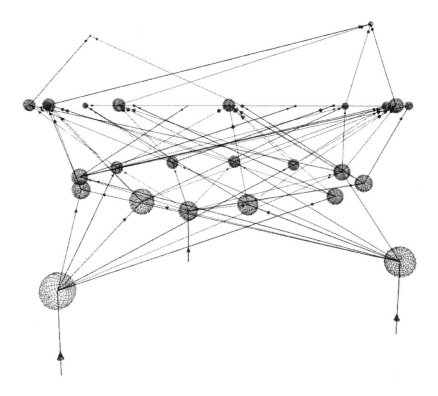

FIGURE 1   Example of a food web resulting from the evolutionary model described in section 2. The radii of the spheres which represent the species are proportional to the logarithm of the population sizes.

they are best suited to exploit, rather than on all possible prey, is essential for the production of realistic food-web structures.

If such dynamics are chosen, then evolved webs such as that shown in figure 1 seem to resemble real webs. It is, therefore, now possible to go on to the second stage of comparison with empirical webs and carry out a more detailed assessment of web properties. Table 1 shows such a comparison for a choice of basic parameters of the model: R (the rate at which resources enter the system), λ (the ecological efficiency), c (the competition parameter, see eq. (8)) and b (the effectiveness of predators, see eq. (4) or eq. (7)). The agreement is reasonable, but not perfect. In particular, the ratio of the number of links to the number of species seems to be too low. However, this figure only includes links which have efforts which are greater than 1%. This is an arbitrary cutoff; it is difficult to assess how to make this choice in order to make the best comparison with the data.

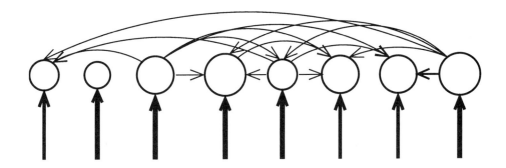

FIGURE 2   Example of a food web resulting from the evolutionary model described in section 2, but with Lotka-Volterra population dynamics.

TABLE 1   Table 1. Results of simulations of the model with $\lambda = 0.1$, $b = 5 \times 10^{-3}$ and for various values of R and c.

R	10^5	10^5	10^5	10^6
c	0.4	0.5	0.6	0.5
No. of species	79	57	55	270
Links per species	2.33	1.91	1.70	2.96
Av. level	2.38	2.35	2.28	3.07
Av. max. level	3.69	3.9	3.91	4.4
Basal species (%)	8	9	9	11
Intermediate species (%)	90	89	90	89
Top species (%)	2	2	1	1
Mean overlap level 1	0.22	0.34	0.37	0.27
Mean overlap level 2	0.08	0.12	0.13	0.15
Mean overlap level 3	0.07	0.09	0.09	0.12

The web structure in this model is in a very real sense emergent—species are not assigned to a particular level and their predator-prey interactions are not assigned beforehand in any way. Therefore, measurements of various web attributes are not influenced by preconceptions and have the potential to yield novel insights. For example, when the distribution of link strengths is measured (Quince et al. 2005b), it is found that there is a large number of weak links, as shown in figure 3. This is in line with various recent suggestions (Berlow et al. 2004). The stability of food webs may also be defined in a way which is closer to that favored by empiricists, since numerical experiments may be carried out in order to investigate the stability of webs to the deletion of species (Quince et al. 2005a).

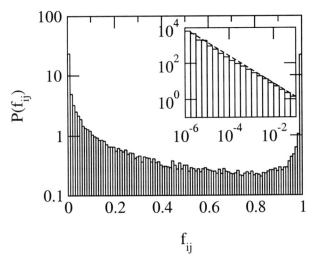

FIGURE 3   Link strength distribution resulting from the evolutionary model. The inset shows the distribution for small link strength on a log-log plot. The straight line indicates a power law with an exponent −0.75.

4   SCOPE OF EVOLUTIONARY MODELS OF FOOD WEBS

In section 1 we have tried to argue that the notion of evolving food webs in the context of computer simulations is a very natural one and in section 2 we have given a concrete example of a model of this kind. The description of this model illustrated the need to introduce dynamics of a novel kind at time scales larger and shorter than those on which population dynamics operate. The freedom associated with the specification of these processes reveals that there is a wide scope for the introduction of a variety of models.

Although there a large number of models that are possible in principle, in practice studies of this kind have only begun in recent years and so the number of distinct models is relatively small. In this section we review those that are published, but we are also aware of relevant work being undertaken by several groups as we write this article.

4.1   THE MODEL OF BASTOLLA, LASSIG ET AL.

The model by Bastolla et al. (2002) and Lassig et al. (2001) is, among all those to be discussed in this section, the one that is closest to our model. It includes dynamics on two time scales. On the shorter time scale, the model has population dynamics of the type shown in eq. (3), using a Lotka-Volterra-type functional response. In addition, the authors include a competition term $-\sum_j \beta_{ij} N_i N_j$ on the right-hand side, with the sum being taken over all competitors of species i (including species i itself). On the evolutionary time scale, speciation events occur. A new species is obtained by modifying the strength of a link or by adding or deleting a link of an existing species. The authors have performed a mean-field calculation of the expected properties of the resulting food webs analytically, finding realistic results for the mean number of species on different trophic levels. However, they have so far not published any complex food webs containing several trophic levels obtained by computer simulations of their model, and it remains to be seen whether this model is capable of generating multi-level webs if Lotka-Volterra dynamics are used. As we have shown, the choice of functional response has a large effect on the resulting web structure.

4.2   AN EVOLVING NETWORK OF MOLECULAR SPECIES

A model for evolving networks of interacting molecular species was introduced by Jain and Krishna (1998, 1999, 2002a, 2002b). Although this model is not a food-web model, it has some similarities with ecological models. In this model, species depend on other species for growth but, in contrast to food-web models, there is no negative impact on a species that facilitates the growth of another species. Furthermore, this model contains no external resources, implying that stable species configurations have a core of species that mutually sustain each other. The population dynamics are $\dot{N}_i = \sum_j c_{ij} N_j - \phi N_i$, with $c_{ij} = 1$ if j catalyzes the growth if i, and zero otherwise. Evolutionary dynamics consist of replacing the species having the smallest population size with a new species that has random couplings $\{c_{ij} \in \{0,1\}\}$ to the existing species. The probability for a nonzero value of c_{ij} to occur is p, which is the parameter of the model. The total number of species is conserved in this model. After some time, a small autocatalytic set emerges, which then expands until it comprises the entire system. The core of the autocatalytic set collapses from time to time and is replaced by a different one, accompanied by crashes in the population size. The immediate application of this model is to the study of the origin of life. However, cores of mutually sustaining species are also found in two ecological models described below, and occasional crashes of the structure are observed in one of these two.

4.3   THE APPROACH OF CATTIN ET AL.

Cattin et al. (2004) suggest a method of constructing food webs that indirectly takes evolution into account. The model is similar to the niche model insofar as species are ordered according to a "niche value." However, motivated by the existence of phylogenetic constraints, the species are made to form a nested hierarchy of consumers in the following way. First, species are ordered according to their "niche value," just as in the niche model, and the number of prey for each consumer is fixed in the same way as in the niche model. Then, trophic links are assigned in a two-stage process for each species i, starting with the species that has the smallest niche value. In stage one, prey species of consumer i are randomly chosen among species with rank less than i. Depending on this randomly chosen prey, j, two cases are possible: (1) prey j has no consumer and, therefore, the next prey of consumer i will again be randomly attributed; and (2) prey j already has one or more consumers and, therefore, consumer i joins the group of species j's consumers, and the next prey of i is then randomly chosen among the set of prey of this group. However, if the number of prey in the group is too small for choosing all remaining prey of consumer i, the remaining prey are again randomly chosen among prey without consumers. The second stage is needed if prey still cannot be attributed; the remaining prey are then chosen from species with ranks greater than or equal to i. By creating nested groups of consumers, this model is capable of faithfully reproducing the structure of real food webs. However, it differs from the other models discussed in this article as it does not explicitly model an evolutionary process.

4.4   MODELS THAT INCLUDE MUTUALISTIC COUPLINGS

Tokita and Yasutomi (2003) introduce an evolutionary model for ecosystems that also includes mutualistic couplings. Population dynamics are of the Lotka-Volterra form, with species that contain less than a fraction δ of all individuals being deleted from the pool; that is, they become extinct. A new species may be introduced in three different ways: with random couplings (an *invasive species*), or with all couplings of the parent species modified by a small random amount, or with only one pair of couplings of the parent species being modified. All three types of rules were investigated one after the other. It is found that species number and average coupling strength increase with a power law in time for the third type of evolutionary dynamics. With the two other types of rules, the species number settles at a small value (around 10), and the average coupling strength remains small. This behavior is explained by the rapid growth of groups with mutualistic mutants, which is only possible with the third type of evolutionary dynamics.

Yoshida (2003) introduces a model that is similar to the one by Tokita and Yasutomi. The main difference is that the couplings are derived from more fundamental properties. These properties are the positions in a 10-dimensional niche space and the ranges of reactivity with respect to positive and negative interactions. The positions in niche space are similar to our features: they are 10 numbers out of 100 possible numbers, but there are 2 such sets of features for each species: those features that are used to determine if species harm each other, and those that are used to determine if species benefit from each other. The ranges of reactivity are random numbers between 1 and 10 and are chosen for each species for each set of features. For instance, if the difference between the nth niche values of the two species is smaller than the reactivity of a species, this increases the probability that the species harm (or benefit) each other. Mutations occur in the space of these properties, and modified couplings are calculated from these.

Dynamics are again of the Lotka-Volterra type. Evolution leads to large stable networks that are dominated by species that harm others. However, from time to time a group of mutualists takes over, eventually destabilizing the system.

Of the above models which relate to food webs obtained through an evolutionary process, none make sufficiently detailed predictions about food-web properties that they may be compared directly with our model described in section 2. It is for this reason that we have discussed the predictions of our model in section 3, and used this section to illustrate the range of models which have been considered to date, and tried to put them in context.

5 DISCUSSION

The construction of model food webs using evolutionary dynamics and their comparison with real food webs is still at an early stage. Here we have reviewed one model in some detail, since it is capable of producing model webs which have a degree of realism that allows them to be compared with real webs. We also discussed a number of other models, although in less detail, since for various reasons they did not lead to stable, evolved predator-prey networks which could be compared to data in the same way.

The model that we have largely focused on here is well developed enough that we can examine it in light of the comments on models of dynamically evolving webs which were made in section 1. By this we mean that the essential character of the model is clear and well established, so the various aspects which are important or largely irrelevant to the construction of realistic webs can be systematically investigated. The possible extensions discussed in section 1 can also be examined. Again, a judgment can be made as to whether or not these extensions would lead to a significant change in the web structure, but ultimately it may be necessary to carry out these modifications to determine their actual importance. For example, as we saw in section 3, the choice of functional response in the population dynamics is important for the creation of realistic evolved webs. Within the context of the model we have identified some factors which the population dynamics should have in order to successfully lead to acceptable webs. One is that predators should not spread their efforts over a large number of different prey, but should focus on one or two of the species that they are best adapted to exploit. In some cases this choice may emerge naturally from the population dynamics, but in other cases it may have to be imposed as a constraint, since the model typically allows a predator to have a large number of prey. Another important factor is the explicit inclusion of terms f_{ij} which specify the fraction of effort that species i puts into preying on species j, and which change in an adaptive way that reflects other changes in the web. Such terms were not included in the first version of the model (Caldarelli et al. 1998), and lead to webs that reached a state in which new species could no longer be introduced.

It seems clear, however, that many choices made in the construction of the model will be unimportant to the final web structure. So while the way species are defined (with no reference to the underlying genetics), and the way that *speciation* occurs (as a single stochastic event which changes one of the phenotypic or behavioral features), are to an extent arbitrary, we believe that other reasonable choices would have resulted in similar webs. Of course, this is an assertion which has to be tested, and this is being undertaken as we write. To begin with, the number of features, the nature of the $m_{\alpha\beta}$ matrix, the precise form of the speciation mechanism, and other specific factors can be varied and the resulting effects on web structure monitored. Later on, attempts could be made to alter some of the more fundamental aspects of the structure of the model. In addition, it is not obvious that extending the model in some of the ways outlined in section 1 would improve the type of webs produced. For instance, reformulating the model as an IBM would take a lot more computer time and might not be worth the effort. The functional responses would have to be translated into interactions between individuals which could be quite complicated, although alternatively it might turn out that this could be modeled by a relatively simple set of mechanisms. The birth-death process would now be stochastic, with no need to enforce *species extinction*. Speciation could take place as before, except that assuming a constant rate of speciation for each individual would make the choice of species undergoing speciation proportional to the population of the species. This was, in fact, the choice made in the original version of the model (Caldarelli et al. 1998), and we do not believe that it is an important choice. Of the possible extensions discussed in section 1, one of the most interesting and probably the most important is the introduction of some spatial aspect. Clearly, making the model spatially explicit would be very costly in terms of computer time, but introducing some notion of biogeography would be valuable.

While some indications of the factors that need to be considered during model building are emerging from the study of the principal model which we have described, whether or not these are necessary remains to be seen. So few different kinds of models have been explored that it is premature to be too dogmatic about the necessary ingredients for evolutionary models. For this reason a large diversity in the types of evolutionary models which are produced is to be welcomed, and it is also for this reason that in our discussion of other approaches, we included models that were not strictly evolutionary models of food webs. These included a model in which the nodes of the evolving network were molecular species (Jain and Krishna 1998, 1999, 2002a, 2002b).

At the same time as the developments we have described in this review were occurring, other researchers were investigating the evolution of other kinds of networks (Albert and Barabási 2002; Dorogovtsev and Mendes 2002). Among these networks are social networks, the Internet, and citation networks. These types of networks typically grow by adding new nodes that become connected to several existing nodes. The emerging structure of the network depends on the growth rule, and there are some models than can even be evaluated analytically.

Characteristic features of such networks are the *small-world* property, which means that the average distance between nodes grows only logarithmically with the number of nodes, and the property of being *scale free*. This means that the number of nodes with a given connectivity is a power law in the connectivity. These growth models are simpler than food-web models based on evolutionary dynamics, since they are usually confined to the evolutionary scale and, in addition, do not include short-term dynamics of a process happening on the network. Certainly, there exist such processes that might determine further growth or survival of nodes, and the progress made with food-web models at combining the different time scales may inspire similar progress in those types of models. On the other hand, the methods developed for studying structure and growth of these other types of networks could inspire new approaches to food webs. In any event, we expect that the number of evolutionary models of food webs which will appear in the next few years will increase significantly, and the greater variety of these models should bring the outlines of the subject into sharper focus.

ACKNOWLEDGMENTS

We wish to thank Carlos Lugo for providing figure 1.

REFERENCES

Abrams, P. A., and L. R. Ginzburg. 2000. "The Nature of Predation: Prey Dependent, Ratio-Dependent or Neither?" *Trends Ecol. & Evol.* 15:337–341.

Akcakaya, H. R., R. Arditi, and L. R. Ginzburg. 1995. "Ratio-Dependent Predation—An Abstraction that Works." *Ecology* 76:995–1004.

Albert, R., and A-L Barabási. 2002. "Statistical Mechanics of Complex Networks." *Rev. Mod. Phys.* 74:47–97.

Arditi, R., and L. R. Ginzburg. 1989. "Coupling in Predator-Prey Dynamics: Ratio-Dependence." *J. Theor. Biol.* 139:311–326.

Arditi, R., L. R. Ginzburg, and N. Perrin. 1992. "Scale-Invariance is a Reasonable Approximation in Predation Models—Reply." *Oikos* 65:336–337.

Bastolla, U., M. Lässig, S. C. Manrubia, and A. Valleriani. 2002. "Dynamics and Topology of Species Networks." In *Biological Evolution and Statistical Physics*, ed. M. Lässig and A. Valleriani. Berlin: Springer-Verlag.

Beddington, J. 1975. "Mutual Interference between Parasites or Predators and Its Effect on Searching Efficiency." *J. Animal Ecol.* 51:597–624.

Berlow, E. L., A-M Neutel, J. E. Cohen, P. De Ruiter, B. Ebenman, M. Emmerson, J. W. Fox, V. A. A. Jansen, J. I. Jones, G. D. Kokkoris, D. O. Logofet, A. J. McKane, J. M. Montoya, and O. Petchey. 2004. "Interaction Strengths in Food Webs: Issues and Opportunities." *J. Animal Ecol.* 73:585–598.

Caldarelli, G., P. G. Higgs, and A. J. McKane. 1998. "Modelling Coevolution in Multispecies Communities." *J. Theor. Biol.* 193:345–358.

Cattin, M. F., L. F. Bersier, C. Banasek-Richter, R. Baltensperger, and J. P. Gabriel. 2004. "Phylogenetic Constraints and Adaptation Explain Food-Web Structure." *Nature* 427:835–839.

Drake, J. A. 1988. "Models of Community Assembly and the Structure of Ecological Landscapes." In *Mathematical Ecology*, ed. T. Hallam, L. Gross and S. Levin, 584–604. Singapore: World Scientific.

Drake, J. A. 1990 "The Mechanics of Community Assembly and Succession." *J. Theor. Biol.* 147:213–233.

Dorogovtsev, S. N., and J. F. F. Mendes. 2002. "Evolution of Networks." *Adv. Phys.* 51:1079–1187.

Drossel, B., and A. J. McKane. 2003. "Modelling Food Webs." In *Handbook of Graphs and Networks: from the Genome to the Internet*, ed. S. Bornholdt and H. G. Schuster. Berlin: Wiley-VCH.

Drossel, B., P. G. Higgs, and A. J. McKane. 2001. "The Influence of Predator-Prey Population Dynamics on the Long-Term Evolution of Food Web Structure." *J. Theor. Biol.* 208:91–107.

Drossel, B., A. J. McKane, and C. Quince. 2004. "The Impact of Nonlinear Functional Responses on the Long-Term Evolution of Food Web Structure." *J. Theor. Biol.* 229:539–548.

Hanski, I. 1991. "The Functional Response of Predators: Worries about Scale." *Trends Ecol. & Evol.* 6:141–142.

Hassell, M. P. 1978. "The Dynamics of Arthropod Predator-Prey Systems." Princeton: Princeton University Press.

Holling, C. S. 1965. "The Functional Response of Predators to Prey Density and Its Role in Mimicry and Population Regulation." *Mem. Ent. Soc. Can.* 45:1–60.

Jain, S., and S. Krishna. 1998. "Autocatalytic Sets and the Growth of Complexity in an Evolutionary Model." *Phys. Rev. Lett.* 81:5684–5687.

Jain, S., and S. Krishna. 1999. "Emergence and Growth of Complex Networks in Adaptive Systems." *Comput. Phys. Commun.* 121:116–121.

Jain, S., and S. Krishna. 2002. "Crashes, Recoveries, and 'Core Shifts' on a Model of Evolving Networks." *Phys. Rev. E* 65(2 pt 2):026103.

Jain, S., and S. Krishna. 2002. "Large Extinctions in an Evolutionary Model: The Role of Innovation and Keystone Species." *Proc. Natl. Acad. Sci.* 99:2055–2060.

Lässig, M., U. Bastolla, S. C. Manrubia, and A. Valleriani. 2001. "Shape of Ecological Networks." *Phys. Rev. Lett.* 86:4418–4421.

Law, R. 1999. "Theoretical Aspects of Community Assembly." In *Advanced Ecological Theory: Principles and Applications*, ed. J. McGlade, 143–171. Oxford: Blackwells.

Law, R., and R. D. Morton. 1996. "Permanence and the Assembly of Ecological Communities." *Ecology* 77:762–775.

Lockwood, J. L., R. D. Powell, P. Nott, and S. L. Pimm. 1997. "Assembling Ecological Communities in Time and Space." *Oikos* 80:549–553.

Maynard Smith, J. 1974. *Models in Ecology.* Cambridge: Cambridge University Press.

McCann, K., A. Hastings, and G. R. Huxel. 1998. "Weak Trophic Interaction and the Balance of Nature." *Nature* 395:794–798.

McKane, A. J. 2004. "Evolving Complex Food Webs." *Eur. Phys. J. B* 38:287–295.

Morton, R. D., and R. Law. 1997. "Regional Species Pools and the Assembly of Local Ecological Communities." *J. Theor. Biol.* 187:321–331.

Pielou, E. C. 1977. *Mathematical Ecology.* New York: Wiley.

Quince, C., P. G. Higgs, and A. J. McKane. 2002. "Food Web Structure and the Evolution of Ecological Communities." In *Biological Evolution and Statistical Physics*, ed. M. Lässig and A. Valleriani. Berlin: Springer-Verlag.

Quince, C., P. G. Higgs, and A. J. McKane. 2005a. "Deleting Species from Model Food Webs." *Oikos* 110:283–296.

Quince, C., P. G. Higgs, and A. J. McKane. 2005b. "Topological Structure and Interaction Strengths in Model Food Webs." *Ecol. Model.* In press.

Roughgarden, J. 1979. *Theory of Population Genetics and Evolutionary Ecology: An Introduction.* New York: MacMillan.

Tokita, K., and A. Yasutomi. 2003. "Emergence of a Complex and Stable Network in a Model Ecosystem with Extinction and Mutation." *Theor. Pop. Biol.* 63:131–146.

Williams, R. J., and N. D. Martinez. 2000. "Simple Rules Yield Complex Food Webs." *Nature* 404:180–183.

Yoshida, K. 2003. "Evolutionary Dynamics of Species Diversity in an Interaction Web System." *Ecol. Model.* 163:131–143.

Phenotypic Plasticity and Species Coexistence: Modeling Food Webs as Complex Adaptive Systems

Scott D. Peacor
Rick L. Riolo
Mercedes Pascual

Many species respond to changes in the density of a second species by modifying phenotypically plastic traits including behavior. Such trait modifications can affect the interaction strength between the responding species and many other species in the system, and, therefore, pair-wise species interaction magnitudes are potentially a function of the density of multiple species. We briefly review recent theoretical studies regarding the potential role of phenotypic plasticity in species interactions (i.e., trait-mediated interactions). While these studies show a stabilizing or destabilizing role of phenotypic plasticity depending on context, we argue that using a complex adaptive systems (CAS) approach can address important features of this problem not possible using traditional theoretical approaches. The adaptive nature of individuals in the short term, and species in the long term, and the feedback between individual level phenotype and the structure and dynamics of the network of interacting species (i.e., the food web), are all hallmarks of CAS. We have constructed an individually-based computational system in which

Ecological Networks: Linking Structure to Dynamics in Food Webs,
edited by Mercedes Pascual and Jennifer A. Dunne, Oxford University Press.
245

species are created by artificial evolution. An individual organism's characteristics determine its ability to survive and reproduce, and offspring receive combinations of characteristics from parents with the possibility of slight alterations (i.e., mutations). Using this computational system, we show that phenotypic plasticity can greatly enhance, by unforeseen mechanisms, the coexistence of two competing species that share a common predator. The discovery of unforeseen processes, especially relating to processes that cross scales of organization, is one strength of using the CAS approach. Our initial studies in this simple system suggest that this approach will help us understand the role of phenotypic plasticity when applied to more complex individuals and communities.

1 INTRODUCTION

Ecological communities are among the most complex systems that scientists attempt to understand, and the food web remains one of the most useful abstractions for disentangling this complexity. A food web integrates direct consumer/resource interactions to form a large network. The food web abstraction makes it clear that species, as well as interacting directly with numerous species, also interact indirectly with vast numbers of species causing interactions that are the basis of concepts such as keystone predation, trophic cascades, and exploitative competition (reviewed in Schoener 1993; Polis and Winemiller 1996). Further, as highlighted in the lead chapter of this book (Pascual and Dunne, this volume), the relationship between the structure of food webs and emergent community properties such as species diversity and stability, invasibility, and effects of perturbations, are fundamental issues in ecology (Pimm 1984; May 2001; Neutel et al. 2002; Fussmann and Heber 2002; Williams et al. 2002; Dunne et al. 2002).

In practice, to address these issues, theorists typically represent food webs by using direct species-pair interactions, which describe how species affect each other's densities, as building blocks are used to construct more complex communities (e.g., Levine 1976; Bender et al. 1984; Holt 1987; Yodzis 1988; Leibold 1989). This tradition, however, skirts a critical element of ecological systems—the existence of phenotypic plasticity: individual animals respond to short-term environmental changes by altering their phenotypes (Agrawal 2001; West-Eberhardt 2003). These plastic responses to environmental conditions might include changes in behavior, morphology or life history (Lima 1998; Tollrian and Harvell 1999). For example, many prey species respond to changes in predator density by modifying their foraging behavior; they may become generally less active, spend increased time in refuges, and move at slower rates. These trait changes affect not only interactions with the predator that they are responding to, but also with other species in the food web. For example, a predator-induced reduction in prey activity could reduce the prey's impact on a food resource, leading to a decrease in prey growth rate and an increase in resource abundance (Huang and Sih 1991; Turner and Mittelbach 1990; Schmitz 1998; reviewed in Werner and Peacor 2003). Thus, predators can affect prey and species that the prey interact with even without actually removing (eating) prey (Wootton 1993; Abrams 1995). These types of indirect interactions are denoted trait-mediated interactions (TMIs), because they are caused by a modification of a trait (e.g., behavior, morphology or life history), rather than the abundance, of the intermediate species (Abrams et al. 1996).

There is strong evidence that TMI are ubiquitous and important in natural systems. The requisite mechanism for TMI, phenotypic plasticity in traits affecting species interactions, is universally represented in different taxa and communities (Lima 1998 reviews hundreds of examples; also Agrawal 2001; West-Eberhardt 2003). Animals have diverse and extraordinary means to detect changes in the density of other species that affect their fitness, and to adjust their behavior (and other traits such as morphology) accordingly. Such modifications will likely affect interactions with other species, given the inherent tradeoffs associated with trait modifications. In particular, changes in traits that reduce predation risk are often associated with a reduction in foraging rates. Consequently, we expect that TMI are widespread (reviewed in Werner and Peacor 2003). Further, there is evidence that the presence of many predators reduces the interaction strength between prey and prey resources by 20–80% (Peacor and Werner 2004), which can cause substantial TMIs (Box 1; Peacor and Werner 2001). There is, therefore, direct evidence that the influence of phenotypic plasticity on species interactions is widespread in nature, and, therefore, the traditional method of linking pairwise density-mediated effects can neglect an important component of species interactions that affect community structure and dynamics.

The inclusion of phenotypic plasticity in food web models introduces an important departure from conventional theory. In the traditional view, chains of species interactions may affect another species density even to the point of extinction, but the basic description of the interactions (edges of network in network theory parlance) connecting the species densities (nodes of the network) typically remains unchanged. In contrast, with the inclusion of phenotypic plasticity, the magnitude of species interactions will change dynamically as a function of changes in species densities and population dynamics.

There is overwhelming evidence that the labile nature of species interaction strength that is a consequence of phenotypic plasticity strongly affects species interaction magnitudes. In the next section, we review theoretical evidence that inclusion of phenotypic plasticity can have a strong influence on community structure and dynamics. The present state of our understanding, however, may neglect important processes, and we argue that using a complex adaptive systems (CAS) approach could provide new tools to explore this problem further. Following a CAS approach, we present a computational tool in which the dynamics and structure of a model community emerge from interacting individuals that adapt to their environment. We then present initial results of a study in

Box 1.

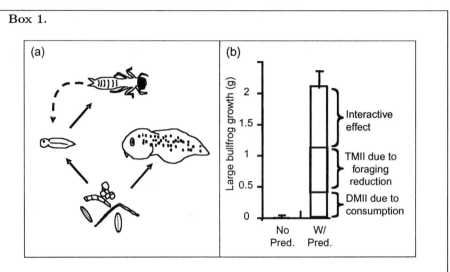

Example study of trait-mediated indirect interactions resulting from a phenotypic response of a consumer to a predator (from Peacor and Werner 2001). (a) The dragonfly predator consumed small bullfrog tadpoles (consumption indicated by straight arrows that point in the direction of energy flow), and induced a change in tadpole behavior (indicated by the dashed curved arrow); small bullfrog tadpoles avoided being preyed upon by spending less time foraging and less time in areas with higher density of resources. (a) In predator absence, the large bullfrogs did not grow, but in predator presence, large bullfrogs grew substantially due to reduced competition for resources from small bullfrog tadpoles. A number of treatments allowed an independent measure of the contribution of the trait-mediated indirect interaction (TMII—due to reduction in small bullfrog foraging) and density-mediated indirect interactions (DMII—due to a reduction in small bullfrog density) of the predator on large bullfrogs. The net indirect effect of the predator was 4 to 7 times greater than the effects due to density reduction alone. Therefore, effects caused by phenotypic plasticity had a dramatic effect, and are required to accurately describe the dynamics of this system.

which we examine the affect of phenotypic plasticity on the dynamics of two competitors in the presence of a shared predator.

2   THEORETICAL INVESTIGATIONS OF TRAIT-MEDIATED INTERACTIONS

Ives and Dobson (1987) performed one of the first analyses examining the effect of phenotypic plasticity on population dynamics, by including optimization techniques used in behavioral ecology to model prey behavior. In their model, prey modified their behavior dynamically, optimizing their fitness to changes in predator density. The population growth equations used were

$$\frac{dC}{dt} = \lambda C \left(1 - \frac{C}{K}\right) - \nu C - CyF(C, P, \nu),$$

$$\frac{dP}{dt} = cCPF(C, P, \nu) - mP,$$

$$F(C, P, \nu) = e^{-\varepsilon\nu}\frac{q}{1 + aC}.$$

Here C and P represent prey (consumer) and predator densities, respectively. The growth rate of the prey alone is described by the logistic equation with intrinsic growth rate λ and carrying capacity K. $F(C, P, \nu)$ is the per capita predation rate of a single predator on a single prey species, which in the absence of a prey response is a Hollings type-II functional response in which the parameter q scales the predation rate, and the parameter a sets the rate at which predators become saturated by prey. The phenotypic response is described by the variable ν, which measures the magnitude of the prey's response to predator presence; higher values of ν cause an exponentially decreasing predation risk, but at the cost of a reduced growth rate proportional to ν. The coefficient e determines the "efficiency" of the antipredator behavior: at higher values of ε there are higher reductions in predation risk at the same cost. The coefficient c is the conversion efficiency of prey into predator growth rate, and m describes the density-independent mortality rate. Phenotypic changes were assumed to occur much faster than changes in density, and thus ν was determined by finding the value that optimized the growth rate of prey species. The results showed that with increasing efficiency (i.e., higher values of ε) two effects damped the predator-prey oscillatory dynamics, suggesting that the phenotypic response stabilizes dynamics. First, the phenotypic response caused the growth rate of the prey to decrease even though prey biomass is not being converted to predator biomass. This reduced the potential for the predator density to overshoot the equilibrium density. Second, the phenotypic response led to increased prey density, and, therefore, the self-regulation of the prey was more pronounced.

Several authors have argued that features not included in the Ives and Dobson model can affect the prediction that phenotypic plasticity will dampen predator-prey oscillations, and, indeed, if included phenotypic plasticity can be destabilizing. Luttbeg and Schmitz (2000; see also Abrams 1992) argue that how prey perceive predation risk can determine the timing and magnitude of phenotypic responses and consequently population dynamics. If phenotypic responses are based on experience, and estimates of predation risk are imperfect, then phenotypic responses will not reliably retard predator population growth as in the model of Ives and Dobson. Indeed, in the model of Luttbeg and Schmitz, increased average individual fitness resulted from phenotypic responses that led to more rapid changes in prey population size. Using a model based on the Nicholson-Bailey predator-prey model, Kopp and Gabriel (manuscript in preparation) further found that phenotypic plasticity can affect stability, and can be stabilizing or destabilizing, depending on the magnitude of the constitutive (non-plastic) defense of the prey.

Two other mechanisms associated with phenotypic plasticity have been recognized as being potentially stabilizing when systems with greater than two species are considered (reviewed in Bolker et al. 2003). First, much theoretical attention has been given to the one-predator-two-prey configuration. A trait-mediated indirect interaction is considered to arise if the density of one prey affects the predation rate on the second prey. In such systems, such switching behavior can stabilize communities and generate indirect mutualisms between prey, because adaptive predators switch their attention away from the more rare species (Murdoch and Oaten 1975; Holt 1984). However, the inclusion of more biological realism in this configuration can be destabilizing (see Bolker et al. for references). Such features include optimal foraging by the predator, partial preferences of prey, and competition between prey species. Second, in two-predator-one-prey systems, if prey responses to predators (e.g., reduced activity) are predator specific, then the inclusion of an adaptive response of the prey increases persistence and stability. However, if the adaptive response of the prey is similar to both predators, then the response is destabilizing to the three-species system. While the results of a two-predator-one-prey system are more straightforward than those of a one-predator-two-prey system, this is possibly because there are fewer studies of the former system (Bolker et al. 2003).

Theoretical studies indicate that in addition to affecting population dynamics, phenotypic plasticity can have profound effects on species interaction strengths, functional relationships, and system structure (e.g., Abrams 1987, 1992, 1995). For example, if phenotypic responses of prey to predators are included in models, a predator is predicted, under some circumstances, to have a net *positive* effect on individual prey or prey population growth rate (Abrams and Rowe 1996; Peacor 2002), and consequently on all trophic levels. This is in direct contrast to predictions of the celebrated "trophic cascade" in traditional models that omit phenotypic plasticity (Hairston et al. 1960; Oksanen et al. 1981; Carpenter et al. 1987).

Theoretical studies thus indicate that phenotypic plasticity can have large effects on system structure and dynamics, but there are no general rules concerning its influence (Bolker et al. 2003). And while the models and approaches used thus far have been useful, they do have limitations. First, only low-dimensional systems, typically with two or three species, have been addressed. Second, we know little about what functional forms are appropriate for representing phenotypic plasticity used at the population level in these models. Third, the mechanisms underlying the stabilizing affect of phenotypic plasticity arise only in systems where the dynamics of predator and prey density are tightly coupled.

More generally, there are properties of natural systems and their interactions that may be important for understanding the effects of phenotypic plasticity that cannot be sufficiently explored with traditional analytic approaches to theory.

For example, within-population variation of individual traits or the local nature of interactions may affect the ecological and evolutionary dynamics of the system in ways that are lost because of the "mean field" nature of population-level models (Wilson 1998). More fundamentally, it is very difficult to use traditional analytic approaches to model feedback across organizational scales, that is, to explore how the structure and dynamics of the community affects the expression of phenotypic traits of individuals, which in turn affect the structures and dynamics of the community. To do so with those approaches typically requires *a priori* assumptions about the temporal scales of responses at the individual level, about the ability of individuals to optimize their behavior, and about the distribution of within-population variability.

3   A COMPLEX ADAPTIVE SYSTEMS APPROACH TO MODELING FOOD WEBS

Understanding the role of phenotypic plasticity in food web dynamics is integral to addressing a fundamental problem in ecology: how does adaptation of traits at the individual animal level affect properties at the community level? In a broader context, this fundamental problem in ecology is a central question about all *complex adaptive systems (CAS)*; that is, how do structures and properties emerge and change at one level of organization as a result of interactions between adaptive components at lower levels of organization? While there are many definitions of CAS (Holland 1995; Gell-Mann 1995; Morowitz 2002; Bar-Yam 1997), Levin (1998) presented a more succinct definition that outlines three key elements: (a) sustained diversity and individuality of components; (b) localized interactions among components; and (c) an autonomous process that selects a subset for replication or enhancement from among those components, based on the results of local interactions. These elements underpin other characteristics associated with CAS, including continual adaptation and the generation of novelty, path-dependent histories, the emergence of hierarchical and other non-random structures, and feedback across organizational scales. Clearly, ecological systems are prototypical CAS (Levin 1998), and the basic characteristics of CAS are exactly those that are key to understanding the effects of phenotypic plasticity on the structure and dynamics of food webs.

The difficulties noted above that arise in applications of traditional analytic approaches to studying the effects of phenotypic plasticity on food web dynamics, thus reflect the same difficulties found in the study of any CAS with those approaches (cf. Holland 1995; Bar-Yam 1997; Wolfram 2002). This suggests that employing conceptual frameworks and computational approaches that are now commonly used to study a variety of CAS could help elucidate the role of phenotypic plasticity in ecological communities. In particular, *agent-based computational models (ABMs)* are often used to gain insights into CAS dynamics (Axelrod 1997; Conte et al. 1997; Michael and Willer 2002; Vicsek 2002). In such models, individual variation of components, their behaviors, and interactions are all represented explicitly, and the structure and dynamics at higher levels emerge as a consequence of interactions of component parts. Importantly, the agent traits and behaviors are the result of the adaptation of the components as they interact with each other and their environment, on both short (e.g., individual learning) and long (evolutionary) time scales (Epstein and Axtell 1996; Belew and Mitchell 1996; Vriend 2000; Gimblett 2002).

The "bottom-up" approach to studying CAS by constructing ABMs is closely related to the use of *individual-based models (IBMs)* in ecology (Huston et al. 1988; DeAngelis and Gross 1992; Judson 1994; Grimm and Railsback 2005). However, most IBMs used to model ecological systems do not include individual adaptation or evolutionary processes (Railsback 2001; Grimm and Railsback 2005). Instead, those models focus on other CAS aspects of ecological systems that require an IBM approach, that is, individual variability, discrete individuals, and local interactions.

There have been some examples of using CAS to study ecological problems (Hraber et al. 1997; Strand et al. 2002; Wilke and Adami 2002). For instance, Hartvigsen and Levin (1997) examined the influence of adaptation on the population and community dynamics of plant-herbivore interactions. The phenotypes of individual plants and herbivores were determined by parameter (i.e., gene) strings. Mutation and recombination of these parameters accompanied reproduction, and, therefore, both species "evolved." The general finding was that small-scale details of phenotypically variable individual-level interactions can lead to evolutionary dynamics that affect large-scale populations and community dynamics. In another study, Muller-Landau et al. (2003) show that a spatially explicit model that incorporates an algorithm to find evolutionary stable strategies of seed dispersal is required to comprehensively elucidate the ecological and evolutionary dynamics of long-distance seed dispersal, and in particular to investigate the consequences of potential time lags introduced by pest infestation.

4   A COMPUTATIONAL SYSTEM: DIGITAL ORGANISMS IN A VIRTUAL ECOSYSTEM

We developed a computational system, Digital Organisms in a Virtual Ecosystem (DOVE), to address ecological questions from a CAS point of view, so that we can better understand how structures and properties emerge and change at one level of organization as a result of interactions between adaptive and evolving components at lower levels of organization. In particular, DOVE was developed to examine the origin and role of phenotypic plasticity in ecological communities. DOVE is a system for implementing individual-based models of food webs consisting of adaptive organisms subject to evolution by natural selection. We create populations of model animals (digital organisms, i.e., predators, omnivores, herbivores) that evolve and interact in a virtual ecosystem. This ecosystem is a spatial landscape with multiple habitats that contain resources distributed heterogeneously (Box 2). Animals can be created with varying degrees of perception of features of the environment (e.g., predator presence) and their own internal state (e.g., hunger level), and their behavior is a function of this perception. Parameters that determine behavioral responses vary between individuals within a population, and successful combinations are reliably passed on to offspring with some small chance of mutation. In this way, optimal behaviors can evolve. Because phenotypic plasticity generally will evolve in animal populations that can perceive different environment states, DOVE allows the study of the dynamics of multiple interacting animal populations that can, but need not, exhibit phenotypic plasticity. Thus DOVE models food webs as complex adaptive systems, where phenotypic plasticity and trait-mediated interactions may lead to emergent properties (processes or patterns) at the population level arising from interactions among individual organisms that adapt to the dynamic environment.

In this section we describe the basic operation of DOVE. In the next section, we describe computational experiments with DOVE in which we examine the consequences of phenotypic plasticity in relatively simple systems, that is, in which two consumer species inhabit two habitats, one of which has higher predation risk than the second, but which also has greater resources. Although DOVE is a very flexible computational system, we discuss the parameters and mechanisms relevant to these basic configurations to simplify the following description of DOVE.

4.1   THE WORLD AND ITS INHABITANTS

The world is composed of two 30 by 15 square grids, denoted low-risk habitat and high-risk habitat, due to the difference in predation risk (see below). Each cell contains resources and may be occupied by multiple animals. Resources grow according to the discrete logistic growth equation with the same growth parameters within a habitat. In the low-risk habitat the intrinsic growth rate is equal to 1.05, and it has a carrying capacity of 1000. The growth rate in the high-risk habitat is higher, 1.1, in order to establish a trade-off between resource abundance and predation risk between the two habitats.

4.2   ANIMAL BEHAVIOR

An individual animal performs one of three actions at each time step: it can remain inactive, eat, or switch habitats. Remaining inactive involves staying on the occupied cell without eating. Eating involves randomly moving to one of nine nearest neighbor cells (the cell it is in, or a cell adjacent or diagonal) and eating 50% of the available resources on the newly occupied cell. Available resources are defined as the density of resources on the cell minus a finite amount in a refuge from herbivory (20 percent of the carrying capacity). If an animal occupies a cell

Box 2. Schematic Diagram of the Dove Computational System.

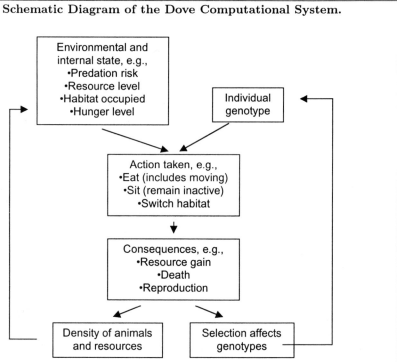

At each time step, individuals perform an action which is based on their perception of the environmental and internal state, and the parameters (denoted "genes") that map this perception into state-dependent actions. Different actions, such as eating or remaining inactive, will lead to different consequences. The consequences in turn determine the distribution of genotypes in each animal population through the process of selection (Box 3), and the density of animals and resources. These two factors then feedback to affect the next actions performed by the individuals.

The left side of the figure is typical of an individual-based model, in which individual actions affect species densities, which feed back to affect individual actions. The right side captures the evolutionary component of DOVE, in which the distribution of genotypes changes over conservative time steps due to selection. If animals can perceive and respond differently to different environmental states, they should respond adaptively to changes in the environment; i.e., there is the potential for phenotypic plasticity to evolve if it is adaptive.

on the edge of a habitat, it can move to cells on the opposite edge (i.e., each habitat is a torus). If an animal switches habitats, it moves to a random cell in the other habitat.

4.3 PREDATION RISK

We use the simplest possible representation of changing predation risk, in which a predation risk sequence alternates between predator absence and predator presence every 10 ± 1 (mean $\pm$ variance) time steps. In predator presence, predation risk is the same magnitude for each cell within a habitat, but varies between habitats (see below). We use this defined predation risk sequence, rather than a dynamic predator density that changes as a function of prey density, for two reasons. First, we are interested in identifying mechanisms that affect the stability of the system that are independent from feedback mechanisms between changes in prey and predator density. Second, in natural communities, predator density may be driven largely by factors other than particular prey densities. For example, in the pond community studied by the authors (Peacor and Werner 2001), the density of *Anax sp.* larval dragonfly predators at the local level is not determined by the local density of any particular larval frog (tadpole) prey density, because *Anax sp.* dragonflies are migratory and because larval dragonflies consume many different prey species.

5 GROWTH, REPRODUCTION, AND DEATH

The energy state of each individual animal is defined by its internal resource level (IRL). Animals are initiated with zero IRL, and this value increases by the amount of resources acquired (see section 4.2, Animal Behavior) when it consumes resources. When sufficient resources are acquired, 60 in the example case below, the animal is marked as reproduction-capable, and will reproduce with a second individual from the same population that is reproduction-capable (if one exists). Reproduction results in the IRL of both parents being set to 0, and one offspring placed in a random cell in the habitat of one of the parents (chosen randomly). There is no loss of internal resources, for example, due to metabolism, in the runs experiments presented here.

Behavior also affects the magnitude of predation risk. An inactive animal has a two percent probability of being preyed upon in predator presence, if it is in the low-risk habitat, while the probability is ten times this if it is active (i.e., either eating or switching habitats). If the animal is in a high-risk habitat, both of these probabilities are doubled. This probability distribution reflects patterns observed in natural systems, in which foraging activity increases vulnerability, and predation risk is strongly dependent on habitat.

In addition to death by predation, consumers can also die due to "background" causes that are independent of the explicitly represented features of the food web. There is a three percent probability that a consumer will die at each time step (independent of IRL, behavior, or habitat).

5.1 GENES, ENVIRONMENT, AND BEHAVIOR

Behavior of the individual organisms is a function of the individuals' genes and perception of the environmental state. In the experiments described here, each consumer can perceive which habitat (high vs. low risk) it is in. To study the effects of phenotypic plasticity, we compare populations of consumers that can perceive predator presence/absence (*P-consumers*) to those that cannot (*NP-consumers*).

P-consumers can perceive predator presence/absence perfectly. Thus for P-consumers there are four distinct environment states to which they must respond, that is, the four combinations of predator present/absent and high/low risk habitat. For each of these four states, a consumer has a set of three values that determine that consumer's probability of remaining inactive ("sitting"), eating, or switching in the given environmental state (see Box 3). Each of the individual parameters is denoted a "gene" following the terminology of the genetic algorithm literature (cf. Holland 1992). We denote each group of three genes corresponding to one environmental state, which must sum to 1, a *gene-set*. Thus each P-consumer individual has 12 gene values in total (three genes for each of four gene-sets). Since each individual P-consumer can potentially have different behavior probabilities in predator presence and absence, it can exhibit phenotypic plasticity. However a P-consumer "need not" exhibit phenotypic plasticity: if it has the same values for the action probabilities in the gene-sets for predator presence and absence, then it will behave the same whether the predator is perceived or not.

NP-consumers, which cannot perceive the presence of the predator, have only two gene-sets, one for each habitat. In this case, a plastic behavioral response to predation risk is not possible.

5.2 GENES AND EVOLUTION

It is selection upon individual consumer genes that allows a population's behavior to evolve. For each individual, the initial values for the genes are chosen at random, and then normalized such that each gene-set sums to one. When two animals reproduce, their young receive a portion of each parent's genes using a single-point crossover routine (Box 3). Further, there is a two percent probability that each gene value will mutate, that is, change in a positive or negative direction by adding a value chosen from a mean-zero normal distribution with variance of 0.1. If genes mutate or a crossover event occurs within a gene-set, then the values of the genes in the gene-set are renormalized so that they again sum to 1. Over time, successful strategies may thus evolve as a result of this optimization technique known as a *genetic algorithm*, an adaptive process that is an abstract model of Darwinian evolution by variation and selection of heritable traits (Holland 1992). Genetic algorithms are routinely used in engineering and computer science (Goldberg 1989; Mitchell 1996) and more recently in biology

(Hraber et al. 1997; Hartvigson and Levin 1997; Johst et al. 1999; Drossel et al. 2001; Lassig et al. 2001; Strand et al. 2002; Wilke and Adami 2002). Note that the algorithm used here is unlike most engineering counterparts (Goldberg 1989), in that there is no explicit exogenous objective function that defines an optimal behavior. Note also that parameters that describe the consequence of a particular action (e.g., resource acquired or probability of being preyed upon) are the same for all individuals within a species. Only the parameters that determine the probability of performing a particular action (the genes) are variable between individuals. Therefore, selection acts entirely on differences among individuals in the probability of performing defined actions.

Because genes are initiated at random, low initial fitness often leads to extinction. Therefore, we use an "anti-extinction" period at the beginning of each model run to increase the probability that a population will evolve a viable strategy before going extinct (Strand et al. 2002). In particular, for the first 5000 time steps of each run, if a population's density falls below 50, the population density is doubled by duplicating each existing individual.

5.3 MODEL EVENT SCHEDULING

Each run consists of initializing the population(s) and environment, followed by a number of "steps" of the model's basic events. For each model step, events occur in the following order:

a Reproduction of all animals (randomly paired) that are reproduction-able
b Perform anti-extinction routine if required
c Resource growth
d Each animal performs one action (i.e., remain inactive, eat or switch) according to its genes. The order of individual actions is dictated by a list of all animals (from both consumer species) that is re-shuffled each model step
e Determine if each individual dies due to background causes or predation (if predation risk is present) and remove killed animals from the world
f Update predation risk state
g Record measures of interest (population density, behavioral choices, genetic composition)

Typically, model runs are executed for 20,000 or more steps (50,000 in most of the experiments described in this paper), in order to generate enough turnover of animals to allow adequate evolution of behavioral strategies.

6 EFFECT OF PLASTICITY ON THE COEXISTENCE OF TWO COMPETING CONSUMERS

We used the DOVE computational system to examine the effect of phenotypic plasticity on the coexistence of two competing consumers subject to a common predator. We addressed the question: if two consumer species optimize their behavior in the presence of variable predation risk, how will phenotypic plasticity, that is, the ability to sense and respond to the predator presence, affect the coexistence of the competing consumers? The basic approach was to allow two competing consumers to evolve together in an environment with variable predator threat, comparing two cases: (a) the two competing consumers can (but need not) evolve to perceive and respond differently to predator presence and absence (P-consumers), and (b) the competing consumers cannot perceive predators (NP-consumers). Thus, in the P-consumers case both species have the potential to evolve phenotypic plasticity, that is, different behaviors as a function of predator presence, while in the NP-consumers case, the same behavior must be exhibited in predator presence and absence.

Results for each case (experimental condition) are based on model runs of 50,000 time steps, repeated 20 times with different random number generator seeds for each run. Coexistence time of the competing pairs was defined as the number of time steps after time step 5000 (at which the anti-extinction procedure ended) until the number of one of the consumer species reached zero. To gain insight into results of these runs, we also examined the results of running the model under a number of other experimental conditions, only some of which are reported here. For example, we also ran experiments in which there was just one consumer, either with (P) or without (NP) perception of predators. In all cases described here, nearly identical species' behavior eventually evolved into the different replicates of the same treatments.

Note that while selection acts on the values that describe the probability of remaining inactive, eating, or switching habitats (Box 3), other model parameter values are fixed and were chosen to satisfy several criteria. First, a relatively high turnover rate (birth and death) of consumers was desired to expedite selection. Second, an intermediate level of herbivory was used, such that the resource level was not restricted to the upper (carrying capacity) or lower (refuge level) ranges, in order to insure resource density responses to fluctuations in consumer density. Third, predation and background death rates were chosen so that predation rate was a large, but not the sole, source of consumer mortality. Fourth, species persisted in each circumstance in the absence of competition.

7 RESULTS

We found that species coexistence was much higher for the case in which two species with the potential for phenotypic plasticity (P-consumers) evolved together, relative to the case in which phenotypical plasticity was not possible (NP-consumers). This result is robust to a series of different scenarios examined, including: (S1) the scenario described in the previous section with two habitats (high-risk and low-risk, in which resource growth rate is higher in the former), two predation risk levels, and in which both consumers had identical fixed parameters; (S2) a more complex scenario similar to (S1) but with four predation risk levels; and (S3) a scenario sharing properties with (S1) but in which there was only one habitat and the two consumers had different fixed parameters that gave one consumer an advantage at gaining resources (higher proportion of resources acquired when they ate) at the cost of higher predation risk (higher probability of being preyed upon when active). Further, we examined the robustness of the stability difference by varying a parameter of one consumer, such as the parameter that determines the amounts of resources acquired when eating, in order to vary the relative competitive advantage of the two species. In each of the three scenarios, there was a wide range of parameter space over which species with the potential for phenotypic plasticity coexisted indefinitely—that is, there was no extinction in 20 replicated runs. But when phenotypic plasticity was not possible there were no parameter values in which even a single pair coexisted.

To ascertain whether or not phenotypic plasticity is responsible for the enhanced species co-existence in runs with P-consumers, we examined the behavior and composition of the populations' gene values over time. We found that phenotypic plasticity evolved (behavior was different in predator presence than predator absence) in all runs in which the species (P-consumers) perceived predator presence and could respond differentially. While the behavior of individuals within a given population at a particular time step was usually similar, several very distinct behaviors evolved at different times and in different runs. For example, in the first scenario with two habitats and two predation risk levels, species often initially evolved a "switching" strategy, occupying the low-risk habitat in predator presence, and switching to the high-risk habitat in predator absence (presumably to take advantage of higher resources due to the trait-mediated indirect effect of the predator on resources through induced reduction in consumer foraging). After a number of time steps (the number strongly dependent on parameter values), behavior often evolved away from the switching behavior to a stationary strategy in which the species specialized in one habitat. An example of this evolution in behavior is illustrated in figure 1, in which Consumer 1 exhibited the change from switching to stationary strategies, while Consumer 2 evolved quickly to specialize in one habitat. Interestingly, we see that although the evolution by Consumer 1 to specialize in one habitat was presumably adaptive, when all individuals adopted this strategy it had a net negative effect on population density. This highlights how adaptive behavior in the DOVE computational system originates at the individual level, as opposed to models that optimize group level responses, as is often the case with other approaches. Note that in addition to the potential for evolution of plastic response in habitat

Box 3.

Evolutionary algorithm.

Environmental states:

Habitat occupied:	1	1	2	2
Predation risk level:	0	1	0	1

Actions:

1 = sit, 2 = eat, 3 = switch:	1	2	3	1	2	3	1	2	3	1	2	3

Individual genes:

Individual 13:	0	0.6	0.4	1	0	0	0	0.95	0.05	0.1	0.0	0.9
Individual 124:	0	0.55	0.45	1	0	0	0	1	0	0.2	0.0	0.8

Offspring of 13 and 124:

	0	0.55	0.45	0.83	0	0.17	0	1	0	0.1	0.0	0.9

                 ↑ Mutation   ↑ Crossover point

Genes determine behavior: At each time step, an individual animal has probability parameters (denoted *genes*) that determine which of its potential actions it will perform. These actions, however, are dependent on the animal's perception of its environmental state and internal state (e.g., hunger level). In the text example represented here, an individual will sit, eat, or switch habitats, depending on which habitat it occupies, and the predation risk level.

Evolution: The genes of two individuals from the same population are shown. Genes are initially seeded randomly and remain unchanged for an individual, but over time their distribution changes due to selection in a manner similar to that used in genetic algorithms, which are abstract models of Darwinian evolution. Two individuals (*parents*) that have acquired enough resources without being killed, reproduce, each passing a segment of genes to the offspring. There is a small probability of a gene mutation during reproduction, as illustrated; the 6th gene from the left mutated from 0 to 0.2. After a mutation (as indicated) or certain crossover events, we must normalize gene values corresponding to the same environmental state to sum to 1. Offspring with high fitness will pass on their genes. In this manner, the population adapts in a dynamic environment. It is important to note that the dynamics of the system, in addition to the state, affect the result of evolution. Different strategies evolve for the same environmental state if the duration and frequency of the environmental states is changed.

Evolved behavior example: In the example shown, individuals have a strategy that is similar to what we have observed evolve when predators are more efficient at capturing individuals in Habitat 1 and active individuals (eating or switching). The genes cause individuals to migrate fairly quickly to Habitat 1 and eat in predator absence, but to migrate immediately to Habitat 0 and sit (remain inactive) in predator presence. Note that in the example shown, the mutation increases the probability to switch in predator presence to the more risky Habitat 1. This is unlikely to be a viable strategy, and, thus, there will likely be selection against mutations like this.

preference, a plastic response in the percent time spent performing different actions within a habitat also evolved. In particular, species ate nearly 100% of the time in predator absence, but a smaller percentage of time (anywhere from 0% to a high percentage depending on the scenario and parameter values) in predator presence.

In contrast, species that could not perceive predator presence (NP-consumers) obviously could not evolve plasticity to predator presence. However, it is important to note that an adaptive generalist strategy did evolve that was different than a strategy that would evolve if predators were continually absent or present (as verified by additional runs). For example, in scenario S1, species typically evolved one of three strategies in which they: (a) specialized in the high-risk habitat and ate about 75% of the time (with the parameters described in the model description section); (b) specialized in the low-risk habitat and ate about 90% of the time; or (c) occupied both habitats (about 50% in both habitats) with little switching behavior and eating at similar rates as in (a) and (b).

There are a number of plausible mechanisms that could generate the kinds of stabilizing effects of phenotypic plasticity between two consumer species that we observed. We have carried out a number of additional experiments with DOVE in order to test these various alternative mechanisms. Here we briefly outline these mechanisms, while their more detailed description and the evidence to support them is presented elsewhere (Peacor et al. in prep).

We considered two mechanisms related to population density dynamics that could possibly contribute to higher coexistence of P-consumers. First, in all three scenarios the average population density was higher with phenotypic plasticity than without. Therefore, decreased co-existence in the non-plastic cases could be due to stochastic fluctuations around a lower population mean. To test this possible explanation, we performed additional runs with modified parameters (e.g., increased world size and resource carrying capacity) that led to higher densities without phenotypic plasticity, and we found that coexistence times were not significantly increased. Second, without plasticity, the variation in population density through time was greater than in runs with plasticity, due to rapid reductions in density in the presence of the predator. However, as expected we found that this high variation was present in runs in which there was only one NP-consumer species, but in these one-species runs this high variation never led to extinction, indicating competition must be involved in exclusion. In short, while these two mechanisms may have contributed marginally to the increase in coexistence time, they do not account for the large observed differences in coexistence time between scenarios with and without the evolution of phenotypic plasticity.

We believe that two other mechanisms, one specific to the scenarios (S1 and S2) with two habitats, and the other specific to the one habit scenario (S3), are primarily responsible for the increased stability.

First, consider the scenario (S1) with two habitats and the parameter values described in the previous section. Although the time it took P-consumers

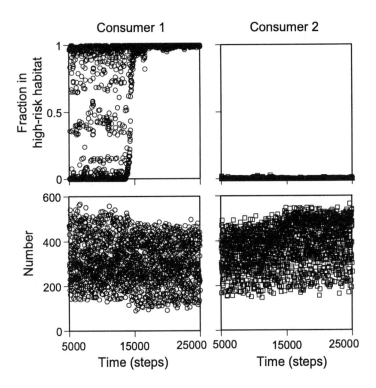

FIGURE 1   Behavior and density of two consumers (denoted Consumer 1—left panels, and Consumer 2—right panels) that evolved in Scenario 1 with two habitats. Both consumer species have identical fixed parameters, but very different "gene" values evolved that determine behavior. Consumer 1 initially switched from the high-risk to the low-risk habitat as the predator came and left. As the figure indicates, however, switching was not immediate for all individuals (gene values indicate that the probability of switching away from the high-risk habitat in predator presence was 65%, while the probability of switching to the high-risk habitat in predator absence was 50%), and, therefore, there was often an intermediate fraction of individuals on the top and bottom. After approximately 15,000 time steps, Consumer 1 evolved a different strategy in which almost all individuals remained in the high-risk habitat (right-hand portion of top left panel). In contrast, Consumer 2 initially evolved and then remained as a specialist in the low-risk habitat (top right panel). The number of individuals was approximately the same for both species (bottom panels). However, Consumer 1 had slightly more individuals initially, but after the change in behavior at approximately 15,000 time steps, its numbers fell and Consumer 2's numbers increased to levels for which Consumer 2 had higher numbers. Thus, while it was adaptive for individuals from Consumer 1 to change strategies, the net effect on the species' number was negative.

to reach a stable state was variable, a final identical stable state was always reached in all replicate runs, a state in which the two P-consumers partitioned the two habitats or, in other words, each P-consumer was a habitat specialist in a different habitat. In contrast, when NP-consumers competed, in the final state one species evolved to occupy both habitats, and it always excluded the second species. While both species often evolved in the first part of the runs to partition the habitat, each becoming a specialist in a particular habitat, in these cases one NP-consumer quickly invaded the other habitat and thereafter occupied both habitats. There was, therefore, a strong difference in the ability to invade a habitat occupied by a specialist, in that NP-Consumers invaded and excluded a competitor that specialized in a second habitat, whereas the P-Consumers did not. By carrying out additional experiments, we have found that this difference arose from differences in the structure of the adaptive landscapes facing the P- versus NP-consumers when they invade and compete with another P- or NP-consumer that specialized in a different habitat, respectively. In particular, we found that the P-consumers evolve to occupy a relatively narrower adaptive peak. We believe that the peak is narrower because with P-consumers, the genes for behavior in predator presence and absence are allowed to evolve independently. As a result, behavior with and without the predator can both be optimized to match a unique state. Thus, an invading P-consumer is easily excluded, as small deviations from the optimal behavior nearly achieved by the resident habitat specialist allow it to exclude any potential invaders. This, in turn, means two P-consumers that specialize in separate habitats will continue to exist indefinitely, each in its separate habitat. In contrast, the adaptive peaks occupied by NP-consumers were broader. This is because small deviations from the optimal phenotype are less detrimental, because the same genes are used to control behavior in the presence and absence of predation risk. This use of single gene values leads to a kind of averaging effect in which various tradeoffs are combined in one gene-value, so any small change in gene-value will result in both benefits and costs. Thus, an invading NP-consumer species can eventually exclude the resident habitat specialist due to stochastic events, because while the habitat specialist will be successful in resisting the invading species the majority of times, eventually the inferior species will be successful since it is only marginally less fit than the resident.

Next consider scenario S3, with only one habitat: again, phenotypic plasticity had a strong stabilizing effect, but we found that a very different mechanism was responsible for this stability. In S3, with phenotypic plasticity, both consumers (C1 and C2) foraged nearly 100% in predator absence, as expected, but in predator presence, C1 ate about 95% of the time while C2 did not eat. Both species thus exhibited plastic response to predator presence, however, because C1 is less vulnerable and gains less energy when it eats than C2, a less pronounced plastic response to predation risk evolved. The result was that, with phenotypic plasticity, the competitive advantage of the two consumers changed as a function of predator presence. In predator absence, both species ate at equivalent rates; however, C2 was a superior forager. In predator presence, C2 did not eat and decreased in density due to background mortality, while C1 population grew since it could forage in the absence of competition from C2. This fluctuating difference in advantage that accompanies the changes in predator presence risk caused intraspecific competition to limit both C1 and C2 growth more strongly than interspecific competition, because the population growth of each occurs primarily when each is the dominant competitor. Such limitation of growth by intraspecific, relative to interspecific, competition is a well known stabilizing mechanism of competitors (Chesson 2000). This scenario with phenotypic plasticity is in stark contrast to the case without phenotypic plasticity, in which both consumers forage at a constant rate in predator presence and absence, and one consumer quickly excluded the second species.

8 DISCUSSION

Because phenotypic plasticity causes dynamic changes in species traits in response to changes in the food web structure and dynamics, it follows that phenotypic plasticity could have profound affects on community properties such as stability. For example, when one species responds to changes in the density of another species, this response modifies both the interaction between the two interacting species, and between the responding species and other species in the community with which it interacts. The conceptual image of a food web network of phenotypically plastic species is thus one with dynamically changing interaction strengths that increase and decrease as a function of the changes in the densities of species. Such changes, especially given their expected nonlinear nature, could strongly affect the dynamics and structure of a community. Understanding the role of phenotypic plasticity in food web dynamics is clearly integral to addressing a fundamental problem in ecology: how does adaptation of traits at the individual animal level affect properties at the community level?

We are in the initial stages of developing a computational system (DOVE) that uses a complex adaptive systems (CAS) approach, that is, an individual-based model (IBM) with adaptive individuals and evolutionary processes, to examine the influence of phenotypic plasticity on food web structure and dynamics. Here we have presented results in which we have applied the computational system to a relatively simple low-dimensional system. This analysis has illustrated that the evolution of phenotypic plasticity can increase the stability of two competitors, and uncovered mechanisms that have not been recognized with traditional approaches. Traditional modeling approaches (reviewed briefly above) have shown that phenotypic plasticity can affect the degree to which the density of interacting species will overshoot equilibrium densities and hence affect stability. By setting the problem up as CAS, additional and unforeseen stabilizing mechanisms emerged. In particular, we have found that the breadth of peaks in the adaptive landscape differed between species with and without plasticity, influencing a species' ability to invade, and consequently, both habitat partitioning and species coexistence. Secondly, we found that phenotypic responses of two competitors to predator presence can magnify temporal differences in intraspecific and interspecific competition, and consequently increase coexistence.

The mechanisms we found were not previously recognized and were not foreseen by us before we carried out the experiments using the CAS approach described in this paper. We believe it will be useful to apply traditional modeling approaches to further investigate these mechanisms, thus using computational and analytical models to complement each other.

However, we also believe the use of a CAS/IBM approach can be very useful (and in some cases necessary) for discovering new mechanisms that are not hypothesized *a priori*. In this sense, CAS/IBM approaches can be seen as a third and complementary way of doing science (Axelrod 1997), combining the explicit formulation of basic assumptions used in deductive (analytic) approaches with the analysis of the outcomes using inductive techniques common to experimentalists. Thus CAS/IBM models can be used to carry out "thought experiments" that are more complicated than those we can actually carry out in our brains, so that we can explore more deeply how particular individual-level traits and mechanisms lead to structural and dynamical regularities at the level of populations and food webs. In addition, this approach allows us to carry out experiments that would be difficult or costly to carry out in the field or lab. Finally, the CAS/IBM approach enables us to discover mechanisms (regularities) expressible at aggregate levels (i.e., in traditional analytic models) that are very difficult or impossible to infer from the micro-level processes that underlie a CAS (Holland 1995; Wolfram 2002). In particular, the approach is particularly well suited to address the interplay of behavioral, ecological, and evolutionary processes across levels of organizations.

Our longer term goal is to examine how the evolution of phenotypic plasticity affects community structure and dynamics in more complex settings. In particular, we plan to use DOVE to assemble larger networks with more species and examine the resulting structures. While the size of networks that we can examine will depend on the (growing) computer power available, we even now can explore networks with around 10 to 12 species, which is beyond what is typically amenable to analytical approaches and still a challenge for numerical explorations of systems of differential equations if many of the parameters specifying plastic responses must be varied. We also plan to represent species' behavioral strategies with individual-level response equations rather than the matrix representation (Box 1) presented here, in order to allow individual behaviors to respond differentially to continuous environmental factors (such as predator density) and to allow the examination of functional responses associated with phenotypic plasticity. Further, we will examine the consequences of phenotypic responses to multiple environmental and internal inputs, such as predation risk, resource level, and hunger level. The basic approach, however, will remain the same; the behavior of multiple species will be dictated by simple environmental context-dependent rules described by individual-specific values subject to selection. Our preliminary studies suggest that this approach can uncover unanticipated and unforeseen mechanisms by which the adaptive nature of individual phenotype can affect the structure and dynamics of ecological communities. It remains to be seen whether the stabilizing effect of phenotypic plasticity observed in the simple system examined here can be generalized to higher-dimensional systems.

ACKNOWLEDGMENTS

We thank Katrina Button, Andrew McAdam and Jennifer Dunne for helpful comments on this paper, and Fernando Diaz, Erik Goodman, John Holland, and Earl Werner for discussions during project development. Thanks to CSCS for computation resources and support for Rick L. Riolo. Scott D. Peacor wishes to acknowledge support for this research from the Michigan Agricultural Experimental Station and the Cooperative Institute for Limnology and Ecosystems Research. This work was supported by NSF grants DEB-0089809 to Earl Werner and Scott D. Peacor. This is Great Lakes Environmental Research Laboratory contribution number 1359. Mercedes Pascual acknowledges support from the James S. McDonnell Foundation through a Centennial Fellowship.

REFERENCES

Abrams, P. A. 1987. "Indirect Interactions between Species that Share a Predator: Varieties on a Theme." In *Predation: Direct and Indirect Impacts on Aquatic Communities*, ed. W. C. Kerfoot and A. Sih, 38–54. Hanover: University Press of New England.

Abrams, P. A. 1992. "Adaptive Foraging by Predators as a Cause of Predator Prey Cycles." *Evol. Ecol.* 6:56–72.

Abrams, P. A. 1995. "Implications of Dynamically Variable Traits for Identifying, Classifying and Measuring Direct and Indirect Effects in Ecological Communities." *Amer. Natur.* 146:112–134.

Abrams, P. A., and L. Rowe. 1996. "The Effects of Predation on the Age and Size of Maturity of Prey." *Evolution* 50:1052–1061.

Abrams, P. A., B. A. Menge, G. G. Mittelbach, D. Spiller, and P. Yodzis. 1996. "The Role of Indirect Effects in Food Webs." In *Food Webs: Dynamics and Structure*, ed. G. Polis and K. Winemiller, 371–395. New York, NY: Chapman and Hall.

Agrawal, A. A. 2001. "Phenotypic Plasticity in the Interactions and Evolution of Species." *Science* 294:321–326.

Axelrod, R. 1997. "Advancing the Art of Simulation in the Social Sciences." In *Simulating Social Phenomena* ed. C. Rosario, R. Hegselmann, and P. Terna, 21–40. Berlin: Springer.

Bar-Yam, Y. 1997. *Dynamics of Complex Systems*. Reading, MA: Addison-Wesley.

Belew, R. K., and M. Mitchell, eds. 1996. *Adaptive Individuals in Evolving Populations*. Santa Fe Institute Studies in the Science of Complexity Series, proc. vol. XXVI. Reading, MA: Addison-Wesley.

Bender, E. A., T. J. Case, and M. E. Gilpin. 1984. "Perturbation Experiments in Community Ecology: Theory and Practice." *Ecology* 65:1–13.

Bolker, B., M. Holyoak, V. Krivan, L. Rowe, and O. Schmitz. 2003. "Connecting Theoretical and Empirical Studies of Trait-Mediated Interactions." *Ecology* 84:1101–1114.

Carpenter, S. R., J. F. Kitchell, J. R. Hodgson, P. A. Cochran, J. J. Elser, M. M. Elser, D. M. Lodge, D. Kretchmer, X. He, and C. N. Vonende. 1987. "Regulation of Lake Primary Productivity by Food Web Structure." *Ecology* 68:1863–1876.

Chesson, P. 2000. "Mechanisms of Maintenance of Species Diversity." *Ann. Rev. Ecol. & System.* 31:343–66.

Conte, R., R. Hegselmann, and P. Terna, eds. 1997. *Simulating Social Phenomena*. Springer-Verlag.

DeAngelis, D., and L. Gross, L. 1992. *Individual Based Models and Approaches in Ecology: Populations, Communities, and Ecosystems*. Chapman and Hill.

Drossel, B., P. G. Higgs, and A. J. McKane. 2001. "The Influence of Predator-Prey Population Dynamics on the Long-Term Evolution of Food Web Structure." *J. Theor. Biol.* 208:91–107.

Dunne, J. A., R. J. Williams, and N. D. Martinez. 2002. "Food-Web Structure and Network Theory: The Role of Connectance and Size." *PNAS* 99(20):12917–12922.

Epstein, J. M., and R. Axtell. *Growing Artificial Societies: Social Science from the Bottom Up*. Cambridge, MA: The MIT Press.

Fussmann, G. F., and G. Heber. 2002. "Food Web Complexity and Population Dynamics." *Ecol. Lett.* 5:394–401.

Gell-Mann, M. 1995. *The Quark and the Jaguar: Adventures in the Simple and the Complex*. New York: Henry Holt and Company.

Gimblett, H. R., ed. 2002. *Integrating Geographic Information Systems and Agent-Based Modeling Techniques for Simulating Social and Ecological Processes*. Santa Fe Institute Studies in the Sciences of Complexity. New York: Oxford University Press.

Goldberg, D. E. 1989. *Genetic Algorithms in Search, Optimization and Machine Learning*. Reading, MA: Addison-Wesley.

Grimm, V., and S. F. Railsback. 2005. *Individual-Based Modeling and Ecology.* Princeton, NJ: Princeton University Press.

Hairston, N. G., F. E. Smith, and L. B. Slobodkin. 1960. "Community Structure, Population Control, and Competition." *Amer. Natur.* 94:421–425.

Hartvigsen, G., and S. Levin. 1997. "Evolution and Spatial Structure Interact to Influence Plant-Herbivore Population and Community Dynamics." *Proc. Roy. Soc. Lond. B* 264:1677–1685.

Holland, J. H. 1992. *Adaptation in Natural and Artificial Systems.* 2d ed. Cambridge, MA: MIT Press.

Holland, J. H. 1995. *Hidden Order: How Adaptation Builds Complexity.* Cambridge, MA: The MIT Press.

Holt, R. D. 1984. "Spatial Heterogeneity, Indirect Interactions, and the Coexistence of Prey Species." *Amer. Natur.* 124:377–406.

Holt, R.D. 1987. "Population Dynamics and Evolutionary Processes: The Manifold Roles of Habitat Selection." *Evol. Ecol.* 1:331–347.

Hraber, P. T., T. Jones, and S. Forrest. 1997. "The Ecology of Echo." *Artificial Life* 3:165–190.

Huang, C. F., and A. Sih. 1991. "Experimental Studies on Direct and Indirect Interactions in a 3 Trophic-Level Stream System." *Oecologia* 85:530–536.

Huston, M., D. DeAngelis, and W. Post. 1988. "New Computer Models Unify Ecological Theory." *BioScience* 38:682–691.

Ives, A. R., and A. P. Dobson. 1987. "Antipredator Behavior and the Population-Dynamics of Simple Predator-Prey Systems." *Amer. Natur.* 130:431–447.

Johst, K., M. Doebeli, and R. Brandle. "Evolution of Complex Dynamics in Spatially Structured Populations." *Proc. R. Soc. Lond. B* 266:1146–1154.

Judson, O. P. 1994. "The Rise of the Individual-Based Model in Ecology." *Trends in Ecol. & Evol.* 9(1):9–14.

Kopp and Gabriel. 2005 "The Dynamic Effects of an Inducible Defense in the Nicholson-Bailey Model." Unpublished manuscript (in preparation).

Lassig, M., U. Bastolla, S. C. Manrubia, and A. Valleriani. 2001. "Shape of Ecological Networks." *Phys. Rev. Lett.* 86:4418–4421.

Leibold, M. A. 1989. "Resource Edibility and the Effects of Predators and Productivity on the Outcome of Trophic Interactions." *Amer. Natur.* 134:922–949.

Levin, S. A. 1998. "Ecosystems and the Biosphere as Complex Adaptive Systems." *Ecosystems* 1:431–436.

Levine, S. H. 1976. "Competitive Interactions in Ecosystems." *Amer. Natur.* 110:903–910.

Lima, S. L. 1998. "Stress and Decision Making Under the Risk of Predation: Recent Developments from Behavioral, Reproductive, and Ecological Perspectives." *Stress & Behav.* 27:215–290.

Luttbeg, B., and O. J. Schmitz. 2000. "Predator and Prey Models with Flexible Individual Behavior and Imperfect Information." *Amer. Natur.* 155:669–683.

May, R. M. 2001. *Stability and Complexity in Model Ecosystems.* With a New Introduction by the Author. Princeton Landmarks in Biology. Princeton, NJ: Princeton University Press.

Michael, M. W., and R. Willer. 2002. "From Factors to Actors: Computational Sociology and Agent-Based Modeling." *Ann. Rev. Sociol.* 28:143–166.

Mitchell, M. 1996. *An Introduction to Genetic Algorithms.* The MIT press.

Morowitz, H. J. 2002. *The Emergence of Everything: How The World Became Complex.* New York: Oxford University Press.

Muller-Landau, H. C., S. A. Levin, and J. E. Keymer. 2003. "Theoretical Perspectives on Evolution of Long-Distance Dispersal and the Example of Specialized Pests." *Ecology* 84:1957–1967.

Neutel, A., J. A. P. Heesterbeek, and P. C. de Ruiter. 2002. "Stability in Real Food Webs: Weak Links in Long Loops." *Science* 296:1120–1123.

Oaten, A., and W. W. Murdoch. 1975. "Switching, Functional Responses, and Stability in Predator-Prey Systems." *Amer. Natur.* 109:299–318.

Oksanen, L., S. D. Fretwell, J. Arruda, and P. Niemela. 1981. "Exploitation Ecosystems in Gradients of Primary Productivity." *Amer. Natur.* 118:240–261.

Pascual, M., and J. A. Dunne. "From Small to Large Ecological Networks in a Dynamic World." This issue.

Peacor, S. D. 2002. "Positive Effects of Predators on Prey Growth Rate through Induced Modifications of Prey Behavior." *Ecol. Lett.* 5:77–85.

Peacor, S. D., and E. E. Werner. 2001. "The Contribution of Trait-Mediated Indirect Effects to the Net Effects of a Predator." *PNAS* 98:3904–3908.

Peacor, S. D., and E. E. Werner. 2004. "How Dependent are Species-Pair Interaction Strengths on Other Species in the Food Web?" *Ecology* 85(10):2754–2763.

Pimm, S. L., J. H. Lawton, and J. E. Cohen. 1991. "Food Web Patterns and Their Consequences." *Nature* 350:665–674.

Polis, G. A., and K. O. Winemiller. 1996. *Food Webs: Integration of Patterns and Dynamics.* New York, NY: Chapman and Hall.

Railsback, S. F. 2001. "Concepts from Complex Adaptive Systems as a Framework for Individual-Based Modelling." *Ecol. Model.* 139:47–62.

Schmitz, O. J. 1998. "Direct and Indirect Effects of Predation and Predation Risk in Old-Field Interaction Webs." *Amer. Natur.* 151:327–342.

Schoener, T. W. 1993. "On the Relative Importance of Direct versus Indirect Effects in Ecological Communities." In *Mutualisms and Community Organization*, ed. H. Kawanabe, J. E. Cohen, and K. Iwasaki, 365–411. New York: Oxford University Press.

Strand, E., G. Huse, and J. Giske. 2002. "Artificial Evolution of Life History and Behavior." *Amer. Natur.* 159:624–644.

Tollrian, R. and C. D. Harvell, eds. 1999. *The Ecology and Evolution of Inducible Defenses.* Princeton, NJ: Princeton University Press.

Turner, A. M., and G. G. Mittelbach. 1990. "Predator Avoidance and Community Structure: Interactions Among Piscivores, Planktivores, and Plankton." *Ecology* 71:2241–2254.

Vicsek, T. 2002. "Complexity: The Bigger Picture." *Nature* 418:131.

Vriend, N. 2000. "An Illustration of the Essential Difference between Individual and Social Learning, and Its Consequence for Computational Analyses." *J. Econ. Dynamics & Control* 24:1–19.

Werner, E. E., and S. D. Peacor. 2003. "A Review of Trait-Mediated Indirect Interactions." *Ecology* 84:1083–1100.

West-Eberhard, M. J. 2003. *Developmental Plasticity and Evolution.* New York: Oxford University Press.

Wilke, C. O., and C. Adami. 2002. "The Biology of Digital Organisms." *Trends in Ecol. & Evol.* 177:528–532.

Wilson, W. G. 1998. "Resolving Differences between Deterministic Population Models and Individual-Based Simulations." *Amer. Natur.* 151:6–134.

Williams, R. J., E. L. Berlow, J. Dunne, A. Barabasi, and N. Martinez. 2002. "Two Degrees of Separation in Complex Food Webs." *PNAS* 99(2):12913–12916.

Wolfram, S. 2002. *A New Kind of Science.* Champaign, IL: Wolfram Media, Inc.

Wootton, J. T. 1993. "Indirect Effects and Habitat Use in an Intertidal Community: Interaction Chains and Interaction Modifications." *Amer. Natur.* 141:71–89.

Yodzis, P. 1988. "The Indeterminacy of Ecological Interactions, as Perceived by Perturbation Experiments." *Ecology* 72:1964–1972.

Exploring the Evolution of Ecosystems with Digital Organisms

Claus O. Wilke
Stephanie S. Chow

We study the evolution of ecosystems in self-replicating computer programs (digital organisms). In previous work, we have shown that in a well-stirred, multiple resource environment, resource limitations are sufficient to induce adaptive radiation. At intermediate resource availability, the digital organisms evolved from a clonal population into a number of distinct, coexisting, and persistent genotypic and phenotypic clusters (species). Here we examine the phenotypic changes that occur as populations diversify and separate into distinct species. We find that there are multiple ways in which new species can arise, including through acquisition or loss of the ability to utilize certain resources, or through conversion, whereby a pathway to utilize one resource is transformed into a pathway to utilize another resource. In future work, we will be able to study the evolution of multilayered food webs as well, by allowing digital organisms to feed on each other instead of only on basal resources.

Ecological Networks: Linking Structure to Dynamics in Food Webs,
edited by Mercedes Pascual and Jennifer A. Dunne, Oxford University Press.   271

1   INTRODUCTION

When we study food webs, or networks of ecological interactions in general, we must not only ask how these networks function and how they react to perturbations, but also how they come about. While the first two questions have received substantial attention in the ecology literature, by comparison the third has been rather poorly studied. To answer the third question satisfactorily, we have to combine knowledge from ecology with that from evolutionary biology, two fields that have traditionally been studied quite separately (the widespread existence of departments for ecology and evolutionary biology in American universities notwithstanding). These two fields are difficult to combine because ecologists frequently make the convenient assumption that species don't change (thus excluding evolution), while evolutionary biologists often study the adaptation of a single species to a particular niche (thus excluding more complex ecological effects).

To study the evolution and emergence of new ecosystems experimentally is a daunting task in any case, because adaptive radiation operates on very long time scales. We can only hope to carry out such experiments with microbes, and even those experiments are tedious and time-consuming. Several groups have been studying the evolution of microbial ecologies in vitro, and have observed adaptive radiation, resource partitioning, stable polymorphisms, and the emergence of strains that cross-feed on other strains' waste products (Rosenzweig et al. 1994; Turner et al. 1996; Rainey and Travisano 1998; Bohannan and Lenski 2000; Kassen et al. 2000; Rainey et al. 2000; Greig et al. 2002; Friesen et al. 2004).

An alternative approach to wet experiments is to study the emergence of ecosystems *in silico*. Examples include studies on the *ab-initio* formation of foodwebs (Caldarelli et al. 1998; Drossel et al. 2001), on evolutionary branching (Geritz et al. 1997; Geritz et al. 1998; Dieckmann and Doebeli 1999; Doebeli and Dieckmann 2000), or on speciation under spatial resource heterogeneity (Day 2000; Kawata 2002; Doebeli and Dieckmann 2003). One problem with *in silico* studies is that the properties organisms can acquire through evolution are restricted to the set of features built into the simulation software. This restriction applies in principle to all *in silico* studies of evolution. However, by building a simulation software that offers a rich environment and a wide range of different phenotypes for the organisms to explore, we can hope that our *in silico* experiments will reveal new and interesting aspects of the evolution of ecosystems. A new paradigm for experimental evolution *in silico* has been developed in recent years: the experimentation with self-replicating and evolving computer programs, so-called *digital organisms* (Wilke and Adami 2002). Since digital organisms are programs written in a full-blown computer language, they have infinite possibilities of acquiring new and unexpected phenotypes, and thus are good candidates for studying complex evolutionary scenarios.

Early work with digital organisms has focused on single-niche environments (Ray 1992; Lenski et al. 1999; Adami et al. 2000; Yedid and Bell 2001; Wilke et al. 2001; Yeidid and Bell 2002; Lenski et al. 2003), but recent progress has made it possible to study ecological questions with digital organisms as well (Cooper and Ofria 2002; Ofria and Wilke 2004; Chow et al. 2004). In particular, we have recently investigated under what conditions digital organisms speciate in an unstructured environment (Chow et al. 2004). We have found that populations diversify rapidly when several limiting resources are moderately abundant, but do not diversify if all but a single limiting resource are either overabundant or scarce. In the present chapter, we extend this work, and analyze the phenotypic changes that occur when populations speciate.

2   THE AVIDA PLATFORM

All the experiments we present here were carried out with the Avida software platform.c Avida (Ofria and Wilke 2004) is the most-widely used system for studying the evolutionary biology of self-replicating computer programs, also called digital organisms (see Box 1). Avida has a number of features that result in rich and interesting evolution experiments. In particular, in a simple digital world in which organisms do nothing but self-replicate, selection will drive the organisms to become shorter and shorter, so that they can replicate as quickly

---

Box 1: Digital organisms

Digital organisms are self-replicating computer programs that mutate and evolve (Wilke and Adami 2002). They live in a simulated environment, a virtual computer, that provides them with the resources they need to survive and reproduce. Their genomes are sequences of computer instructions written in a special language that can be executed by the virtual computer. A successful digital organism will locate suitable computer memory, store a copy of its genome, and then notify the environment that the offspring is ready to start a life of its own. Typically, the total amount of memory in the virtual computer is limited, so that after a while, newborn digital organisms start to replace already existing ones. This constant replacement of older organisms provides a selective pressure: those organisms that manage to replicate faster than others will prevail, while the slower replicators die out.

Mutations arise because the organisms have to copy their genomes instruction for instruction—much like a DNA polymerase, which copies base after base—and the copy process is not error-free. Occasionally, while copying their genome, the organisms will write a random instruction instead of the instruction they intended to copy. This process is controlled by the simulated environment, not by the organisms themselves, and the rate at which it occurs is set by the researcher. Most of these incorrectly copied instructions (i.e., mutations) will be deleterious. However, occasionally a mutation improves an organism's replicatory ability, and this organism will then be able to outcompete the other organisms in the population.

continued on next page

Box 1 continued

The first simulation system in which computer programs were properly evolving and adapting in the manner outlined in the preceding paragraphs was Tierra, developed by Tom Ray in the early 1990s (Ray 1992). Later, Titus Brown, Charles Ofria, and Chris Adami developed an alternative system, which they called Avida (Adami 1998; Ofria and Wilke 2004). The main difference between Tierra and Avida is that organisms in Avida are more clearly separated from each other, which simplifies the analysis and interpretation of experimental results in many cases. Also, organisms in Avida can carry out computations (tasks) on numbers provided by the environment, and in this process use up resources which increase the organisms' replication speed (see figure below). A third system, called Physis, is currently being developed by Attila Egri-Nagy and coworkers (Egri-Nagy and Nehaniv 2003).

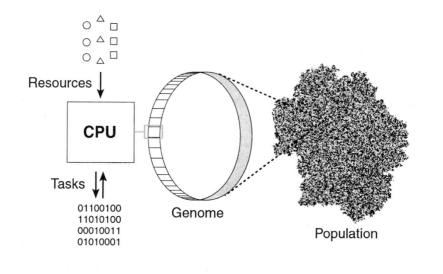

FIGURE: Schematic drawing of Avida world.

as possible (Ray 1992; Yedid and Bell 2002). To counterbalance this pressure, the Avida software provides an additional element, CPU speed (Ofria and Wilke 2004). Different digital organisms can run at different CPU speeds, so that they execute different numbers of instructions within the same amount of simulated time. Then, an organism that, for example, has to copy twice as many instructions, but runs at more than twice the speed of another organism, will have a selective advantage. There are two mechanisms that influence CPU speed in Avida. First, by default the CPU speed is set to a constant multiplied by the genome length of the organism. This setting is meant to prevent the otherwise strong selective pressure to reduce genome size, and allows organisms to grow in size and complexity. Second, organisms can further increase their CPU speed beyond this basal level by completing logic functions, which they can carry out on numbers that the environment provides. If an organism successfully calculates a particular logic function, it receives a bonus in the form of additional CPU speed, which can more than counterbalance the additional instructions that need to be stored in the genome (and thus copied) and executed to carry out this calculation. These logic functions (which are often also called *tasks*) serve as a computational metabolism for the digital organisms, and we refer to those parts of a digital organism's genome that are necessary to carry out logic functions as *metabolic genes* (in comparison to the *replicatory gene*, which copies the genome and creates the offspring organism). When organisms are allowed to evolve a computational metabolism in Avida, they grow quickly in size and complexity as they learn to carry out a number of different logic functions (Adami et al. 2000; Lenski et al. 2003).

In the original Avida implementation, bonuses were always infinitely abundant (Adami 1998). Whether an organism received a bonus depended only on the successful completion of a logic function, but not on how many other organisms had previously carried out the same function. With an infinite bonus supply, the Avida world provides only a single niche, because an organism that can execute more logic functions or can execute them more efficiently than the current population can always invade. Single-niche Avida has been used successfully to study a number of questions in evolutionary biology (Lenski et al. 1999; Adami et al. 2000; Wilke et al. 2001; Lenski et al. 2003), but is not of much use for ecological questions. Recently, a new bonus system has been introduced into Avida. With this new bonus system, the number of bonuses that are available is finite (Ofria and Wilke 2004). Each logic function is coupled to a simulated resource, the amount of which is reduced whenever an organism executes the associated logic function, and a bonus is given to the organism only if the resource is not fully depleted. The researcher can set up resources to flow into and out of the system at specific rates. If the inflow of a specific resource is low, then only a limited number of organisms can derive a benefit from carrying out the associated logic function. The researcher can also adjust how many times the organisms can make use of a single resource during one gestation cycle. If resources are unlimited, then it is often useful to put a cap on the number of times an organism can access a resource in a single generation, to prevent the evolution of organisms that try to gain infinite CPU speed by using infinitely many resource units. If resources are limited, such a cap is not needed. In all results presented here, unless noted otherwise, no cap on resource usage was implemented.

To summarize, there are three types of resources in Avida. First, there is the basal resource, which is available to all organisms and allows them to replicate even if they cannot compute any logic functions. Second, there are computational resources associated with the logic functions. There are plans to remove the basal resource in future versions of Avida, so that organisms can survive only when they can utilize computational resources, but currently, the basal resource is always available to every organism, and is always unlimited. Finally, there is space. The memory of the virtual computer in which the Avida organisms live can hold only a finite number of organisms, which are replaced at random when new organisms are born.

3   METHODS

The work we present here is a direct extension of Chow et al. (2004), and we refer the reader to this article for a detailed description of our experimental conditions. In short, we set up the Avida world such that there are nine depletable resources associated with nine simple one- and two-input logic functions. The resources flow into the system at a constant rate, and out at a rate proportional to their concentration. Thus, the resource flow through the system mimics a chemostat. Each experiment is seeded with a single ancestral organism of length $L = 100$. Organisms are not allowed to change their genome length. The per-site copy mutation rate is 0.005, which implies a genome mutation rate of 0.5. All experiments are carried out for 400,000 updates. (One update is an arbitrary time unit used in the Avida system. Five to ten updates typically correspond to one generation.)

Here, we analyze the change of community structure and resource utilization patterns for several experiments of Chow et al. (2004). This analysis is carried out in the following way: for a given experiment, we determine the most abundant genotype for each species (the species representative, in the terminology of Chow et al. (2004)) in the final, evolved community, and follow the line of descent from each species representative back to the original ancestor with which the experiment has been seeded. Then, we determine the resource utilization pattern for each organism along the lines of descent. For visualization purposes, we record only whether an organism uses or does not use a given resource, even though organisms can use resources to varying degrees (see previous section).

4   PATTERNS OF GENOTYPIC DIFFERENTIATION AND SPECIATION

In Chow et al. (2004), we used a clustering algorithm on the phylogenetic tree of the evolved community to determine whether speciation had occurred, and how many species were present in the final community. (This phylogenetic tree is fully known in Avida, and need not be inferred from sequence data.) However, visual inspection of the evolved phylogenetic tree can usually serve the same purpose, because separate species show up as extremely deep, coexisting branches in the phylogenetic tree. To visualize phylogenetic trees, we plot phylogenetic depth as a function of time. Phylogenetic depth is the number of genotypically distinct

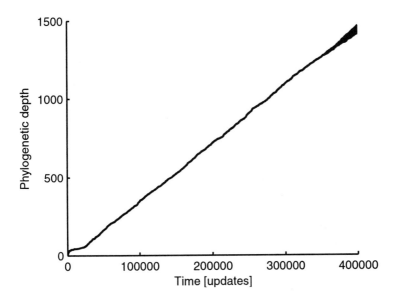

FIGURE 1   Phylogenetic depth versus time in an environment that does not promote speciation. Each line traces an organism in the final population back to the original unevolved ancestor.

ancestors that separate a given organism from the original ancestor with which the experiment was seeded.

Figure 1 shows a typical phylogenetic depth profile for a community of digital organisms that has not speciated (all depletable resources, that is, all resources coupled to logic functions, are overabundant in this example). When we go from the final community backwards in time, we see that all lines of descent quickly coalesce in the most recent common ancestor of the final community. This most recent common ancestor lived around update 350,000, approximately 50,000 updates before completion of the experiment. From the most recent common ancestor, a single line of descent leads back to the seeding ancestor. In contrast, figure 2 shows a typical phylogenetic depth pattern for a community of digital organisms that has speciated (the depletable resources are moderately abundant in this example). We can clearly distinguish four deep branches that separate early (around update 100,000). Close to the end of the experiment (around update 300,000 to 350,000), each of these deep branches fans out in the same way as the single branch in figure 1 does. The interpretation of this observation is simple: in a community that has speciated, all organisms within a single species share a fairly recent common ancestor, but the most recent common ancestor of all species lies in the distant past.

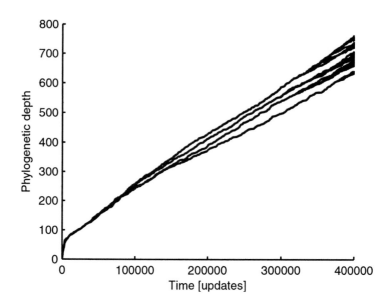

FIGURE 2   Phylogenetic depth versus time in an environment that does promote speciation. Each line traces an organism in the final population back to the original unevolved ancestor.

Here, we are interested in understanding in detail what happens when populations speciate. Is speciation triggered when a mutant organism acquires the ability to utilize a new resource? Can the resource utilization pattern within the community change in the absence of a speciation event? Can speciation occur through the loss of function? We will see in the following that there a many ways in which speciation can come about, and that community structure and resource utilization patterns can change in a variety of different ways.

Our first example is that of a community evolved at medium resource availability from an unevolved, hand-written organism that can self-replicate but cannot utilize any of the computational resources. The evolved community consists of four species, and among them, all nine computational resources are being used. Figure 3 shows the pattern of resource utilization along the lines of descent of the four species representatives. These lines of descent coalesce as we go back in time, and, therefore, part of the resource utilization patterns are shared among species. In fact, we can identify the speciation events from the points in time at which resource utilization patterns start to diverge. For clarity, the four species in figure 3 are ordered in chronological order of emergence.

All four species share their history for the first 70,000 updates. We see that the resource utilization pattern changes quickly and drastically during the first

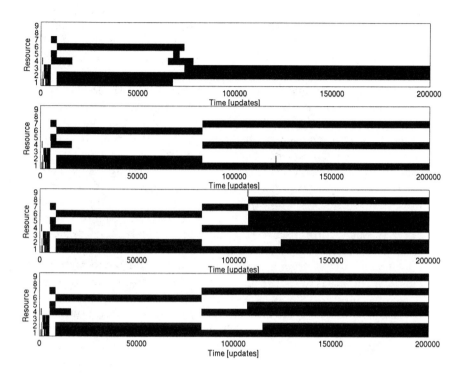

FIGURE 3   Resource utilization pattern as a function of time for a community of four species in an experiment seeded with an unevolved ancestor. Black areas indicate that a given resource is used at the particular point in time. Each subplot corresponds to the line of descent of one of the four species in the evolved community.

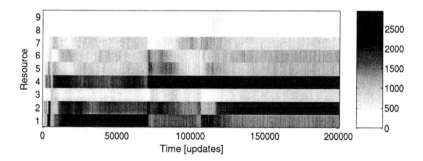

FIGURE 4   Resource utilization pattern of the complete community as a function of time, for the same experiment as shown in figure 3. Gray levels indicate how many organisms in the community utilize a given resource.

10,000 to 15,000 updates, as the species acquire the ability to utilize different resources. Then, the species settle on using resources 1, 2, and 6, until at approximately 70,000 updates, the topmost species splits off from the other species. From update 70,000 to about 80,000, there is considerable change in resource utilization pattern of the species, but no further speciation event. Finally, around update 100,000, the third and fourth species split off from the second species.

From figure 3, we can get the impression that during the first 25,000 updates, the community was rapidly switching the resource utilization pattern, without undergoing any speciation events. This impression is misleading. It is important to recall that figure 3 shows only the resource utilization pattern along the lines of descent from the final species back to the seeding ancestor. The figure does not show the resource usage of those organisms that did not end up on the line of descent. We can get a feeling for their activity from plotting the total resource usage of the whole community as a function of time. Figure 4 shows that resources 1 through 7 are discovered almost immediately, and are being used continuously until the end of the experiment. This observation implies that during the long time interval from approximately 15,000 updates to approximately 70,000 updates, in which the single ancestral species in figure 3 does not undergo any change, it is probably stably coexisting with one or more other species. The other coexisting species use the remaining resources. However, ultimately these species die out and are replaced by newly evolving species, either because these species find a new, more efficient way to utilize a certain combination of resources (this is what probably caused the speciation event close to update 70,000) or because they discover new resources that have not been used previously (this is what probably caused the speciation event close to update 100,000).

At update 200,000, the four final species overlap in their resource use, but species 1 uses the most of resources 2 and 3; species 2 uses the most of resources 4 and 7; species 3 uses the most of resources 5 and 8; and species 4 uses the most of resources 1, 4, and 9 (fig. 5). There is no significant change in resource usage from update 200,000 to update 400,000 (not shown).

Our second example shows the diversification of an evolved organism. This organism had evolved in an environment where all resources were infinitely abundant, but where each resource could be used only once per generation, and had acquired the capability of carrying out all nine logic functions. Thus, while highly evolved, the organism was ill-adapted to an environment in which the computational resources were only moderately abundant, but could be harvested multiple times.

As in the previous example, by the end of the experiment, at update 400,000, the original ancestor has diversified into four distinct species. However, unlike the previous example, and despite the ability of the original ancestor to use all nine resources, three of the four final species are using only a single resource each, and only the fourth species is using several resources at once.

Figure 6 shows the pattern of resource utilization for these four species during the first 200,000 updates of the experiment. We see that the original ancestor

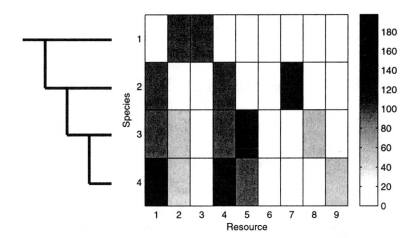

FIGURE 5 Matrix of resource utilization for the four species of figure 3 at update 200,000. Gray levels indicate the degree to which the different species use the resources within one generation (in arbitrary units). In two cases (species 3/resource 6, species 4/resource 7), resource usage is so low that the coloring is indistinguishable from the background white. The branching pattern to the left of the utilization matrix indicates the relationship between the species based on their phylogeny (relative branch lengths not to scale).

starts out consuming all nine resources, but has already lost the first resource (resource 7) 15,000 updates into the experiment, and loses another one (resource 3) shortly thereafter. The first speciation event occurs approximately at update 40,000, when the three bottom species start specializing on resource 8, while the top species loses the ability to utilize this resource. From update 40,000 to approximately 80,000, we see some changes in the resource utilization pattern of the bottom three species that are not accompanied by any lasting speciation events. Then, the bottom two species split from the second species. The bottom two species continue to undergo changes in their resource utilization pattern, and split approximately at update 160,000. At this time, the third species specializes on resource 6, and the fourth species specializes on resource 8. From update 200,000 to 400,000, there is still some change in the resource usage pattern, even though all four species have formed. Species 1 learns to use resource 5 in addition to 1, 2, and 4, and species 3 switches from resource 5 to resource 6 (not shown).

5 CONCLUSIONS

Even though sympatric speciation has often mystified biologists, it seems that it is inevitable in asexually reproducing organisms that have access to several

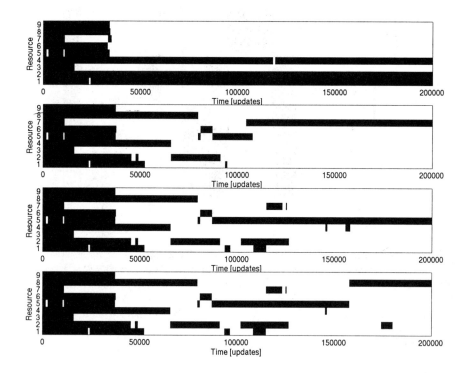

FIGURE 6   Resource utilization pattern as a function of time for a community of four species in an experiment seeded with a generalist ancestor. Black areas indicate that a given resource is used at the particular point in time, and white areas indicate the opposite. Each subplot corresponds to the line of descent of one of the four species in the evolved community.

moderately abundant resources. In previous work, we have studied the pattern of speciation in asexually replicating digital organisms as we change the availability of computational resources, but keep the basal resource and the space limitation constant (Chow et al. 2004). We have found a hump-shaped pattern of species richness. When the abundance of computational resources was either very low or very high, we did not observe significant speciation, and the evolved communities consisted of a single species. If computational resources were moderately abundant, on the other hand, then speciation occurred in almost all cases, and the evolved communities consisted of up to six stably-coexisting species (on nine resources). We can conclude from this work that speciation occurs readily in asexually reproducing digital organisms if the limiting resources are sufficiently available such that no other factor (for example space) becomes the dominant limitation to population growth (Chow et al. 2004).

Here, we have studied what phenotypic changes take place as organisms speciate. We have seen that speciation is often initiated when a mutant acquires the capability of utilizing a new resource that was previously untouched. Since this resource is abundant when the mutant first appears, the mutant initially derives a large selective advantage and can gain a foothold in the population. What happens next depends on whether the new mutant continues to use all resources it has been using previously, or gives up on some of them. Naively, we might think that the organisms should simply continue to acquire the capability of using new resources, without ever giving up old resources, until the community is dominated by a single generalist that can use all resources. However, this reasoning neglects the fact that resource usage comes at a cost. In the case of digital organisms, this cost is manifest in increased genome size (to store the additional instructions needed to use the resources) and increased number of instructions that have to be executed. If all organisms use a moderately abundant resource, then this resource will be drawn down to a level at which it cannot provide sufficient benefit to counteract the cost of its usage. At this point, organisms will be better off losing the genes necessary to take advantage of that resource. This reasoning shows that a single species is unlikely to evolve into a stable equilibrium in a system with limited resources. The only way in which the communities can equilibrate is by splitting into separate species that specialize on different resources. In this case, the different species stabilize each other through frequency-dependent selection, and the overall growth rate of the complete community is maximized. (Note that frequency-dependent selection in the evolved ecosystems has been shown in Chow et al. [2004].)

We have presented indirect evidence that early in the experiments, there is substantial turnover as species come about, only to be replaced by other species shortly thereafter. Towards the end of the experiments, we see this activity less and less. Instead, the evolved communities seem to settle into fairly stable patterns of coexisting species. We can understand this development with tradeoffs among resource usage and increasingly optimized digital organisms. It is known that certain logic functions in Avida are closely related to certain other logic functions (Ofria and Wilke 2004). Therefore, there must be more advantageous and less advantageous resource combinations. For example, it may be more cost-efficient for an organism to use resources 1 and 2 than to use 1 and 3. In this case, if through happenstance an early species evolves that uses resources 1 and 3, it is bound to be replaced later on by a more efficient species that uses resources 1 and 2. Note, however, that we don't have evidence for a single, optimal state towards which all ecosystems evolve. We see wide variability in the evolved, apparently stable diversified communities (Chow et al. 2004).

The experiments we have presented here constitute only the very first step towards a full-blown analysis of the evolution of ecosystems with artificial-life systems. In our current setup as consumers of basal resources, all digital organisms occupy the same trophic layer. The next step towards increased realism will be to introduce the possibility that digital organisms can consume other digital organisms, rather than only basal resources. This possibility will be simple to implement: all we have to do is add a command that lets one digital organism kill another one, and in that process absorb the bonus CPU speed that the prey organism has accumulated so far. To make the interaction between predator and prey more interesting, we can also add the possibility that predators can test prey for certain properties (such as the consumption of particular resources) before they kill them. In this way, predators can evolve to consume only specific types of organisms, rather than any organism indiscriminately. With these changes in the simulation, we will have a system in which proper multilevel food webs can evolve, and we will be able to study the conditions under which certain types of food-web structures emerge, and the range of different structures that emerge in a given set of conditions.

ACKNOWLEDGMENTS

This work was supported in part by the NSF under Contract No. DEB-9981397.

REFERENCES

Adami, C. 1998. *Introduction to Artificial Life*. New York: Springer.

Adami, C., C. Ofria, and T. C. Collier. 2000. "Evolution of Biological Complexity." *Proc. Natl. Acad. Sci. USA* 97:4463–4468.

Bohannan, B. J. M., and R. E. Lenski. 2000. "Linking Genetic Change to Community Evolution: Insights from Studies of Bacteria and Bacteriophage." *Ecol. Lett.* 3:362–377.

Caldarelli, G., P. G. Higgs, and A. J. McKane. 1998. "Modelling Coevolution in Multispecies Communities." *J. Theor. Biol.* 193:345–358.

Chow, S. S., C. O. Wilke, C. Ofria, R. E. Lenski, and C. Adami. 2004. "Adaptive Radiation from Resource Competition in Digital Organisms." *Science* 305:84–86.

Cooper, T., and C. Ofria. 2002. "Evolution of Stable Ecosystems in Populations of Digital Organisms." In *Artificial Life VIII*, ed. R. K. Standish, M. A. Bedau, and H. A. Abbass, 227–232. Proceedings of the Eighth International Conference on Artificial Life. Cambridge, MA: MIT Press.

Day, T. 2000. "Competition and the Effect of Spatial Resource Heterogeneity on Evolutionary Diversification." *Amer. Natur.* 155:790–803.

Dieckmann, U., and M. Doebeli. 1999. "On the Origin of Species by Sympatric Speciation." *Nature* 400(6742):354–357.

Doebeli, M., and U. Dieckmann. 2000. "Evolutionary Branching and Sympatric Speciation caused by Different Types of Ecological Interactions." *Amer. Natur.* 156:S77–S101.

Doebeli, M., and U. Dieckmann. 2003. "Speciation Along Environmental Gradients." *Nature* 421:259–264.

Drossel, B., P. G. Higgs, and A. J. McKane. 2001. "The Influence of Predator-Prey Population Dynamics on the Long-Term Evolution of Food Web Structure." *J. Theor. Biol.* 208:91–107.

Egri-Nagy, A., and C. L. Nehaniv. 2003. "Evolvability of the Genotype-Phenotype Relation in Populations of Self-Replicating Digital Organisms in a Tierra-like System." *Lect. Notes Artif. Int.* 2801:238–247.

Friesen, M. L., G. Saxer, M. Travisano, and M. Doebeli. 2004. "Experimental Evidence for Sympatric Ecological Diversification due to Frequency-Dependent Competition in *Escherichia coli*." *Evolution* 58:245–260.

Geritz, S. A. H., E. Kisdi, G. Meszena, and J. A. J. Metz. 1998. "Evolutionarily Singular Strategies and the Adaptive Growth and Branching of the Evolutionary Tree." *Evol. Ecol.* 12:35–57.

Geritz, S. A. H., J. A. J. Metz, E. Kisdi, and G. Meszena. 1997. "Dynamics of Adaptation and Evolutionary Branching." *Phys. Rev. Lett.* 78:2024–2027.

Greig, D., E. J. Louis, R. H. Borts, and M. Travisano. 2002. "Hybrid Speciation in Experimental Populations of Yeast." *Science* 298:1773–1775.

Kassen, R., A. Buckling, G. Bell, and P. B. Rainey. 2000. "Diversity Peaks at Intermediate Productivity in a Laboratory Microcosm." *Nature* 406:508–512.

Kawata, M. 2002. "Invasion of Vacant Niches and Subsequent Sympatric Speciation." *Proc. Roy. Soc. Lond. B* 269:55–63.

Lenski, R. E., C. Ofria, T. C. Collier, and C. Adami. 1999. "Genome Complexity, Robustness and Genetic Interactions in Digital Organisms." *Nature* 400:661–664.

Lenski, R. E., C. Ofria, R. T. Pennock, and C. Adami. 2003. "The Evolutionary Origin of Complex Features." *Nature* 423:129–144.

Ofria, C., and C.O. Wilke. 2004. "Avida: A Software Platform for Research in Computational Evolutionary Biology." *Artificial Life* 10:191–229.

Rainey, P. B., and M. Travisano. 1998. "Adaptive Radiation in a Heterogeneous Environment." *Nature* 394:69–72.

Rainey, P. B., A. Buckling, R. Kassen, and M. Travisano. 2000. "The Emergence and Maintenance of Diversity: Insights from Experimental Bacterial Populations. *Trends Ecol. & Evol.* 15:243–247.

Ray, T. S. 1992. "An Approach to the Synthesis of Life." In *Artificial Life II*, ed. C. G. Langton, C. Taylor, J. D. Farmer, and S. Rasmussen, 371. Santa Fe Institute Studies in the Sciences of Complexity. Reading, MA: Addison-Wesley.

Rosenzweig, R. F., R. R. Sharp, D. S. Treves, and J. Adams. 1994. "Microbial Evolution in a Simple Unstructured Environment—Genetic Differentiation in *Escherichia coli*." *Genetics* 137:903–917.

Turner, P. E., V. Souza, and R. E. Lenski. 1996. "Tests of Ecological Mechanisms Promoting the Stable Coexistence of Two Bacterial Genotypes." *Ecology* 77:2119–2129.

Wilke, C. O., and C. Adami. 2002. "The Biology of Digital Organisms." *Trends Ecol. Evol.* 17:528–532.

Wilke, C. O., J. L. Wang, C. Ofria, R. E. Lenski, and C. Adami. 2001. "Evolution of Digital Organisms at High Mutation Rates leads to Survival of the Flattest." *Nature* 412(6844):331–333.

Yedid, G., and G. Bell. 2001. "Microevolution in an Electronic Microcosm." *Amer. Natur.* 157(5):465–487.

Yedid, G., and G. Bell. 2002. "Macroevolution Simulated with Autonomously Replicating Computer Programs." *Nature* 420(6917):810–812.

Network Evolution: Exploring the Change and Adaptation of Complex Ecological Systems Over Deep Time

Neo D. Martinez

1 INTRODUCTION

One of the most exciting new directions in research on food webs and ecological networks is network evolution or, in other words, the development of the structure and function of ecological networks over time scales long enough for node selection and speciation to occur (Caldarelli et al. 1998; Drossel and McKane 2002; Worden 2003; Yoshida 2003; Cattin et al. 2004; Rossberg 2004; Anderson and Jensen 2005). Most food-web studies focus on shorter-term snapshots of network structure and time series of network dynamics. The difference between long-term evolution over "deep time" and short-term structure and dynamics is a familiar one. Hutchinson distinguished the two by suggesting that we "view the natural world as the ecological theater, serving as a stage for the evolutionary play" (Hutchinson 1965). Following this suggestion, research on network evolution goes well beyond contemporary topology, theatrical population booms and busts, and more staid equilibria. That is, network evolution encompasses a broad

Ecological Networks: Linking Structure to Dynamics in Food Webs,
edited by Mercedes Pascual and Jennifer A. Dunne, Oxford University Press. **287**

spectrum of studies, from the dynamics of intra-node stocks and inter-node flows to the evolutionary play of node survival, adaptation, and speciation over time.

Though the distinction between network dynamics and network evolution may be familiar, much research occupies a kind of gray area in between these two ends of a spectrum. For example, in adaptive prey preference and feeding behavior models (Post et al. 2000; Brose et al. 2003; Kondoh 2003), network topology changes and species can be lost but the most fundamental traits of nodes do not change. Similarly, network structure changes during the assembly of local communities from regional species pools. Such assembly (Holt et al. 1999; Fox and McGrady-Steed 2002) and also disassembly (Holt et al. 1999; Fox and McGrady-Steed 2002) informs network evolution by exploring how networks respond to species gain and loss but again, without adaptive change in intrinsic properties of nodes.

The difference between network evolution and network dynamics is that network evolution describes changes in the more fundamental properties of nodes, such as which species potentially interact with a node, the functional form of these interactions, and even species' metabolic, assimilation, and maximum ingestion rates. These properties typically remain constant in the context of network dynamics even though species' biomasses, feeding rates, and presence or absence can change. The invasion by, and extirpation of nodes, which drives network and community assembly and disassembly, belongs to the intermediate scale between shorter-term network dynamics and longer-term evolution. While these changes occur simultaneously in nature, they are still somewhat distinguishable by their frequencies or rates of variation that decrease from food-web dynamics through assembly/disassembly to evolution.

In their chapter of this book, Drossel and McKane (chapter 9) review recent research on network evolution. This chapter discusses a range of other outstanding questions about network evolution, focusing on the effects of different types of evolutionary rules including the networks that emerge from implementation of the rules. Then, different approaches for evaluating emergent networks are discussed. With these basic concepts in place, the chapter subsequently describes more specific modeling approaches that implement rules in ways that facilitate the evaluation of their results. The chapter concludes by pointing toward the development of new evaluation criteria and by connecting the study of evolving networks to other areas of ecology.

2 EXPLORING NETWORK EVOLUTION

The first question one might ask is, "What sorts of rules for network evolution lead to the networks that we see in nature?" Typically, "rules" represent formalizations of biological mechanisms thought to be responsible for network evolution, but could represent any formalized process for constructing networks. An important caveat is that the empirical networks we see in nature should not be taken completely literally due to limitations of the data (May 1983; Paine 1988; Martinez 1991; Polis 1991; Cohen et al. 1993; Martinez 1993b; Borer et al. 2002; Dunne chapter 2). In particular, there is a frequent mismatch between the larger spatiotemporal scale of evolution and the much smaller spatiotemporal scales at which food webs are currently being documented (Martinez and Lawton 1995; Martinez and Dunne 1998). Still, overall generalities that appear to survive methodological variation (Winemiller 1990; Martinez 1991; Goldwasser and Roughgarden 1993; Martinez 1993b; Martinez et al. 1999; Thompson 2000; Banašek-Richter et al. 2004) and empirical variation (Martinez 1992, 1993a, 1994; Warren 1994; Williams and Martinez 2000; Dunne et al. 2002b; Williams et al. 2002) are more likely to represent robust regularities that do warrant attention. For example, in no case with which I am familiar has *connectance*, defined as the fraction of all possible network links realized (Martinez 1992), exceeded 50%; food-web connectance is almost always much closer to 10%. Rules that create networks exceeding 50% should be treated skeptically. Other generalities go beyond statistical patterns in the data to indicate sets of rules that successfully predict the complete structure of food webs (Williams and Martinez 2000; Dunne et al. 2004), including rules that explicitly attempt to capture evolutionary aspects of food-web structure (Cattin et al. 2004). Such relatively comprehensive models generate networks that can be evaluated using increasingly many network properties (Dunne et al. 2002a; Garlaschelli et al. 2003; Cattin et al. 2004; Dunne et al. 2004) instead of restricting evaluations to one or two properties such as connectance or mean food-chain length. Perhaps the most exciting empirical development relevant to network evolution is paleofoodweb research that reconstructs food webs over "deep" time, stretching back over a half billion years before the present (Dunne and Erwin unpublished data). As this research progresses to the point of describing series of food webs from different times and habitats, it will provide both new tests of hypotheses (e.g., trophic specialization increases with time following major extinction events) and new empirical patterns that should inspire yet more hypotheses and theories.

Two of the most central questions that network studies aim to answer are, "How do networks come into being?" and, "How do networks vary over time scales within which species originate, invade, and go extinct?" Answers to these questions seek to uncover the evolutionary plot concerning the *players*, or nodes, and their mechanistic interactions that are responsible for network evolution. Such interactions are likely to involve well-known evolutionary mechanisms such as natural selection, less well-articulated constraints on network structure (e.g., plants must exist before herbivores can evolve), and newly emerging dynamical constraints that restrict the number and types of species that can persist and interact as nodes within a variable environment. A productive strategy for answering these questions would be to uncover a minimal set of rules that create networks consistent with empirical patterns. An important issue involves the environment within which species and their interactions evolve, such as environmental and demographic stochasticity (Cushing et al. 2003) and spatial heterogeneity.

3 RULES FOR NETWORK EVOLUTION

There is a wide range of mechanisms whose formalizations can provide important and useful rules for implementing network evolution. One of the most fundamental mechanisms with respect to species and the individuals that comprise their populations concerns the metabolic rates determining energetic maintenance cost of staying alive long enough to reproduce. Much if not most of the significance wrapped around the structure of the archetypal ecological networks called food webs is the need for this network to supply species with their metabolic requirements. Formalization of metabolic rules needs to take into account the degrees of constraint and flexibility associated with the body sizes and metabolic types of: (a) autotrophs including vascular plants, phytoplankton, and microbial chemoautotrophs; (b) animals including ectotherms, endotherms, invertebrates and vertebrates; and (c) microbes including bacteria, fungi, and protozoa. Closely associated with metabolic rates are the maximum rates of ingestion and assimilation that indicate potentially important variability among different types of animals and microbes (Yodzis and Innes 1992). One may want to relate these rates to trophic generality and specialization, trophic level, and omnivory. It is also important to consider rules that determine whether species are strong interactors such as keystone species or weak interactors whose removal less dramatically impacts their communities (Berlow et al. 2004). The emergence of such species would appear to depend on generating variation in resource consumption efficiencies and preferential predation upon the most efficient consumers.

Other rules relate more directly to the interactions themselves as opposed to the interactors, though the two are not completely separable. For example, there are many different types of functional responses that determine how consumption varies with the abundance of resources and other species in ecological networks (Gentleman et al. 2003; Skalski and Gilliam 2001). Few if any interactions are thought to be "Holling type I" interactions, which lack thresholds and in which feeding rates continually increase without satiation, regardless of food availability. Some interactions with thresholds are thought to be "Holling type II" responses in which feeding rates quickly accelerate as rare resources become more abundant. "Holling type III" responses are characterized by a much slower acceleration of feeding on rare resources, and are thought to occur in the presence of either limited refuges for resources or more intelligent consumers. Intelligent consumers avoid expending energy to hunt and feed unless resources are at a high enough density to provide more energy than is needed for consumption. Other "Beddington-Holt" functional responses address whether consumers are likely to interfere with each other's feeding rates as their densities increase (Skalski and Gilliam 2001). While the variation of these responses is known to greatly alter the dynamics of both low-dimensional (species < 5) and high-dimensional ($10 \leq$ species ≤ 30) networks (Williams and Martinez 2004b,c), their evolutionary stability and other evolutionary consequences are not well understood. Another important issue regarding interactions and network development concerns the hypothesis, based on non-ecological networks, that new nodes preferentially attach to existing highly connected nodes when joining established networks (Albert et al. 2000; Newman 2003). If relevant for ecology, this may suggest that speciation is more likely to either generate prey species that are eaten by generalists or generate predator species that eat prey with the most consumers. However, the ecological probability, dynamic effects, and evolutionary stability of such issues as attachment preference are poorly known (Dunne et al. 2002b).

Several other rules that may be implemented during network evolution are common to both nodes and interactions. For example, might there be tradeoffs between different properties such as the generality of nodes and the functional responses of their interactions, or between the assimilation efficiency of an interaction and the maximal ingestion rate of the node possessing that interaction? Still more general questions concern the directionality of evolution, i.e., whether or not any system's or component's traits tend to evolve in a consistent direction, either increasing or decreasing. For example, do networks evolve based on selection for nodes with high assimilation efficiencies or does such evolution lack any such directional selection? Such rules could emerge from network evolution rather than being implemented during network evolution as emphasized in this section. Still, implementing well-chosen rules can locate selection at genetic, individual, population, or network levels. How do networks differ when selection operates at these different levels? A similar question of scale involves exploring the effectiveness of immediate payoffs (e.g., quick biomass gain) compared to nodes that achieve long-term success at the evolutionary scale.

Many other questions concern environmental effects on network evolution. How is evolution altered by a stochastic versus stable environment? When evolution occurs in a stochastic environment, how does dominance by low frequency variation, known as *red noise*, alter network evolution in comparison to dominance by high frequency variation, known as *blue noise*? Also, what are the evolutionary effects of episodic versus gradual environmental change? How are resulting networks and their evolution altered when played out in a spatially explicit environment? Such spatial considerations could be examined at an individual level or at a meta-network level in which networks are connected in a landscape by inter-network migrations of different species.

This focus on the network is quite different from the usual focus of evolutionary theory on individuals, populations, and species. Still, these entities evolve within a network that Darwin described as a "tangled bank" of interactions. How does the evolution of nodes such as populations and species within a network differ from evolution within simpler contexts? Evolution of nodes within networks allows the consequences of a myriad of indirect interactions to influence the outcome of each species' evolution (e.g., Worden 2003). Additionally, networks or graphs can provide structure to evolutionary interactions such that nodes only directly interact with nodes that are connected with a link or edge to other nodes (Lieberman et al. 2005). Such evolution within or "on" networks is quite interesting apart from the evolution "of" networks emphasized here.

4 EVALUATING NETWORKS USING EVOLUTIONARY RULES

Beyond explorations of how the many factors mentioned above alter the evolution of networks, one needs criteria for evaluating those effects. Of course, matching networks generated using rules with empirical data is the most obvious criterion to employ. Less obvious are the aspects of network data on which to focus, and what priority to ascribe to the numerous quantifiable aspects of network structure. Of course, aspects of network nodes (e.g., total number and distribution among trophic levels) and interactions (e.g., link density and degree distribution) warrant attention, but also notable are statistical (e.g., regressions) and more mechanistic models of larger sets of empirical data. An important example is a recent comparison between the niche model (Williams and Martinez 2000; Camacho et al. 2002a,b; Dunne et al. 2004) and the nested hierarchy model (Cattin et al. 2004). The nested hierarchy model actually predicts less accurately a suite of increasingly common network properties (Martinez and Cushing Box A). However, unlike the niche model, the nested hierarchy model uses evolutionary rules to make networks that are non-interval, similar to most reasonably-sized food webs (Bersier et al. Box B). How is one supposed to balance such disparate success and failure (Stouffer et al. 2005)? There are also more general criteria to be addressed such as the robustness to disturbance and evolutionary trajectory of networks over deep time.

One place to start an empirically based evaluation of ecological network evolution is to focus on the network nodes, ideally meant to represent species or populations. What kinds of nodes emerge during network evolution? For example, when allowed to evolve, do metabolic rates assume empirically observed values? Do they assume uni-modal, bi-modal, or multi-modal distributions, as appears to be the case in nature given the presence of several metabolic types? Do the body sizes implied by "evolved" metabolic rates assume empirically observed ratios between predators and prey, herbivores and plants, or parasites and hosts? Does generality (number of species a node consumes) or vulnerability (number of species that consume a node), assume empirically observed distributions? For example, generality appears much more variable and widely distributed than does vulnerability (Williams and Martinez 2000). However, narrow vulnerability distributions may relax when consumption is coupled with mutualism such as pollination (Memmott 1999; Bascompte et al. 2003; Vazquez 2005).

If such patterns are robust, they may have profound evolutionary implications. What about the distribution of nodes among trophic levels and degrees of om-

Box 1.

A variety of criteria can be used to evaluate ecological networks in terms of their evolution and assembly, as indicated by the following questions:

1. How does network evolution differ from network assembly?
2. Does network evolution attain adaptive peaks?
3. How does the evolved network vary with time and over various spatial scales?
4. Are there consistent transient or asymptotic states?
5. How do the spatial scales of observations or mechanisms (e.g., local interaction with few species vs. regional interaction with many species) alter evolved networks in terms of bioenergetics or speciation?
6. How are non-trophic species traits (e.g., dispersal ability, mobility, tracking efficiency, and tracking ability) impacted by trophic network evolution?
7. How does the evolution of agent-based models differ from the evolution of ODE-based models of population biomass?

nivory (Williams and Martinez 2004a)? If we turn our focus to interactions rather than nodes, do we see empirically observed distributions of type II and type III functional responses, or degrees of predatory interference (Skalski and Gilliam 2001; Sarnelle 2003) when functional responses are allowed to evolve in terms of recently introduced control parameters (Williams 2004b,c)?

Beyond focus on the node and interaction components of networks, one can evaluate ecological network evolution based on the network as a whole. For example, different habitats generally have different networks. Pelagic networks are characterized by extremely small planktonic autotrophs and larger, though still small, herbivores such as zooplankton. Terrestrial networks are characterized by relatively large autotrophs including vascular plants and trees, along with much smaller herbivores such as insects, with the significant exceptions of less speciose larger-bodied mammalian herbivores such as ungulates (Hairston et al. 1960). During network evolution, do such characterizations represent different basins of attraction? While it is tempting to evaluate this question based on network architecture, recent claims that, for example, marine food webs are structurally quite different from non-marine food webs (Link 2002) appear to be overstated (Dunne et al. 2004) and there is little evidence that overall food-web structure reveals the profound differences seen in the distributions of body sizes among different trophic groups. However, perhaps research on paleofoodwebs will reveal structural variation over time rather than among habitats that will provide more useful, empirically-based criteria for evaluating network evolution.

Once empirically based criteria are chosen to evaluate network evolution, another question arises as to which data to use for comparison. Early food-web data were criticized for numerous methodological inconsistencies both among and within webs (Winemiller 1989; Martinez 1991; Polis 1991). Some nodes would be resolved to subspecies while others would aggregate entire kingdoms (e.g., microbes). The early data were also criticized for unacceptably low sampling effort that misrepresented empirical food-web structure (Tavares-Cromar and Williams 1996; Goldwasser 1997; Martinez et al. 1999; Thompson 2000). More recent data have reduced but not eliminated such inconsistencies and low sampling effort. However, such problems still require focused attention when empirical networks are used to evaluate network evolution. One needs to balance the abundance of older webs with more problems against the paucity of newer webs with fewer problems. One aspect of empirical webs that can help guide their use is the purpose for which they were constructed. For example, smaller less-complete webs that were constructed to illustrate the distribution of interaction strengths among a few species (Paine 1992) should be applied in different ways than larger, more complete webs that were constructed to illustrate overall network structure (Martinez 1991).

Beyond using network data to evaluate network evolution, one can compare hypotheses of ecological network evolution against models of network structure. For example, there are several statistical models that characterize the variation of food-web structure with scale in terms of species diversity (Martinez 1992, 1993a, 1994), space (Brose et al. 2004), and taxonomic resolution and trophic similarity (Martinez 1991, 1993b; Solow and Beet 1998; Thompson and Townsend 2000; Williams and Martinez 2000). As evolution adds and eliminates species, one would want to know whether or not the variation of network structure follows the variation specified by these statistical models. Similarly, other network models are based on degree distribution (Barabasi and Albert 1999) and assortative linking (Newman 2003). One would like to know how variation characterized by these statistical models compares to variation seen both during network evolution and within a collection of networks that result from some evolution-based rules. More mechanistic models characterize variability due to randomness (Erdös and Renyi 1960; Bollobás 1985; Williams and Martinez 2000), trophic hierarchy (Cohen 1990, Williams and Martinez 2000), niche structure (Williams and Martinez 2000; Cattin et al. 2004), phylogenetic structure (Cattin et al. 2004) and preferential attachment (Newman 2001). Again, how do evolved networks compare to these network models? If hypothesized mechanisms for network evolution give rise to networks that are generally consistent with accurate statistical and more mechanistic models of network structure, then such hypothesized evolutionary processes can be claimed to be generally consistent with the data. On the other hand, consistent differences between initial networks and those that result from longer-term evolution may point towards patterns to search for among paleofood-webs extracted from the fossil record. For example, specialists that eat only one or a very few species may be very rare during early simulated network evolution, increasing in abundance in later evolutionary stages. While current food-web data seem to have many specialists, paleofoodwebs may have lacked specialists both early in their history and shortly after major extinction events. Perhaps significant numbers of specialist species only occur well after such events. Such an empirical finding could provide important support for evolutionary models that match this pattern and also illustrate as-yet unknown general differences between current and ancient ecological networks.

In addition to specific criteria for evaluating hypotheses of network evolution such as empirical patterns and statistical and mechanistic models of current food-web structure, one can use much more general criteria. One example includes robustness that occurs when food webs stop changing during evolution. What sorts of webs achieve such robustness? Beyond robustness to evolution, one would like to know which evolutionary mechanisms give rise to networks that resist invasions by other species or that contain keystone species whose exclusion precipitates dramatic changes to the rest of the network. More subtly, does network evolution generally give rise to particular kinds of qualitative population dynamics?

Other general sets of criteria involve the concept of directionality at the node, link, and network levels. Is there any directional progression of, for example, the degree or omnivory of nodes (Williams and Martinez 2004), the compartmentalization of links (Girvan and Newman 2002; Krause et al. 2003), or the amount of connectance (Martinez 1992; Dunne et al. 2004)? Network evolution dynamics could also be directional. For example, network evolution might progress towards species that have particular assimilation efficiencies, or particular levels of generality and vulnerability. Whole networks could progress through different qualitative types of dynamics from chaotic fluctuations to equilibria.

5 IMPLEMENTING EVOLUTIONARY RULES AND EVALUATING RESULTS

Most of the discussion of approaches to implementing rules and evaluating the results, based on the criteria described above, focuses on different aspects of simulations. However, network evolution can also be explored by other approaches such as analytical modeling (Camacho et al. 2002b; Cameron 2002; Stouffer et al. 2005) and microcosm experiments (Bohannan and Lenski 2000; Morin 2004). Still, such approaches and computer simulations can all pursue a general approach of building and exploring models that synthesize a wide range of fundamental components of ecological networks. For example, bioenergetic models of trophic interactions can be coupled with models of shared resource consumption among plants and detrital models of deposition of biomass, microbial consumption of non-living biomass, and mineralization of nutrients. Another approach could be to link abstract analytical models (Harte and Kinzig 1993) with more detailed computational models (Brose et al. 2003; Martinez et al. Chapter 6) and agent-based models (Pfister and Peacor 2003). Successful linking would show that synthetic models can be phenomenologically accurate at both individual and system levels. Agent-based models could also be used to explore the accuracy of functional responses that model feeding behavior of larger undifferentiated stocks of biomass containing many individuals. This could test whether, and to what degree, there is consistency between differential equation approaches and agent-based models, especially with respect to functional responses and their consistency with allometric scaling (Economo et al. 2005), optimal feeding, and other behavior by agents.

Ideally, one might like to explore hybrid dynamics-evolution models which incorporate time scales relevant to both ecological and evolutionary dynamics. While much of such dynamics involves nutrient flows and living and dead biomass, it is also important to develop non-trophic and quasi-trophic interaction models involving ecosystem-engineering interactions. Such interactions are common to ecosystems, biofilms, and pollination and dispersal systems in which strong mutualisms overlap and interact in different than strictly trophic interactions (Martinez 1995). For example, while pollinators and seed dispersers often benefit from consuming plant nectar, pollen, and fruit, plants can benefit from using animals' metabolic energy to disperse propagules (Bascompte et al. 2003). Additionally, early colonizers of biofilms can both exclude competitors and provide food for later colonizers. Additional interactions include mimicry, camouflage, and phenotypic plasticity in models of network evolution. Pursuing these approaches will require increased sophistication of models (e.g., adding detrital pathways, engineering, and selection). Such sophistication may be made more tractable by focusing on simpler or earlier systems (e.g., bacteria and plants) that provide insight into how many more species evolve within more complex networks. Explorations of how the first ecosystems functioned and what they looked like, or how population cycles and size structure influence ecosystem dynamics, could provide the foundation for explaining the rich complexity apparent today.

6 CONCLUSION

While much of the focus of this chapter is to develop theories, approaches, and criteria for modeling ecological network evolution, such modeling should also lead to suggestions or specifications of observational (e.g., in the fossil record) and manipulative (e.g., in microcosms) experiments that rigorously test models. Such tests should attend to both means and variation of models and empirical data. Evaluations are typically more compelling when they involve parameters within models that can be empirically measured. In developing models, dialog with field biologists is critical in order to assess which parameters and patterns appear most important in the natural world. As an iterative process, such dialogue with empiricists will facilitate comparison between high-quality snapshots of real ecosystems and longer-term development of ecosystems over deep time. Beyond the inherent interest in network evolution in its own right, exploring network evolution with approaches developed for ecological networks should also greatly elucidate the rich, yet still opaque, relationship between the ecology and evolution of complex natural ecosystems.

ACKNOWLEDGMENTS

The ideas presented here are largely the result of a discussion section on network evolution during a conference entitled, "From Structure to Dynamics in Complex Ecological Networks" at the Santa Fe Institute, sponsored by the McDonnell Foundation, the Pacific Ecoinformatics and Computational Ecology Lab, and the Santa Fe Institute. The following discussion-section participants are greatly appreciated for their contributions: Jordi Bascompte, Ulrich Brose, Barbara Drossel, Gregor Fussman, Sanjay Jain, Michio Kondoh, Per Lundberg, Alan McKane, Carlos Melián, José Montoya, Claus Wilke, Richard Williams. Special thanks to Jordi Bascompte, Mercedes Pascual, and Jennifer Dunne for suggestions that significantly improved the manuscript. NSF Biological Databases and Informatics grant DBI-0234980 and Information Technology Research grant ITR-0326460 supports NDM and the Pacific Ecoinformatics and Computational Ecology Lab.

REFERENCES

Anderson, P. E., and H. J. Jensen. 2005. "Network Properties, Species Abundance and Evolution in a Model of Evolutionary Ecology." *J. Theor. Biol.* 232:551–558.

Banašek-Richter, C., M.-F. Cattin, and L.-F. Bersier. 2004. "Sampling Effects and the Robustness of Quantitative and Qualitative Food-Web Descriptors." *J. Theor. Biol.* 226:23–32.

Barabasi, A.-L., and R. Albert. 1999. "Emergence of Scaling in Random Networks." *Science* 286:509-512.

Bascompte, J., P. Jordano, C. J. Melian, and J. M. Olesen. 2003. "The Nested Assembly of Plant-Animal Mutualistic Networks." *PNAS* 100:9383–9387.

Berlow, E. L., A.-M. Neutel, J. E. Cohen, P. C. de Ruiter, B. Ebenman, M. Emmerson, J. W. Fox, V. A. A. Jansen, J. Iwan Jones, G. D. Kokkoris, D. O. Logofet, A. J. McKane, J. M. Montoya, and O. Petchey. 2004. "Interaction Strengths in Food Webs: Issues and Opportunities." *J Animal Ecol.* 73:585–598.

Bohannan, B. J. M., and R. E. Lenski. 2000. "The Relative Importance of Competition and Predation Varies with Productivity in a Model Community." *Amer. Natur.* 156:329–340.

Bollobás, B. 1985. *Random Graphs.* London: Academic Press.

Borer, E. T., K. Anderson, C. A. Blanchette, B. Broitman, S. D. Cooper, and B. S. Halpern. 2002. "Topological Approaches to Food Web Analyses: A Few Modifications May Improve Our Insights." *Oikos* 99:397–401.

Brose, U., R. J. Williams, and N. D. Martinez. 2003. "Comment on 'Foraging Adaptation and the Relationship Between Food-Web Complexity and Stability.'" *Science* 301:918b.

Brose, U., A. Ostling, K. Harrison, and N. D. Martinez. 2004. "Unified Spatial Scaling of Species and Their Trophic Interactions." *Nature* 423:167–171.

Caldarelli, G., P. G. Higgs, and A. J. McKane. 1998. "Modelling Coevolution in Multispecies Communities." *J. Theor. Biol.* 193:345–358.

Camacho, J., R. Guimera, and L. A. N. Amaral. 2002a. "Robust Patterns in Food Web Structure." *Phys. Rev. Lett.* 88(22):228102.

Camacho, J., R. Guimerà, and L. A. N. Amaral. 2002b. "Analytical Solution of a Model for Complex Food Webs." *Phys. Rev. E* 65:030901.

Cameron, T. 2002. "2002: The Year of the 'Diversity-Ecosystem Function' Debate." *Trends Ecol. & Evol.* 17:495–496.

Cattin, M.-F., L.-F. Bersier, C. Banašek-Richter, R. Baltensperger, and J.-P. Gabriel. 2004. "Phylogenetic Constraints and Adaptation Explain Food-Web Structure." *Nature* 427:835–839.

Cohen, J. E., F. Briand, and C. M Newman. 1990. *Community Food Webs: Data and Theory.* New York: Springer-Verlag.

Cohen, J. E., R. A. Beaver, S. H. Cousins, D. L. DeAngelis, L. Goldwasser, K. L. Heong, R. D. Holt, A. J. Kohn, J. H. Lawton, N. Martinez, R. O'Malley, L. M. Page, B. C. Patten, S. L. Pimm, G. A. Polis, M. Rejmanek, T. W. Schoener, K. Schoely, W. G. Sprules, J. M. Teal, R. E. Ulanowicz, P. H. Warren, H. M. Wilbur, and P. Yodzis. 1993. "Improving Food Webs." *Ecology* 74:252–258.

Cushing, J. M., R. F. Constantino, B. Dennis, R. A. Desharnais, and S. M. Hanson. 2003. *Chaos in Ecology: Experimental Nonlinear Dynamics.* San Diego, CA: Academic Press.

de Ruiter, P. C., A.-M. Neutel, and J. C. Moore. 1995. "Energetics, Patterns of Interaction Strengths and Stability in Real Ecosystems." *Science* 269:1257–1260.

Drossel, B., and A. J. McKane. 2002. "Modelling Food Webs." In *Handbook of Graphs and Networks*, ed. S. Bornholdt and H. G. Schuster. Berlin: Wiley-VHC.

Dunne, J. A., R. J. Williams, and M. D. Martinez. 2002a. "Network Structure and Biodiversity Loss in Food Webs: Robustness Increases with Connectance." *Ecol. Lett.* 5:558–567.

Dunne, J. A., R. J. Williams, and N. D. Martinez. 2002b. "Food-Web Structure and Network Theory: The Role of Connectance and Size." *PNAS* 99:12917–12922.

Dunne, J. A., R. J. Williams, and N. D. Martinez. 2004. "Network Structure and Robustness of Marine Food Webs." *Marine Ecol.—Prog. Ser.* 273:291–302.

Economo, E. P., A. J. Kerkhoff, and B. J. Enquist. 2005. "Allometric Growth, Life History Invariants, and Population Energetics." *Ecol. Lett.* 8:353—360.

Erdös, P., and A. Renyi. 1960. "On the Evolution of Random Graphs." *Publ. Math. Inst. Hung. Acad. Sci.* 5:17–61.

Fox, J. W., and J. McGrady-Steed. 2002. "Stability and Complexity in Microcosm Communities." *J. Animal Ecol.* 71:749–756.

Garlaschelli, D., G. Caldarelli, and L. Pietronero. 2003. "Universal Scaling Relations in Food Webs." *Nature* 423:165–168.

Gentleman, W., A. Leising, B. Frost, S. Strom, and J. Murray. 2003. "Functional Responses for Zooplankton Feeding on Multiple Resources: A Review of Assumptions and Biological Dynamics." *Deep Sea Research Part II: Topical Studies in Oceanography* 50:2847–2875.

Girvan, M., and M. E. J. Newman. 2002. "Community Structure in Social and Biological Networks." *PNAS* 99:7821–7826.

Goldwasser, L., and J. Roughgarden. 1997. "Sampling Effects and the Estimation of Food-Web Properties." *Ecology* 78:41–54.

Goldwasser, L., and J. Roughgarden. 1993. "Construction and Analysis of a Large Caribbean Food Web." *Ecology* 74:1216–1233.

Hairston, N. G., F. E. Smith, and L. B. Slobodkin. 1960. "Community Structure, Population Control, and Competition." *Amer. Natur.* 94:421–425.

Harte, J., and A. P. Kinzig. 1993. "Mutualism and Competition between Plants and Decomposers—Implications for Nutrient Allocation in Ecosystems." *Amer. Natur.* 141:829–846.

Holt, R. D., J. H. Lawton, G. A. Polis, and N. D. Martinez. 1999. "Trophic Rank and the Species-Area Relationship." *Ecology* 80:1495–1504.

Hutchinson, G. E. 1965. *The Ecological Theater and the Evolutionary Play.* New Haven: Yale University Press.

Kondoh, M. 2003. "Foraging Adaptation and the Relationship between Food-Web Complexity and Stability." *Science* 299:1388–1391.

Krause, A. E., K. A. Frank, D. M. Mason, R. E. Ulanowicz, and W. W. Taylor. 2003. "Compartments Revealed in Food-Web Structure." *Nature* 426:282–285.

Lieberman, E., C. Hauert, and M. A. Nowak. 2005. "Evolutionary Dynamics on Graphs." *Nature* 433:312–316.

Link, J. 2002. "Does Food Web Theory Work for Marine Ecosystems?" *Marine Ecol. Prog. Ser.* 230:1–9.

Martinez, N. D. 1991. "Artifacts or Attributes—Effects of Resolution on the Little-Rock Lake Food Web." *Ecol. Monogr.* 61:367–392.

Martinez, N. D. 1992. "Constant Connectance in Community Food Webs." *Amer. Natur.* 139:1208–1218.

Martinez, N. D. 1993a. "Effect of Scale on Food Web Structure." *Science* 260:242–243.

Martinez, N. D. 1993b. "Effects of Resolution on Food Web Structure." *Oikos* 66:403–412.

Martinez, N. D. 1994. "Scale-Dependent Constraints on Food-Web Structure." *Amer. Natur.* 144:935–953.

Martinez, N. D. 1995. "Unifying Ecological Subdisciplines with Ecosystem Food Webs." In *Linking Species and Ecosystems*, ed. C. G. Jones and J. H. Lawton, 166–175. New York, NY: Chapman & Hall.

Martinez, N. D., and J. A. Dunne. 1998. "Time, Space, and Beyond: Scale Isues in Food-Web Research." In *Ecological Scale: Theory and Applications*, ed. D. L. Peterson and T. Parker, 206–226. New York: Cornell University Press.

Martinez, N. D., and J. H. Lawton. 1995. "Scale and Food-Web Structure—From Local to Global." *Oikos* 73:148–154.

Martinez, N. D., B. A. Hawkins, H. A. Dawah, and B. P. Feifarek. 1999. "Effects of Sampling Effort on Characterization of Food-Web Structure." *Ecology* 80:1044–1055.

May, R. M. 1983. "The Structure of Food Webs." *Nature* 301:566–568.

Memmott, J. 1999. "The Structure of a Plant-Pollinator Food Web." *Ecol. Lett.* 2:276–280.

Mikkelson, G. M. 1993. "How do Food Webs Fall Apart? A Study of Changes in Trophic Structure During Relaxation on Habitat Fragments." *Oikos* 67:539–547.

Morin, P. J., and Jill McGrady-Steed. 2004. "Biodiversity and Ecosystem Functioning in Aquatic Microbial Systems: A New Analysis of Temporal Variation and Species Richness-Predictability Relations." *Oikos* 104:458–466.

Newman, M. E. J. 2001. "Clustering and Preferential Attachment in Growing Networks." *Phys. Rev. E* (Statistical, Nonlinear, and Soft Matter Physics) 64:025102.

Newman, M. E. J. 2003. "The Structure and Function of Complex Networks." *SIAM Rev.* 45:167–256.

Paine, R. T. 1988. "Food Webs—Road Maps of Interaction or Grist for Theoretical Development?" *Ecology* 69:1648–1654.

Paine, R. T. 1992. "Food-Web Analysis through Field Measurement of Per-Capita Interaction Strength." *Nature* 355:73–75.

Pfister, C. A., and S. D. Peacor. 2003. "Variable Performance of Individuals: The Role of Population Density and Endogenously Formed Landscape Heterogeneity." *J. Animal Ecol.* 72:725–735.

Polis, G. A. 1991. "Complex Trophic Interactions in Deserts—An Empirical Critique of Food-Web Theory." *Amer. Natur.* 138:123–155.

Post, D. M., M. E. Conners, and D. S. Goldberg. 2000. "Prey Preference by a Top Predator and the Stability of Linked Food Chains." *Ecology* 81:8–14.

Rossberg, A. G. 2004. "An Explanatory Model for Food-Web Structure and Evolution." arXiv.org E-print Archive, Cornell University Library. http://arXiv.org/abs/q-bio.PE/0410030.

Sarnelle, O. 2003. "Nonlinear Effects of An Aquatic Consumer: Causes and Consequences." *Amer. Natur.* 161:478–496.

Skalski, G. T., and J. F. Gilliam. 2001. "Functional Responses with Predator Interference: Viable Alternatives to the Holling Type II Model." *Ecology* 82:3083–3092.

Solow, A. R., and A. R. Beet. 1998. "On Lumping Species in Food Webs." *Ecology* 79:2013–2018.

Stouffer, D. B., J. Camacho, R. Guimera, C. A. Ng, and L. A. N. Amaral. 2005. "Quantitative Patterns in the Structure of Model and Empirical Food Webs." *Ecology* 86:1301–1311.

Tavares-Cromar, A. F., and D. D. Williams. 1996. "The Importance of Temporal Resolution in Food Web Analysis: Evidence from a Detritus-Based Stream." *Ecol. Monogr.* 66:91–113.

Thompson, R. M., and C. R. Townsend. 2000. "Is Resolution the Solution?: The Effect of Taxonomic Resolution on the Calculated Properties of Three Stream Food Webs." *Freshwater Biol.* 44:413-422.

Vazquez, D. P. 2005. "Degree Distribution in Plant-Animal Mutualistic Networks: Forbidden Links or Random Interactions?" *Oikos* 108:421–426.

Warren, P. H. 1994. "Making Connections in Food Webs." *Trends in Ecol. & Evol.* 9:136–141.

Williams, R. J., and N. D. Martinez. 2000. "Simple Rules Yield Complex Food Webs." *Nature* 404:180–183.

Williams, R. J., and N. D. Martinez. 2004a. "Limits to Trophic Levels and Omnivory in Complex Food Webs: Theory and Data." *Amer. Natur.* 163:458–468.

Williams, R. J., and N. D. Martinez. 2004b. "Diversity, Complexity, and Persistence in Large Model Ecosystems." Working Paper 04-07-022, Santa Fe Institute, Santa Fe, NM.

Williams, R. J., and N. D. Martinez. 2004c. "Stabilization of Chaotic and Non-Permanent Food-Web Dynamics." *Europ. Phys. J. B* 38:297–303.

Williams, R. J., E. L. Berlow, J. A. Dunne, A. L. Barabasi, and N. D. Martinez. 2002. "Two Degrees of Separation in Complex Food Webs." *PNAS* 99:12913–12916.

Winemiller, K. O. 1990. "Spatial and Temporal Variation in Tropical Fish Trophic Networks." *Ecol. Monogr.* 60:331–367.

Worden, L. 2003. "Evolution, Constraint, Cooperation, and Community Structure in Simple Models." Ph.D. diss., Princeton University, Princeton, NJ.

Yodzis, P., and S. Innes. 1992. "Body Size and Consumer-Resource Dynamics." *Amer. Natur.* 139:1151–1175.

Yoshida, K. 2003. "Evolutionary Dynamics of Species Diversity in an Interaction Web System." *Ecol. Model.* 163:131–143.

Stability and Robustness of Ecological Networks

"In short, there is no comfortable theorem assuring that increasing diversity and complexity beget enhanced community stability; rather, as a mathematical generality the opposite is true. The task, therefore, is to elucidate the devious strategies which make for stability in enduring natural systems."

— Bob May,
Stability and Complexity in Model Ecosystems, 1973

Ecological Network Meltdown from Habitat Loss and Fragmentation

Ricard V. Solé
José M. Montoya

Human-induced habitat alteration is the major cause leading to biodiversity loss and eventually to ecosystem collapse. The response of communities to habitat loss depends on both species' characteristics and the extent to which species interact. In this sense, habitat destruction may yield qualitatively new consequences when considering species that are embedded in an intricate web of ecological relationships. Here we review the theory linking habitat loss and food webs, from single predator-prey interactions to large food webs. We present a new multitrophic model for simulating the dynamics of multispecies ecosystems in fragmented landscapes. The model shows that as habitat destruction increases, predators are lost first, then herbivores and finally producers. But the interaction between trophic levels yields new outcomes not predicted when a single trophic level is observed alone, as trophic cascades. Our analysis of multispecies ecosystems indicates that the effects of habitat loss and those of network structure interact in new but predictable ways, and, therefore,

Ecological Networks: Linking Structure to Dynamics in Food Webs, edited by Mercedes Pascual and Jennifer A. Dunne, Oxford University Press.

their integration is necessary and useful for a deep understanding of the effects of habitat alteration on ecosystems.

1 THRESHOLDS IN HABITAT FRAGMENTATION

Current rates of habitat destruction are extremely high. Habitat loss strongly enhances the detrimental effects triggered by species introductions, pollution, climate change, and hunting. Habitat loss is strongly tied to specific patterns of patch destruction, leading to habitat fragmentation (fig. 1).Once fragmented into isolated patches, small discrete populations are much more vulnerable to extinction due to environmental fluctuations as well as demographic and genetic factors (such as inbreeding depression). The physical changes associated with habitat loss and fragmentation include reduction of total area and productivity of native areas, isolation of forest remnants and changes in physical conditions of the remnant fragments.

These anthropogenic changes trigger further community responses that sometimes end in a biotic collapse. The sequence of biotic decay (Wilcove 1987) includes the initial exclusion of some species, deleterious effects of isolation, and ecological imbalances. The latter involve nonlinearities and cascade effects through the ecological webs. The loss of a key species can promote the subsequent loss of its predators, parasites or mutualists. One dramatic example (among others) is the so-called *ecological meltdown* observed in predator-free forest fragments. The loss of predators generates strong imbalances shown by the disproportionate increase in the densities of prey and severe reductions of seedlings and saplings of canopy trees (Terborgh et al. 2001). Conversely, the re-introduction of a previously removed predator can sometimes restore the previous biodiversity.

Human-induced habitat alteration is the major cause of biodiversity loss (Wilcox and Murphy 1985; Barbault and Sastrapradja 1995). Some authors have estimated that extinction rates have increased one-thousand-fold during the last 300 years, a rate comparable in magnitude to one of the five big mass-extinction events (Lawton and May 1994). Actually, some relevant differences between both events should be highlighted. In particular, previous mass-extinction events seem to be largely initiated by loss of diversity at the bottom layer of the networks (i.e., affecting primary producers). The current event is damaging webs at different levels. Primary producers are certainly being affected by habitat loss and degradation, but keystone species (particularly large mammals) are being decimated too. Active, widespread hunting is reducing the populations of many species to red numbers. In this way, bottom-up and top-down cascades are unleashed. Habitat loss and degradation are often associated with tropical ecosystems, but they are widespread phenomena in all ecosystems. Even in the Antarctic, benthic ecosystems suffer strong disturbances due to iceberg movement damaging the sea floor (Teixidó 2003). Since these communities are known to grow very

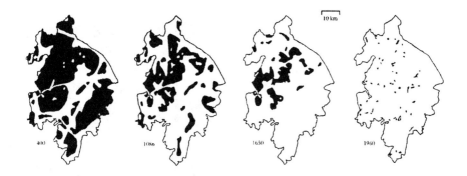

FIGURE 1   Temporal sequence of habitat loss and fragmentation in forests of Warwick-shire, England. The sequence starts near A.D. 400 and ends in A.D. 1960. Extensive removal has taken place ending up with a scattered set of small patches.

slowly, the increasing levels of ice breaking associated with global warming are severely damaging these communities.

Given the magnitude and consequences of habitat destruction, it is imper-ative that we gain enough insight into understanding the effects of habitat loss on species survival, in order to predict its further consequences. Since economic tradeoffs are at play, scientists are often faced with the question of how much habitat can be destroyed before a certain species becomes extinct. Data on the consequences of habitat loss are becoming available, both from using field obser-vations (Andrén 1994, 1996), and from experiments (Holt et al. 1995; Robinson et al. 1992; Debinski and Holt 2000). While this approach is fundamental, it is difficult to obtain sufficient information about the long-term consequences of habitat loss due to the large spatial and temporal scales at which this process takes place. For example, we can record the loss of species a few years after a human alteration, but there may be even more time lags (Tilman et al. 1994). Other species still present may become extinct at a later time, causing us to underestimate the effects of habitat destruction. In order to fully understand the consequences of habitat loss, models play a relevant role, particularly in forecasting the effects of landscape degradation on web structure and stability.

Theory suggests that the response of communities to habitat loss depends on species' characteristics and on the extent to which species interact. Larger-bodied and rare species are usually the first losers in most ecosystems around the world (Woodroffe and Ginsberg 1998; Purvis et al. 2000). Similarly, food-web theory predicts that habitat loss and fragmentation reduces population densities of top predators (Holyoak 2000), and, therefore, species from higher trophic levels are more frequently lost than species from lower levels (see Petchey et al. 2004, for a review). In a related context, the consequences of species loss are highly mediated by the position of a species within the interaction network (Pimm 1991;

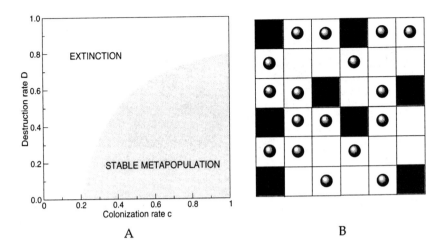

FIGURE 2 (A) Parameter space for the (spatially implicit) Levin's model, using a local extinction rate $e = 0.2$. Two domains are obtained from the transition curve $D_c = 1 - e/c$. This critical boundary separates the domain where a stable metapopulation exists from the extinction phase, where no metapopulation is able to persist. A discrete, spatial view of Levin's model would involve a set of patches (B) occupying a given domain. Here empty, destroyed and occupied sites are indicated as white, black and gray balls, respectively. The standard model assumes that all empty patches have the same probability of being colonized, whereas a more realistic scenario would consider colonization as a local phenomenon: only nearest occupied patches can colonize a given empty site.

Solé and Montoya 2001; Dunne et al. 2002). The disappearance of prey species that are attacked by numerous specialized predators, for example, has larger consequences than the loss of prey species with fewer specialized predators.

In this chapter we have tried to synthesize this theory, highlighting new avenues linking habitat loss and food webs. Habitat destruction may yield qualitatively new consequences when considering species that are embedded in an intricate web of ecological relationships. Our belief is that advances in predicting the effects of habitat loss and fragmentation, and ultimately our ability to make realistic estimations of extinction rates, will require us to consider the network of biotic interactions among species.

2 HABITAT DESTRUCTION IN LEVIN'S MODEL

The simplest, first approximation is obtained by exploring the consequences of habitat loss in a metapopulation. A metapopulation can be defined as a set of geographically distinct local populations maintained by a dynamical balance between colonization and extinction events. Let us start this section by revisiting the Levin's (1969) model, which captures the global dynamics of a metapopulation:

$$\frac{dx}{dt} = cx(1 - x) - ex, \tag{1}$$

where x is the fraction of patches occupied, and c and e are the colonization and extinction rates, respectively. This model has a non-trivial solution given by $x^* = 1 - e/c$. The colonization rate has to be larger than the extinction rate for the metapopulation to persist.

One can easily introduce habitat loss into the framework of model (1). If a fraction D of sites are permanently destroyed, this reduces the fraction of vacant sites that can potentially be occupied. Model (1) becomes:

$$\frac{dx}{dt} = cx(1 - D - x) - ex. \tag{2}$$

This model has two equilibrium points (to be obtained from $dx/dt = 0$): $x^* = 0$ (extinct population) and

$$x^* = 1 - D - \frac{e}{c}. \tag{3}$$

This equilibrium point (3) decays linearly with habitat loss, becoming zero when $1 - D - e/c = 0$. This condition gives a critical destruction level D_c:

$$D_c = 1 - \frac{e}{c} \tag{4}$$

indicating that a non-trivial dependence exists between available sites and the species-specific extinction and colonization properties. Once we cross this threshold (i.e., if $D > D_c$) the population inevitably becomes extinct. This situation is illustrated in figure 2(a), where the critical line separating the two qualitative types of metapopulation is shown. Essentially, as we approach the critical value D_c by increasing the amount of habitat destroyed, the frequency of populated patches decays linearly, becoming zero at the boundary.

The main lesson to be extracted from this model is that (perhaps against our intuition) no sustainable metapopulation is possible once we reach a critical amount of habitat loss, even though a fraction of $1 - D$ patches is still habitable. The interaction between available areas and demographic parameters (here reduced to two local, average rates) leads to a threshold behavior.

Although Levin's model is a good approximation when dealing with metapopulations with long-range colonization, a more realistic view involves using a spatially extended domain of available space and resources that can be colonized through local rules of death and dispersal. As more realism is introduced into the model, new important features become apparent. An especially interesting one is the effect of using a spatially explicit framework (Bascompte and Solé 1996),

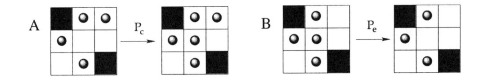

FIGURE 3 Local updating rules in a stochastic cellular automaton implementation of Levin's model. Here space is made explicit and the two basic events are colonization of empty sites (A) and extinction of occupied sites (B). These events have well-defined probabilities P_c and P_e, respectively.

such as the lattice shown in figure 2(b). The empty sites in the lattice (white patches) can only get colonized if at least one of the (four or eight) neighboring patches is already occupied.

The rules for the model are now local and stochastic and easily defined in probabilistic terms. In figure 3, we summarize the two basic rules to be applied. The first is colonization: an empty site will be colonized with a probability P_c proportional to the number of its occupied neighbors (fig. 3(a)). Extinction is simpler: a given occupied site becomes empty with some probability P_e, independently of the state of its neighbors. This spatial constraint strongly limits the dynamics of spatial occupation of a given landscape, and was shown to modify the predictions obtained from Levin's model, where all patches are equally available and thus global mixing is taking place. In particular, the predicted threshold for extinction becomes lower as the range of local colonization is reduced. Besides, spatial landscapes with random destruction of patches (here D would be the probability of a patch being non-available) display sharp thresholds: after a given D_c is reached, the largest connected patch suddenly becomes broken into many small subpatches. Such a transition is sharp and strongly limits colonization even at high dispersal rates.

3 SMALL ECOLOGICAL WEBS AND HABITAT LOSS

One particularly important instance of ecosystem decay due to the increase in habitat degradation within a network structure is provided by the loss of top predators from fragmented habitats. It has been known from different available examples that after strong habitat reduction (either as a consequence of fragmentation or simple loss) top predators are likely to become extinct, unleashing a cascade of changes. This sequence of changes is well documented from the study of artificial islands created by the flooding of forests due to the building of dams.

One of the best known examples of ecosystem changes due to the creation of an artificial island is provided by Barro Colorado. The island was formed during the construction of the Panama Canal when the dam on the Chagres river was

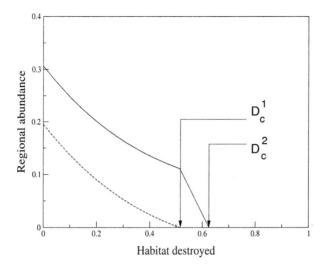

FIGURE 4 Equilibrium populations in the prey-predator metapopulation model when top-down control ($\mu > 0$) is considered. The regional abundance of Prey (solid line) and predators (broken line) is displayed against habitat loss D. Here we used: $\mu = 0.2, c_x = 0.4, c_y = 0.9, e_x = 0.15, e_y = 0.1$. The two arrows indicate the presence of two different extinction thresholds (see text).

built and the water level rose, flooding the area that is now Lake Gatun. One direct consequence of the formation of Barro Colorado was the extinction of its most prominent top predator, the jaguar:

> *When jaguars and pumas disappeared from Barro Colorado (...) because the forest was no longer extensive to support them, the prey species soon increased tenfold. Effects of this shift in balance now appear to be rippling downward through the food chain. Coatis, agoutis, and pacas feed on large seeds that fall from the rain-forest canopy. When they become superabundant, they reduce the reproductive ability of the particular tree species that produce these seeds. Other species whose seeds are too small to be of interest of the animals benefit from the lessened competition (...). Over a period of years, the composition of the forest shift in their favor.*
> —Wilson 1992, pp.165.

How can we introduce the importance of biotic (e.g., trophic) interactions in habitat fragmentation models? A good illustration of the importance of ecological interactions in enhancing the effects of habitat destruction is provided by considering a prey-predator metapopulation model (see Bascompte and Solé 1998, and references therein). It is a simple, but non-trivial extension of Levin's model. Here the predators can only live in patches occupied by prey. If x and y denote the fraction of patches occupied by prey and predator, respectively, the following model describes the trophic interaction:

$$\frac{dx}{dt} = c_x x(1 - x - D) - e_x x - \mu y, \tag{5}$$

$$\frac{dy}{dt} = c_y y(x - y) - e_y y. \tag{6}$$

Where, again, c and e are extinction and colonization rates, respectively, new terms have to be considered. Prey mortality is described by two terms. They have a mortality rate e_x at patches where only prey are present (a fraction $x - y$ of sites). Additionally, it is further increased to $e_x + \mu$ in those patches where both prey and predators are present (a fraction y). Then, total mortality for prey is $e_x(x - y) + (e_x + \mu)y = e_x x + \mu y$. Available sites for predators are now the non-destroyed, empty sites occupied by prey (i.e., $x - y$), since predators are specialists, and cannot live without their prey. The model can describe different trophic interactions: donor control ($\mu = 0$) and top-down control ($\mu > 0$). This is important for two reasons. First, theoretical models of trophic interactions have generally assumed scenarios in which natural enemies have a significant impact on prey populations (top-down control, but see Cohen et al. 1990). Second, donor control may operate in half of the parasitoid-host interactions, which in turn contain roughly half of the world's multicellular species (Hawkins 1992).

The basic outcome of the model (see Bascompte and Solé 1998 for details) is shown in figure 4. Here the equilibrium populations of both predators (dashed lines) and prey (continuous line) are shown. The key result here is that two different thresholds of habitat destruction are shown to exist (arrows in fig. 4). Below the first, $D_c^{(1)}$, prey and predators coexist in the landscape. However, once the first threshold is reached, predators no longer exist. By further increasing habitat loss, prey also become extinct at the second threshold $D_c^{(2)}$.

Two relevant conclusions can be obtained from this study. The first is that extinction takes place first for the predator, then for the prey. If predators are specialists, we can expect a well-defined order in the pattern of extinctions, starting with the highest trophic level species and going down through the trophic chain. This may change if the predator is a generalist (Swihart et al. 2001). This means that habitat destruction is not only going to reduce biodiversity, but it will also reduce the length of the food chain.

A second, counterintuitive result also displayed in figure 4 is that the response of prey to habitat loss depends on the fraction of habitat which has been destroyed. There are two different patterns of decay separated by $D_c^{(1)}$, the extinction threshold for the predator. When predators are present, prey abundance decays more *slowly* than when predators are extinct. Such a result comes from the response to habitat loss, which is a tradeoff between different trends depending on the trophic position. Thus, while habitat loss has a negative effect

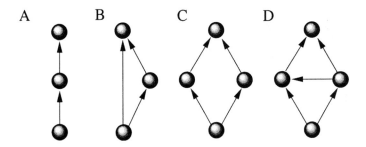

FIGURE 5   Basic food web modules analyzed in (Melian and Bascompte 2002). From bottom to top, plant, herbivores, and top predators, with arrows indicating feeding links pointing from resources to consumers. (a) simple food chain; (b) omnivory (i.e., predator feeds both on herbivore and plant species); (c) apparent competition (predator feeding on two herbivores), (d) intraguild predation (one intermediate species feeds on the other).

on predators, prey face two opposite trends. On one hand, their habitat is reduced, which tends to decrease regional abundance. On the other hand, habitat loss largely affects predators, and so, reduces predation pressure, which tends to increase the abundance of prey. Once predators became extinct, the trade-off between the two opposite forces disappears, and prey abundance decreases much faster with additional habitat loss. Thus, the rate of decrease depends on the amount of habitat already destroyed.

Melian and Bascompte (2002) explored this problem in more detail using Pimm's basic modules (fig. 5). They showed that the sensitivity of the top predator to habitat loss depends on the food web module to which it belongs. In particular, omnivorous predators (fig. 5(b)) are more resistant to habitat loss than predators embedded in other types of food web modules. Also, the extinction threshold of predators is lower for top-down control than for donor control,[1] but this difference attenuates with decreasing trophic level. Whether predictions arising from food web modules scale up to more complex communities needs further testing.

All these results based on simplified food webs indicate that the pattern of species interactions is a key ingredient in an ecosystem's responses to habitat loss. The next step requires going beyond small, predefined subgraphs, and exploring the full complexity of rich multispecies assemblages.

---

[1]For example, when changes in population size of a prey may be relatively independent of predator's population, while predator's dynamics is critically dependent on the availability of its prey.

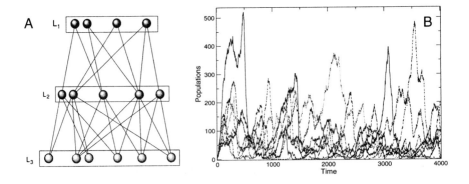

FIGURE 6   Spatially explicit, 3-trophic-level model of many species interactions. The model is a general framework for simulating the dynamics of multispecies ecosystems in fragmented landscapes. Network structure (A) is easily defined through a matrix of weighted interactions connecting producers, herbivores, and predators (here indicated as light, dark, and darker gray balls, respectively). In the version presented here, omnivory is not considered. The model exhibits complex dynamics, with fluctuations in population numbers and species turnover (B).

4   ECOSYSTEM MELTDOWN UNDER FRAGMENTATION

In order to understand the whole impact of habitat loss and fragmentation on complex ecosystems, we need to consider both species diversity and spatial constraints. Predicted diversity losses as habitat destruction increases with time rely on species-area relationships (Pimm and Raven 2000). But the structure of ecological networks has been shown to be no less important (Solé and Montoya 2001; Dunne et al. 2002). Additionally, models should consider the possibility of immigrating species from regional pools as well as species turnover.

The effects of habitat loss in complex, multispecies food webs are explored here by means of a stochastic spatial model comprising three multispecies trophic levels. We present an earlier framework (Solé et al. 2000; McKane et al. 2000) defined by a simple model of species interactions which has been shown to display most of the statistical properties characteristic of diverse communities. These include species-abundance relationships, connectance-species stability curves, species-area patterns, and other relevant dynamical properties (Solé et al. 2002). The model considered here is an example of an interacting particle system (Durrett and Levin 1994). Here a two-dimensional lattice $N = L \times L$ of available sites is considered, together with a pool $\Sigma(S)$ of possible species. Each site can be occupied (or not) by a single individual of a given species. Sites are assumed to be identical, except for the binary distinction of "available" or "destroyed." Interactions among individuals are introduced through a random matrix Ω, and the strength of species interactions is drawn from a uniform distribution. This matrix is fixed and has a predefined connectivity C.

In our setting, we consider three trophic levels. The set of all species in a regional pool, $\Sigma(S)$, is defined as

$$\Sigma(S) = \{0, 1, 2, ..., S\}, \tag{7}$$

where 0 indicates empty space. Here we consider three trophic levels (fig. 6(a)). The previous set is subdivided into three classes plus empty spaces:

$$\Sigma(S) = \{0\} \cup \Sigma_1(S) \cup \Sigma_2(S) \cup \Sigma_3(S) \tag{8}$$

with the three levels having the same numbers of (potential) species per level:

$$\Sigma_1(S) = \{1, 2, ..., [S/3]\}$$
$$\Sigma_2(S) = \{[S/3] + 1, ..., [2S/3]\}$$
$$\Sigma_3(S) = \{[2S/3] + 1, ..., S\}$$

(here $[x]$ indicates integer part). The following four rules are then applied.

IMMIGRATION

An empty site 0 is occupied by a species randomly chosen from the set of (non-empty) species with probability μ, i.e., $A \in \Sigma(S) - \{0\}$:

$$0 \xrightarrow{\mu} A . \tag{9}$$

For simplicity, we use the same immigration rate for all species. Here, we take $\mu = 10^{-4}$.

DEATH

All occupied sites can become empty with some fixed probability e_i:

$$A \xrightarrow{e_i} 0 \tag{10}$$

here the same probability is applied to all species, with $e_i = e = 0.02$.

COLONIZATION OF EMPTY SITES BY PLANTS

Plants can occupy empty sites by colonization, as defined in Levin's model. Here a given species $A \in \Sigma_1$ will occupy a neighboring empty site,

$$0 + A \xrightarrow{c} 2A \tag{11}$$

with some probability of colonization c.

BIOTIC INTERACTIONS

Pairs of individuals belonging to different species will interact through a probabilistic community matrix Ω. The entries of this $S \times S$ matrix are the probabilities of pairwise interactions. Such interactions are defined in terms of rules:
two species $A, B \in \Sigma(S)$ having non-zero entries Ω_{AB} can interact, provided
they are nearest neighbors. Given the three-layer structure of the community
under consideration, three different cases need to be considered. The first set are
competitive interactions among species within $\Sigma_1(S)$. A given individual from
species i will invade a neighboring patch occupied by a different species j

$$P_{ij} = \pi[\Omega_{ij} - \Omega_{ji}] \,, \tag{12}$$

where $\pi[x] = x$ when $x > 0$ and zero otherwise. This probability of an interaction
occurring in the system between species i and j is a measure of the interaction
strength linking these species.

Interactions between levels $1 \leftrightarrow 2$ and $2 \leftrightarrow 3$ are also defined in terms
of probabilistic rules. For example, a predator $B \in \Sigma_3(S)$ will prey on a site
occupied by a species $A \in \Sigma_2(S)$ with probability Ω_{AB}. In this context, the
propagation of predators through space requires the presence of their prey. If
$\Omega_{AB} = 0$, no expansion will take place. The same rule applies to pairs of species
$C \in \Sigma_2(S)$ and $D \in \Sigma_1(S)$.

For a very small $(S = 9)$ pool, the structure of the matrix used in our model
is a block matrix with a well-defined structure:

$$\Omega = \begin{pmatrix} w_{11} & w_{12} & w_{13} & 0 & 0 & 0 & 0 & 0 & 0 \\ w_{21} & w_{22} & w_{23} & 0 & 0 & 0 & 0 & 0 & 0 \\ w_{31} & w_{32} & w_{33} & 0 & 0 & 0 & 0 & 0 & 0 \\ w_{41} & w_{42} & w_{43} & 0 & 0 & 0 & 0 & 0 & 0 \\ w_{51} & w_{52} & w_{53} & 0 & 0 & 0 & 0 & 0 & 0 \\ w_{61} & w_{62} & w_{63} & 0 & 0 & 0 & 0 & 0 & 0 \\ 0 & 0 & 0 & w_{74} & w_{75} & w_{76} & 0 & 0 & 0 \\ 0 & 0 & 0 & w_{84} & w_{85} & w_{86} & 0 & 0 & 0 \\ 0 & 0 & 0 & w_{94} & w_{95} & w_{96} & 0 & 0 & 0 \end{pmatrix} \tag{13}$$

Only some non-zero values are possible: the allowed matrix elements are filled
with (connectance) probability C. Notice that here connectance also includes
competitive interactions among plants, and, therefore, its values will be typically
higher than those observed in real food webs.

The model exhibits complex dynamics, including wide time fluctuations and
(for small numbers of species) properties typically associated with deterministic
chaos. An example of the time series generated by the previous rules is shown in

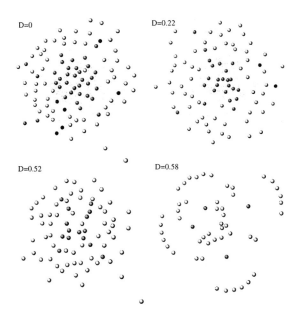

FIGURE 7   An example of the changes occurring in the web structure of the multispecies, three-trophic stochastic model. As habitat destruction increases, predators die off first, followed by a brief increase in herbivores, after which they, too, die off. Eventually, as herbivores are gone, an increase in the number of plants is observed, followed by a final crash for large enough destruction levels. Producers, herbivores, and predators are indicated as light, dark, and darker gray balls, respectively.

figure 6(b). Since we allow immigration (and thus potential invaders) to occur, as well as extinction, species turnover takes place.

In figure 7 we show an example of the network changes that take place in an $S = 300$ system with $C = 0.13, \mu = 10^{-4}, L = 200$. Here we start from a random initial condition and plot the network of trophic interactions (thus excluding competition) at some given step. In these pictures only the largest connected component is shown. For no destruction ($D = 0$) we find a network where the three levels are represented. As habitat loss increases, the network starts losing mainly top predators. The time series of population changes after the loss of some of these predators reveals wide fluctuations (particularly in herbivore populations) over a transient period of time. At $D = 0.22$ only two predators remain and for $D = 0.52$, no predators are present. At this stage, surviving herbivores tend to be generalists. Such tendency rapidly increases as D is further increased. For $D = 0.58$ only four highly connected (i.e., generalist) herbivores are left. A slight increase in habitat loss triggers the extinction of all of them.

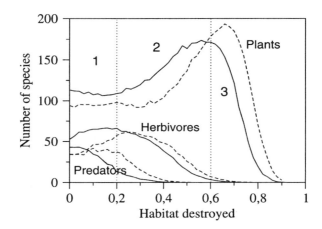

FIGURE 8 Number of species in each trophic level for different levels of habitat destroyed. Here two different probabilities of colonization by plants are used ($p = 0.15$,
straight lines; $p = 0.25$, dashed lines). Three different regions are defined and separated
by dotted lines (numbers in the figure). See text for details. Other parameters: $C = 0.28$
(including competitive interactions among plants).

Beyond the transient dynamics revealed by population changes over time, we
can look at ecological meltdown in a broader perspective. By increasing habitat
destruction in the stochastic model, we analyze how biodiversity is changed in
terms of the number of species present.

Our model shows that as habitat destruction increases, predators are lost
first, then herbivores, and finally producers. But the interaction between trophic
levels yields very interesting outcomes not predicted when a single trophic level is
observed alone. In figure 8 we can observe three different regions. The first region
($D < 0.2$) is characterized by a quick decline in predator species richness and a
consequent increase in herbivores and decrease in plant richness, corresponding
to a typical trophic cascade. For high colonization rates of plants, predators start
declining later, and, more interestingly, the trophic cascade is not observed; that
is, plant richness does not decrease as herbivores release from their predators.
Our results suggest that trophic cascades triggered by habitat loss are more
frequent when plant colonization is low or reduced by other factors (e.g., climate
change). The second region ($20 < D < 60$) is characterized by the decrease in
herbivore richness and the quick drop in plant richness. Indeed, the maximum
values for plant richness are observed within this region, together with predator-
free ecosystems. The third and last region corresponds to the collapse of the
plant assemblage ($D > 60$), because colonization is not enough to compensate for
habitat destruction. These results are robust to both the spatial scale considered
(we have used different lattice sizes) and the richness of the regional pool.

Our model shows that predator-free ecosystems can maintain numerous herbivores and plants. However, we believe this might only be true when interactions among plants within the second trophic level are mainly antagonistic (i.e., trophic, parasitic). When positive interactions between both trophic levels are more frequent than negative ones (i.e., when mutualistic interactions like dispersal and pollination are stronger and more frequent than feeding links), this particular result would not be observed. In this case, plant colonization relies upon herbivore dispersal, and, therefore, the dynamic behavior of plant richness would be similar to that of dispersers, and ecosystem collapse might occur earlier. Further developments in our model will introduce these mutualisms for predicting different biodiversity collapse scenarios depending on the relative importance of antagonistic and mutualistic interactions. Interestingly, mutualisms become more frequent in comparison to feeding links as we move towards the tropics, suggesting critical thresholds might be achieved faster in these ecosystems.

5 DISCUSSION

Habitat loss and fragmentation together with species invasions and climate change, are primarily responsible for the current extinction event. Degraded habitats lead to species loss and the success of exotic species. Climate change is slowly modifying the distribution ranges of many species and, indirectly, the spatiotemporal patterns of ecological networks. The synergies between habitat loss and fragmentation, climate change and network architecture are barely known today, and models can help to forecast possible scenarios of future biodiversity loss.

Previous studies of species removal from food webs (Pimm 1991; Solé and Montoya 2001; Dunne et al. 2002) have been concentrated in ecological networks lacking spatially explicit components. But all of them reveal that some species play a particularly relevant role in maintaining network stability. One important question here is how much are these keystone species affected by habitat loss and fragmentation. Is habitat loss specifically affecting them? This seems to be the case for many vertebrates (Terborgh et al. 2001; Wilson 1992) and is well exemplified by its effects on army ants. Army ants (Holldobler and Wilson 1990) are known to have a great impact in neotropical rainforests. Many species of vertebrates and the invertebrates associated with them will face extinction if the army ants disappear (Boswell et al. 1998). Given their special patterns of spatial search across the forest floor, that involves searching over wide areas, army ants have a huge impact on the spatial and temporal distribution of their ecological partners. Army ant raids facilitate species diversity by altering local ecological succession, creating a spatial mosaic of habitat patches. (Partridge et al. 1996). Additionally, in some rainforests a large number of bird species are specialized in following their raids, feeding on insect prey that are flushed out from leaf litter. The analysis of the effects of habitat fragmentation reveals that their populations

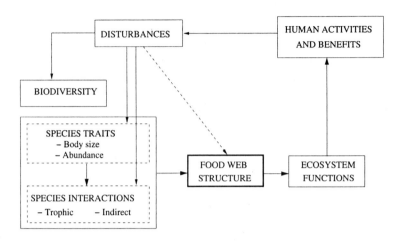

FIGURE 9 Conceptual summary that may guide future models aiming to integrate disturbances and ecological networks. Here food web structure emerges from the interaction between biodiversity (i.e., species richness), species traits (e.g., body size), and species interactions. These food webs support some of the functions and services provided by ecosystems. Humans, as part of the ecosystem, are benefited and can develop different types of activities, some of them beyond the resilience of the ecosystem. These disturbances may affect biodiversity in general, species with particular traits, or species interactions. This can promote further changes in food web structure and functioning, ultimately affecting our activities.

would get extinct once critical destruction thresholds are reached (Boswell et al. 1998).

Our analysis of multispecies ecosystems indicates that the effects of habitat loss combined with those associated with network structure interact in predictable ways. The pattern of species loss affects plant species directly through the loss of available space, but the trend (in terms of the total number of species present) can be reversed if top predators and herbivores benefit from decreasing diversity. Top predators are particularly affected by loss of habitat, as observed in field studies. Our study predicts that the demographic patterns of herbivore increases due to ecosystem meltdown are actually also present at the species level: the decay of diversity at higher levels is followed by an increase in herbivore diversity at intermediate levels of destruction. In this domain, plant diversity slightly decreases. Once habitat loss starts to trigger the decline of herbivores, a rapid increase in plant diversity is observed. This is actually an interesting counterpart to the explosion of herbivore populations following the loss of top predators. Under our approximations, it is predicted that plant diversity will increase when there is no grazing pressure. This is, of course, limited by the important role played by mutualistic interactions associated with species in other trophic levels. Such interactions should be included in future models of community dynamics.

Different extinction patterns have different effects on ecosystems, and they ultimately determine changes in network structure and ecosystem functioning (e.g., Raffaelli 2004). It is, therefore, crucial to identify the route to extinction triggered by different disturbances, addressing whether they primarily affect biodiversity in general (i.e., random extinctions), species with particular traits (body size, temperature tolerance), or species interactions (via spatiotemporal changes). It is time to incorporate more realistic extinction patterns into more realistic models. The integration of multitrophic complex communities and habitat loss presented here is an example of how to achieve this goal.

ACKNOWLEDGMENTS

The authors thank the members of the CSL for useful discussions. This work has been supported by an FIS2004-05422 and REN2003-03989 grant and by the Santa Fe Institute.

REFERENCES

Andrén, H. 1994. "Effects of Habitat Fragmentation on Birds and Mammals in Landscapes with Different Proportions of Suitable Habitat: A Review." *Oikos* 71:355–366.

Andrén, H. 1996. "Population Responses to Habitat Fragmentation—Statistical Power and the Random Sample Hypothesis." *Oikos* 76:235–242.

Barbault, R., and S. D. Sastrapradja. 1995. "Generation, Maintenance, and Loss of Biodiversity." In *Global Biodiversity Assessment*, ed. V. H. Heywood, 193–264. Cambridge, UK: United Nations Development Programme.

Bascompte, J., and R. V. Solé. 1996. "Habitat Fragmentation and Extinction Thresholds in Spatially Explicit Metapopulation Models." *J. Animal Ecol.* 65:465–473.

Bascompte, J., and R. V. Solé. 1998. "Effects of Habitat Destruction in a Prey-Predator Metapopulation Model." *J. Theor. Biol.* 195:383–393.

Boswell, G. P., N. F. Britton, and N. R. Franks. 1998. "Habitat Fragmentation, Percolation Theory and the Conservation of a Keystone Species." *Proc. Roy. Soc. Lond. B* 265:1921–1925.

Debinski, D. M., and R. D. Holt. 2000. "A Survey and Overview of Habitat Fragmentation Experiments." *Conserv. Biol.* 14:342–355.

Dunne, J. A., R. J. Williams, and N. D. Martinez. 2002. "Network Structure and Biodiversity Loss in Food Webs: Robustness Increases with Connectance." *Ecol. Lett.* 5:558–567.

Durrett, R., and S. A. Levin. 1994. "Stochastic Spatial Models: A User's Guide to Ecological Applications." *Phil. Trans. Roy. Soc. Lond. B* 343:329–350.

Hawkins, B. H. 1992. "Parasitoid-Host Food Webs and Donor Control." *Oikos* 65:159–162.

Holldobler, B., and E. O. Wilson. 1990. *The Ants.* New York: Springer-Verlag.

Holt, R. D., G. R. Robinson, and M. S. Gaines. 1995. "Vegetation Dynamics in an Experimentally Fragmented Landscape." *Ecology* 76:1610–1624.

Holyoak, M. 2000. "Habitat Subdivision Causes Changes in Food Web Structure." *Ecol. Lett.* 3:509–515.

Lawton, J. H., and R. M. May, eds. 1994. *Extinction Rates.* Oxford, UK: Oxford University Press.

McKane, A., D. Alonso, and R. V. Solé. 2000. "A Mean Field Stochastic Theory for Species-Rich Assembled Communities." *Phys. Rev. E* 62:8466–8484.

Melian, C. J., and J. Bascompte. 2002. "Food Web Structure and Habitat Loss." *Ecol. Lett.* 5:37–46.

Partridge, L. W., N. F. Britton, and N. R. Franks. 1996. "Army Ant Population Dynamics: The Effects of Habitat Quality and Reserve Size on Population Size and Time to Extinction." *Proc. Roy. Soc. Lond. B* 263:735–741.

Petchey, O. L., A. L. Downing, G. G. Mittelbach, L. Persson, C. F. Steiner, P. H. Warren, and G. Woodward. 2004. "Species Loss and the Structure and Functioning of Multitrophic Aquatic Ecosystems." *Oikos* 104:467–478.

Pimm, S. L. 1991. *The Balance of Nature? Ecological Issues in the Conservation of Species and Communities.* Chicago, IL: Chicago University Press.

Pimm, S. L., and P. H. Raven. 2000. "Biodiversity: Extinction by Numbers." *Nature* 403:843–845.

Purvis, A., J. Gittleman, G. Cowlishaw, and G. M. Mace. 2000. "Predicting Extinction Risk in Declining Species." *Proc. Roy. Soc. Lond. B* 267:1947–1962.

Raffaelli, D. 2004. "How Extinction Patterns Affect Ecosystems." *Science* 306:1141–1142.

Robinson, G. R., R. D. Holt, M. S. Gaines, S. P. Hamburg, M. L. Johnson, H. S. Fitch, and E. A. Martinko. 1992. "Diverse and Contrasting Effects of Habitat Fragmentation." *Science* 257:524–526.

Solé, R. V., and J. M. Montoya. 2001. "Complexity and Fragility in Ecological Networks." *Proc. Roy. Soc. Lond. B* 268:2039–2045.

Solé, R. V., D. Alonso, and A. McKane. 2000. "Connectivity and Scaling in S-Species Model Ecosystems." *Physica A* 286:337–344.

Solé, R. V., D. Alonso, and A. McKane. 2002. "Self-Organized Instability in Complex Ecosystems." *Phil. Trans. Roy. Soc. Lond. B* 357:667–681.

Swihart, R. K., Z. Feng, N. A. Slade, D. M. Mason, and T. M. Gehring. 2001. "Effects of Habitat Destruction and Resource Supplementation in a Predator-Prey Metapopulation Model." *J. Theor. Biol.* 210:287–303.

Teixidó, N. 2003. *Analysing Benthic Communities in the Weddell Sea (Antarctica): A Landscape Approach.* Ph.D. diss., Alfred Wegener Institüt.

Terborgh, J., L. Lopez, P. Nuñez, M. Rao, G. Shohabuddin, G. Orihuda, M. Riveros, R. Ascanio, G. H. Adler, T. D. Lambert, and L. Balbos. 2001. "Ecological Meltdown in Predator-Free Forest Fragments." *Science* 294:1923–1926.

Tilman, D., R. M. May, C. L. Lehman, and M. A. Nowak. 1994. "Habitat Destruction and the Extinction Debt." *Nature* 371:65–66.

Wilcove, D. S. 1987. "From Fragmentation to Extinction." *Natural Areas J.* 7:23–29.

Wilcox, B. A., and D. D. Murphy. 1985. "Conservation Strategy: The Effects of Fragmentation on Extinction." *Amer. Natur.* 125:879–887.

Wilson, E. O. 1992. *The Diversity of Life.* New York: Norton and Company.

Woodroffe, R., and J. R. Ginsberg. 1998. "Edge Effects and the Extinction of Populations Inside Protection Areas." *Science* 280:2126–2128.

Biodiversity Loss and Ecological Network Structure

Jane Memmott
David Alonso
Eric L. Berlow
Andy Dobson
Jennifer A. Dunne
Ricard Solé
Joshua Weitz

1 INTRODUCTION

The world is currently experiencing exceptionally high rates of species extinctions, largely as as result of human activity (Lawton and May 1996). Currently, most conservation research on habitat destruction focuses on the species as the unit of study, looking at either the impact of habitat destruction on individual species, or collections of species from particular habitats. There is, however, increasing recognition that species and species lists are not the only, nor perhaps the best, units for study by conservation biologists. This is because species are linked to other species in a variety of critical ways, for example as predators or prey, or as pollinators or seed dispersers. Consequently, the extinction of one species can lead to secondary extinctions in complex ecological networks (Dunne et al. 2002a; Memmott et al. 2004; Solé and Montoya 2001). The presence of links between species can also lead to community closure after the loss of a species, with the result that this species cannot then be reintroduced (Lundberg et al. 2000). Moreover, in the case of restoration biology, restoration will not be sus-

Ecological Networks: Linking Structure to Dynamics in Food Webs,
edited by Mercedes Pascual and Jennifer A. Dunne, Oxford University Press.
   325

tainable for a given species unless the ecological links with other species are also restored (Palmer et al. 1997).

Food webs have long been used in ecological research and provide a description of the trophic links in a community. In addition to their obvious role as descriptions of community structure, food webs are being used increasingly as the basis for an experimental approach. For example, a food web from Belize (Lewis et al. 2002) was recently use to predict the likely role of indirect effects in structuring the community, these predictions were then tested by a field manipulation (Morris et al. 2004). While food webs may be the most commonly described ecological network, other types of interaction webs are increasingly being investigated that include a variety of trophic and non-trophic interactions such as pollination, seed dispersal, interference competition, habitat or shelter provisioning, and recruitment facilitation or inhibition (e.g., Bascompte et al. 2003; Dicks et al. 2002; Jordano et al. 2003; Menge 1995). Food webs and other ecological networks have not been widely applied to the field of conservation biology, but given the practical advances being made in food web construction (for example, ecoinformatics), the theoretical advances (for example, models of extinction dynamics), and the ongoing threat of biodiversity loss, now is a good time to begin to use ecological networks as a conservation tool.

Conservation biology aims to protect intact native ecosystems and restore degraded ones. Even when these conservation goals focus on protecting one individual threatened or endangered species, there is an increasing recognition that multispecies approaches are essential for success (Chapin et al. 2000; Costanza et al. 1997; Ehrenfeld 2000; Schlapfer et al. 1999). Multispecies considerations are also critical for any ecological monitoring or assessment designed to protect key ecosystem services, such as pollution control, pest control, water filtration, and water clarity (e.g., Carpenter and Kitchell 1988). At the root of these calls for multispecies approaches to conservation and restoration is an appreciation that species *interactions* are important. Ironically, conservation, restoration, or monitoring projects rarely collect data on the structure of these interactions or how they change over time and across gradients of human impacts. In this chapter we will outline five areas of research that we believe could make a significant difference to the speed with which ecological networks can be used in conservation biology.

2   QUESTION 1: THE NATURE OF HABITAT DESTRUCTION

2.1   BACKGROUND

Habitat destruction is a collective term for a variety of environmental troubles, each of which may have different effects on food web structure and, given that they often act concomitantly, may also interact with each other in unpredictable ways. While a frequent outcome of habitat destruction is species loss, whether the different types of habitat destruction lead to different patterns of species loss remains unknown.

2.2 WHAT IS KNOWN

Habitat destruction can be put into four main categories, none of which is mutually exclusive of the others:

a. Habitat removal leading to habitat fragmentation: This is probably the most widely known form of habitat destruction and leads to both habitat loss and habitat isolation. The hostility of the matrix between the remaining habitat fragments is likely to affect different species in different ways. For example, the rodents responsible for the spread of Lyme's disease are not affected by fragmentation whereas their competitors and predators are. This difference in matrix permeability leads to both an increase in the incidence of the rodents and in the prevalence of Lyme's disease in humans (Allan et al. 2003). This example demonstrates how food web structure can mediate unpredictable effects—here woodland fragmentation indirectly caused an increase in the prevalence of a disease affecting man. The effect of habitat fragmentation on food web structure is beginning to be addressed, for example, pioneering work by Gilbert et al. (1998) revealed that predators are both most affected by fragmentation and that they are most rescued by wildlife corridors. The experiment by Gilbert et al. (1998) combined a real (albeit micro) community with rigorous experimental design and teased apart the effects of habitat fragmentation on the different trophic levels. However, the work did not explicitly identify links between species; indeed no published work to date has replicated food webs on fragmented habitats.

b. Hunting: Hunting occurs at all trophic levels; for example canopy trees are removed from tropical rain forests in selective logging programs and vertebrate herbivores and predators are hunted as trophies and for pest control. Insects are rarely hunted, although over-collecting may contribute towards the demise of some butterfly species. Untold millions of fish, sea turtles, sharks, and manatee have been hunted from the Caribbean Ocean and today no coastal sea can be considered pristine (Jackson 2001). In addition to potentially driving the hunted species extinct, hunting will obviously also lead to the extinction of species dependent upon the hunted species, for example, host-specific species of fleas and lice (Stork and Lyal 1993).

c. The introduction of alien species: Aliens pose a significant threat to global biodiversity, second only to habitat loss (Schmitz and Simberloff 1997). They can occur at all trophic levels. Alien plants are a particularly serious threat given their ability to displace native plants, to change the composition of native plant communities, and to alter a range of ecosystem processes such as nutrient cycling and disturbance regimes. Alien herbivores constitute some of worst agricultural pests, damaging native ecosystems and costing many millions of pounds to control. Alien parasitoids, principally biocontrol agents, have been widely reported from natural habitats around the world (Boettner et al. 2000; Funasaki et al. 1988; Munro and Henderson 2002) and are the subject of considerable current concern (Pemberton and Strong 2000). Mammals are one of the most important groups of alien taxa, especially on islands (Courchamp et al. 2003). Alien insect predators can also be very common, for example in Hawaii, 38% of spiders, 23% of predatory beetles, and 14% of Neuroptera are alien species (Bishop Museum 2002). While extensive data exist on the distributions of alien species and their impacts on native species such as competitors, prey species, predators, pollinators, and parasites, and even their impact upon ecosystem properties, there is extraordinarily little data on how aliens are accommodated in food webs. In reality, only two datasets exist: Henneman and Memmott (2001) working in Hawaii and Schonrogge et al. (Schonrogge and Crawley 2000) working in the UK, both describe plant-herbivore-parasitoid networks invaded by alien insects.

How these three factors interact remains unknown. For example, if a community is fragmented, has had its top predators hunted out, and is also being invaded by aliens (a not uncommon combination of events), then do these three effects act independently or do they have a synergistic effect whereby the effect of the three together is worse than would be predicted from the effect of each individual threat? If we are to have powers of either prediction or remediation (restoration) then the data concerning the impact of habitat destruction on food webs are essential.

2.3   WHAT NEEDS TO BE KNOWN

Currently, it is not known whether food webs collapse simply in the opposite order to the way in which they were assembled or whether there is a "first in last out" (FILO) order of collapse. Essentially, these questions ask whether the restoration and recovery of ecosystem function are the mirror image of the processes that lead to decline and collapse. In essence, is there any hysteresis to ecosystem function? Here hysteresis represents the history dependence of the system; thus, if you push on something and it yields, does it spring back completely when released? If it does not, it is exhibiting hysteresis. Or are there situations in which destructive activity focuses on different trophic levels, while recovery always proceeds from the bottom up? Destructive activity varies with trophic position; thus, plants at the base of a food chain will be lost at rates slightly slower than the average; the decomposers, such as worms, soil mites and bacteria will be lost at the slowest rates. In contrast, the species that feed directly on plants will be lost at a faster rate, while the charismatic tigers and eagles that feed at the top of the food chain are lost at the fastest rate (fig. 1). An understanding of how food webs both collapse and reassemble is crucial both in expanding our understanding of how food webs are structured, and also in determining how the

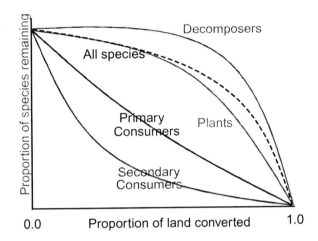

FIGURE 1   The loss of biological diversity and ecosystem services as natural habitat is eroded. The dashed line gives the classic species area decline in abundance such that a 90% loss of area dooms 50% of the original species to extinction. The lines indicate that species at different levels in the food chain will be lost at different rates.

ecosystems services they provide will decline as ecosystem functions break down due to species loss (or decline in abundance). Examples of ecosystem collapse and reassembly should provide insights into how species loss or addition leads to change in food-web structure. An important way to phrase these questions should be from the perspective of ecosystem and food web hysteresis; specifically, in which ways do the processes of food web/ecosystem assembly resemble those of food web/ecosystem collapse?

   The classic example of ecosystem collapse comes from John Terborgh's work on Lago Guri in Venezuela (Lambert et al. 2003; Terborgh et al. 2001; Terborgh et al. 1997). This lake was flooded in the late 1980s to create a huge artificial lake that would drive a huge hydroelectric dam. Flooding a forested river created a series of islands of different sizes that could be studied as their faunas and floras collapsed. Island biogeography theory predicts that larger islands should conserve a greater diversity of species, which in turn should maintain a higher proportion of ecosystem services. Terborgh's work uses island size as an adjunct for time in the process of ecosystem collapse; the smaller islands have lost more species and are thus closer to total faunal and floral "meltdown." The work illustrates that food web collapse is driven by the initial loss of species from the highest trophic levels; this in turn leads to the release from predation of herbivorous species at intermediate trophic levels. The increase in the abundance of the most competitive herbivorous species in turn selects for plants that can withstand herbivory. In the most degraded habitats, leaf cutter ants are only the herbivores that can exploit these plants, which eventually die from defoliation. The general pattern observed here is first a shortening of the web followed by a sequential collapse.

An important contrast to Lago Guri are the studies of Krakatau in the straits between Java and Sumatra (Thornton et al. 2002; Thornton et al. 1988). The volcanic eruption in 1883 completely destroyed all life on Krakatau. However, biologists have visited and surveyed the island, and the surrounding newly emergent islands, with an almost decadal frequency; this has provided a unique set of data on how island communities reassemble themselves. While most studies have focused on these data as ways of testing the laws of island biogeography, they represent a rich source of information that could be exploited by the food web community. When the species living on Krakatau are divided among different trophic levels, we see first a rapid increase in plant species at the base of the web. Herbivorous insects are blown in on the wind, followed by marine and terrestrial bird and bat species. The marine birds feed predominantly on fish, but significantly enhance the soluble nutrients on the island in the guano dumps around their nesting sites. The insectivorous birds and bats feed on the herbivorous insects. Later in the sequence of arrivals are frugivorous birds and bats, which manage to exploit the fruit produced by regurgitated and defecated seeds from the migratory species passing through the islands. When the buildup of trophic diversity on Krakatau is examined, there are some hystereses like similarities to the patterns observed in Lago Guri. There is an initial rapid assemblage of weakly interacting plant and animal species (to about 1920) followed by a slower and more steady increase in diversity at all trophic levels as interactions between species multiply, and intra- and interspecific regulation begin to replace colonization and persistence as the forces structuring the community.

The Krakatau data can be used to compare potential indices of ecosystem resilience and stability. The most widely used index is mean trophic position; this index is calculated by categorizing each species in a trophic level (e.g., plants—level 1; herbivores, level 2, etc.) and then calculating the mean for all the species in the community or food web. The index is most widely used in studies of overexploitation of fisheries, where the mean trophic level of species removed from the fishery is estimated (Pauly et al. 1998). Almost by definition, this will exclude many species from lower trophic levels from the calculation, as few fisheries actively exploit plankton. A logical extension of this calculation is to calculate the variance, or the standard deviation of trophic position. When we undertake this for the Krakatau data we find that the standard deviation provides a more sensitive index of food web change than mean trophic level. In particular, it is intriguing that mean trophic level quickly equilibrates as the Krakatau community assembles, while the standard deviation takes longer to stabilize and then does so once the different trophic levels seem to grow at roughly equal rates. Similar patterns are observed when this approach is applied to the Lago Guri data for community meltdown (Dobson unpublished). In both cases, community change is most rapid when the standard deviation of mean trophic position falls below unity. It would be an interesting exercise to calculate the standard deviation of mean trophic level for other food webs, particularly those that are collapsing. It is probable that food webs with standard deviations of mean trophic level that fall below unity are in the process of collapse.

3   QUESTION 2: HOW DO SPECIES INTERACTIONS MEDIATE SPECIES EXTINCTIONS?

3.1   BACKGROUND

With most, if not all ecosystems facing mild to extreme stress due to a variety of anthropogenic factors, the question of whether species interactions amplify or ameliorate such perturbations takes on practical importance. Increasingly, conservation biologists and ecologists are called on to help set priorities for conservation and restoration programs. Understanding and predicting ecosystem robustness to perturbations should be a fundamental part of making sustainable conservation decisions. Food webs and other types of ecological networks can provide a very useful framework for beginning to evaluate the robustness of ecosystems to species loss, and the tendency of complex species interactions to mediate ecosystem responses. By robustness, we refer to a type of stability that focuses on the persistence of features of interest in a system in response to perturbations, particularly those not normally experienced by the system in its development or history (Jen 2003). Thus, robustness is a useful way to think about ecosystem response to perturbations such as species loss.

This section discusses a particular aspect of robustness: how differences in ecological network structure may mediate the degree to which primary species losses, however they are caused, may be followed by additional secondary extinctions. In a conservation setting, this type of research can help focus effective protection or intervention on less robust habitats, taxa, and ecological processes. This can help direct policy efforts and increased resources towards more fragile systems whose persistence may be relatively more threatened.

3.2   WHAT IS KNOWN

Food webs provide a useful framework for beginning to assess the potential robustness of multispecies assemblages to biodiversity loss, as well as the impact of losing species with different functional or structural roles within ecological networks. One recent approach for assessing ecosystem robustness uses empirical data from ecological networks and makes use of simulation techniques from network structure analysis (e.g., Albert et al. 2000). As shown in several well-resolved food webs (Solé and Montoya 2001) and pollination webs (Memmott et al. 2004), secondary extinctions are much greater when species with many linkages to other species go extinct, compared to random loss of species. These studies suggest that regardless of the ecosystem, the order of species loss and the pattern of linkages among species strongly influence the rate of secondary extinctions. From a topological perspective, increased connectance (links per species[2]) across food webs buffers the impact of species losses: As connectance increases, the robustness of food webs also increases, because, on average, species have more potential prey items (Dunne et al. 2002b).

These types of analyses also need to be done with models that include dynamics among interacting taxa. Results based on network structure may differ from what is seen in a dynamical framework. For example, a 20-species Lotka-Volterra dynamical food web has been used to study random versus ordered species losses in model systems (Ives and Cardinale 2004). By assigning species a "tolerance" to a theoretical stressor (e.g., climate change, but it could as easily represent sensitivity to habitat fragmentation or to pollutants) and then removing species randomly versus based on their level of intolerance to the stressor, it was found that the average tolerance of remaining species is greater under ordered removals versus random removals. Thus, robustness (what the authors call resistance) of the system to future change was greater under ordered loss of species. As with the previous network structure studies, the authors found that the order of species loss, given how species interact with each other as well as their tolerance to a perturbation, is very important for understanding secondary effects. Similarly, the order of species loss was found to have a profound effect on the extinction dynamics of two large plant-pollinator communities (Memmott et al. 2004). Risk is not equal for all species, but instead will be greater for species of high trophic position, rare species, and specialists (e.g., Gilbert et al. 1998). Pollinators are considered to be at higher risk of anthropogenic extinction than plants, due to their higher trophic position and other aspects of their biology (e.g., Kearns et al. 1998; Kevan and Baker 1983). By assigning pollinators a higher risk of extinction in their models, Memmott et al. (2004) ameliorated the collapse of the network.

Perhaps even more importantly, some authors also suggest that because the food web structure and dynamics change with each extinction, whether random or ordered, there is an unpredictability to compensatory dynamics that can change the potential impact of the future loss of a particular species (Ives and Cardinale 2003). In their view, this unpredictability "argues for 'whole-ecosystem' approaches to biodiversity conservation, as seemingly insignificant species may become important after other species go extinct" (Ives and Cardinale 2004). This result is consonant with a finding from a previously described network structure analysis (Dunne et al. 2002b). Dunne and colleagues showed that while ordered removals of the least-connected species often lead to few secondary extinctions, in about a third of the webs examined such ordered removals resulted in relatively high secondary extinctions. Unlike robustness of webs to loss of either most-connected or random species, which is positively correlated with connectance, robustness of food webs to loss of least-connected species is not correlated with any obvious network structure metric (i.e., species richness, connectance, omnivory). Thus, the loss of apparently topologically insignificant species could lead to large negative effects, in a way not obviously predictable from the overall structure of the food webs.

Other models of food-web dynamics that go beyond equilibrium Lotka-Volterra-type dynamics to include more ecologically plausible assumptions about nonlinear, non-equilibrium dynamics, network structure and diversity, and include ecological and evolutionary adaptive dynamics (e.g., Brose et al. 2004; Drossel et al. 2001; Kondoh 2003; Williams and Martinez 2004) will also be useful for exploring issues concerning the robustness of ecosystems to perturbations such as species loss due to habitat loss or climate change. The major challenge will be to bridge the gap between model and empirical ecosystems (whether from lab or field) to usefully predict robustness of ecosystems to different types of perturbation and species loss.

3.3 WHAT NEEDS TO BE KNOWN

Several interrelated research questions emerge out of the broad question of the robustness of multispecies assemblages:

a. Do food webs from different habitat types (e.g., terrestrial vs. aquatic, desert vs. rain forest, fresh-water vs. estuarine vs. marine, boreal vs. equatorial, aboveground vs. belowground) respond to perturbations in predictably different ways?

b. How robust are food webs to different types of perturbation (e.g., habitat damage, loss and fragmentation; species loss/removal and invasion; chemical and climatic stresses)?

c. How robust are food webs to varying perturbation intensity or order? To multiple perturbations? Are impacts likely to be synergistic? Are there thresholds of response?

d. Ecological robustness can be assessed in many different ways (e.g., persistence of species diversity, retention of ecosystem function, low probability of destructive invasions, etc.). Do different types of robustness preclude each other in the form of tradeoffs, or are they complementary?

e. Ecological robustness can be assessed for different organizational levels and ways of grouping taxa (e.g., particular populations vs. a small set of interacting populations vs. a single trophic level vs. a food web for a particular habitat vs. a mosaic of dynamically shifting habitat types with loosely linked food webs). How do particular types of robustness, or the lack thereof, at different levels relate to each other?

f. Are there general characteristics of the network structure, diversity, or dynamics of ecological networks that that can be used to predict robustness to perturbations, or can be used to suggest ways to preserve or increase desired types of robustness?

Research that touches on some of these issues for diverse, multitrophic-level systems is just beginning, and there is much more work to be done. Most of the largely theoretical work concerning complexity-stability in food webs, and the experimental and observational work concerning diversity-ecosystem function (see both briefly reviewed in McCann 2000) helps identify the potential robustness of ecosystems to perturbations. However, both of these research approaches tend to focus on more narrow definitions of internal system stability rather than on system responses to external perturbations. Each approach also lacks the combination of empiricism (characteristic of diversity-ecosystem function research) and multitrophic-level inclusion (characteristic of complexity-stability research) that is necessary to facilitate advances that may help conserve natural systems and answer the robustness research questions posed previously. An example of how the two aspects can be fruitfully combined is research that used a theoretical (but plausible) bioenergetic food-web model, parametrized with data from a 29-taxa version of the Benguela marine food web, to look at the likely impact that culling fur seals would have on fisheries in that system (Yodzis 1998, 2000). We have highlighted another approach that integrates empirical data on ecological network structure with simulations of species loss to assess ecosystem robustness (Solé and Montoya 2001; Dunne et al. 2002b; Memmott et al. 2004). This type of research has been made possible by the fact that over the last decade, more detailed data for food webs and other types of ecological networks have become available, as well as by increased computing power. Wider deployment of methods discussed at the end of this chapter will facilitate rapid compilation of future data. Improved data, combined with better analysis and modeling methods, can support crucial new research on the robustness of ecosystems to a variety of perturbations.

4 QUESTION 3: MINIMUM VIABLE AREAS FOR FOOD WEBS

4.1 BACKGROUND

Ecology has contributed two basic insights into the biology of extinction: larger areas hold more species than smaller areas and larger populations persist longer than smaller ones (Bond 1994). One key observation from field studies concerns the impact of the removal of top predators (and large mammals in general) in unleashing top-down effects. The loss of such species can promote subsequent loss of further species and eventually lead to a so-called "ecological meltdown." Such a phenomenon has been reported from predator-free forest fragments. The loss of predators generates strong imbalances shown by the disproportionate increase in the densities of prey and severe reductions of seedlings and saplings of canopy trees (Terborgh et al. 2001). An important issue here is that when looking at the space available for a given predator species, it is often assumed that it is the whole space that counts, while actually the (average) space occupied by its prey is what counts. In this context, models reveal that the fragmentation thresholds experienced by a given species depending on a given resource are lower than one would expect by just considering the effective spatial area available (Bascompte and Solé 1998). In other words, the patchiness defined by the spatial distribution of the resource strongly constrains the metapopulation dynamics of those species exploiting it and thus their effective fragmentation thresholds.

4.2 WHAT IS KNOWN

The relevance of topology in enhancing the effects of habitat loss and fragmentation can be observed by modeling different types of community structure. Melian and Bascompte (2002) have studied small food webs involving different types of architectures. They considered four trophic web modules: simple food chain, omnivory, apparent competition, and intraguild predation. Each module contained three trophic levels with three or four species. Again, it was shown that extinction thresholds are not only determined by life history traits, competitive-colonization abilities, and landscape properties, but also by the complexity of the food web. In particular, omnivory was shown to confer the higher persistence to the top species, while interactions between the two intermediate species decrease its patch occupancy.

Beyond the few-species model, other approaches have been taken using a number of community structures with different internal organizations, for example hierarchical communities of competitors at one extreme (Stone 1995; Tilman et al. 1994) and neutral and quasi-neutral communities on the other (Solé et al. 2002). These have shown that community organization is relevant in determining the effects of habitat loss and its spatial patterning.

4.3 WHAT NEEDS TO BE KNOWN

Future research should address these topics by considering the food web structure and, in particular, should use approaches that seem to capture the fundamental traits of community organization. One particularly difficult problem here deals with the possibility of considering all kinds of community interactions (not just trophic links). How parasites and mutualistic interactions might modify previous predictions or alter the final outcome of ecosystem melting is largely unknown. Finally, when dealing with the sixth extinction at its largest scale, it is also important to provide useful tools for conservation efforts that deal with multiple scales and also include the intrinsic variability of climate and resources as an input (something largely ignored in current models of habitat destruction).

5   QUESTION 4: INTEGRATING HABITAT DESTRUCTION INTO FOOD WEB THEORY

5.1   BACKGROUND

In the past few years there has been a revival of interest in the theoretical analysis of food webs, motivated in part by advances in the theory of networks (Albert and Barabasi 2002; Strogatz 2001). While a comprehensive food web theory or theories is still very much in development, as illustrated in other chapters of this book, progress has been made in developing models that describe the static, statistical properties of food webs. The most successful of these theories is the niche model (Williams and Martinez 2000), which assumes that predation occurs in an approximately hierarchical fashion. Despite producing artificial webs in extraordinarily good accord with analysis of the largest available food webs, the niche model is not a panacea. The niche model does not purport to be a dynamic model, nor, despite the temptation, is it possible to equate niche value with a clearly measurable organismal trait such as body size. Further, there is no mechanism for maintenance of any particular food web pattern, either ecologically or over evolutionary time scales. All of these issues would be interesting courses for further theoretical investigation.

5.2   WHAT IS KNOWN

A standard approach to the study of food webs is the development of dynamical models describing changes in population densities in multitrophic systems (Drossel et al. 2001; Montoya and Solé 2003). A similar and seemingly worthwhile variation would be to develop individual-based models in which the food web and its organization constitute "emergent" phenomena resulting from competitive, predatory, and/or mutualistic spatially explicit interactions. All of these models may be viewed, with some historical deference, as the next generation of food web models based on Lotka-Volterra-type interactions (Cohen et al. 1990; Pimm and Lawton 1977; Post and Pimm 1983). The theoretical motivation for dynamic modeling is clear; food webs evolve on a variety of time scales, from the loss or introduction of a few species due to exogenous/endogenous causes, to wide-scale displacement of organisms because of climactic changes and/or anthropogenic disturbance. Only a dynamic, spatially explicit model can hope to explain the underlying ecological reasons for the observed patterns. Indeed, habitat degradation is occurring at unprecedented rates; therefore, it seems irresponsible, even reckless, to wait the necessary time to agree on a comprehensive food web theory. While progress is made on general principles underlying a dynamic, spatially explicit, and possibly stochastic food web theory, modeling should attempt to offer practical recommendations and partial answers to the crucial questions in conservation and restoration ecology. As such, we outline below what we believe to be the most promising avenues for integrating reasonable notions of habitat degradation, an inherently spatial phenomenon, into food web models.

5.3   WHAT NEEDS TO BE KNOWN

a. Non-random destruction of habitat: Complex ecological networks such as food webs arise from explicitly spatial and often heterogeneous interactions (Holt 2002). The degradation of habitat is heterogeneous in its spatial distribution. Theoretical models of habitat destruction rarely incorporate such complexities, relying instead on simplified notions of disturbance amenable to computational and mathematical analysis. Some examples of theoretical approximations of disturbance include homogeneous site loss (Neuhauser 1998; Tilman 1994), homogeneous barriers (Nakagiri and Tanaka 2004), as well as spatially correlated degradation in models with exogenous and endogenous heterogeneity (Bolker 2003). Spatially realistic models of degradation in community/regional level food web models would be well served by borrowing from the field of metapopulation biology, in which habitat fragmentation is the norm and the methodology to deal with it has developed in kind. Examples of useful concepts include the use of incidence functions to describe the flux of immigrants between disconnected patches as well as the scaling of extinction probability with patch area (Hanski 1994). The development of inexpensive and web-available remote sensing landscape data also bodes well for opportunities to integrate explicit land-use changes, habitat fragmentation, and vegetation changes into spatial models. This is relatively uncharted territory for theory and seems to require close affiliation between theory and empiricists working on local conservation and/or ecological studies. Finally, it is worth noting that not all habitat destruction is damaging. The pitch pines of the Pine Barrens are infamous for relying on forest fires to gain an advantage over other species that have diminished ability to re-sprout and recover after fires. Taking a broader perspective, there is also evidence that at intermediate levels of disturbance the benefits of spatially limited refuges outweigh the costs to the community as a whole (Bolker 2003; Weitz and Rothman 2003). In a sense, fragmentation can lead to a spectrum of solutions to the colonization/competition tradeoff, but in a completely connected world the best competitor wins. Deciding what degree of degradation a food web can sustain without collapsing is an important problem that theory is only beginning to confront (Melian and Bascompte 2002).

b. Species-area and link-area laws: Food webs are typically defined at a fixed scale, but the composition of local communities is determined by local as well as regional processes taking place at multiple scales. There is a widely held, though controversial, belief that local interactions and regional co-occurrence are responsible for the purported power-law scaling of species richness S with area A, $S = cA^z$ (Connor and McCoy 1979; Crawley and Harral 2001; Hubbell 2001). A recent theory proposes a means to extend the species-area law to a link-area law (Brose et al. 2004), that is, a higher-order indicator of the organization of communities across scales. This theory generates many new questions for research. For example, to what extent will such a link-area law hold in a fragmented habitat? Available empirical data might permit computational removal experiments so as to examine the robustness of the link-area laws to artificial destruction of habitat. Also, is it possible to integrate the niche model into link-area laws, or at least to parametrize the niche model and test its predictions with this scaling theory?

c. Theory for reserves: The flip side of developing theory for habitat degradation is the ongoing interest in the design of reserves (Bascompte et al. 2002; Lubchenco et al. 2003; Williams et al. 2004). The basic scientific question in reserve design may be likened to that of a Lagrange optimization problem from elementary calculus, wherein a student is asked to design a rectangular enclosure capable of holding the maximum number of cattle given a fixed amount of available fence. The answer is, of course, to construct a square enclosure. Unlike this simple example, the tradeoffs in reserve design are myriad and far more complex (Williams et al. 2004) and the "value" to be maximized is often contentious and under dispute. How many reserves should be set aside? Is it better for them to be contiguous or separated? What is the minimum useful reserve size and intra-reserve distance? How should the reserves be allocated dynamically in time (Costello and Polasky 2004)? Answering any/all of these questions is not strictly a matter of judging long-term sustainability of populations that depend on life-history, dispersal functions, evolutionary changes in organismal size, as well as on multispecies interactions. Benefits to tourism, public use constraints, and industrial access must also be considered. Some reserve design models incorporate explicit population dynamics, others do not. However, as pointed out in a review by Gerber et al. (2003), marine food webs are rarely considered. Doing so would certainly add complexity to an already difficult problem. But in the long run, understanding the future of a single population requires at least some understanding of its interdependencies within the larger web.

6   QUESTION 5: WHAT DATA DO WE NEED IF THIS RESEARCH AREA IS GOING TO GROW?

6.1 BACKGROUND

Data about the structure of species interactions will be invaluable in understanding key unanswered questions about how these interactions either respond to or mediate the effects of human perturbations (e.g., Ives and Cardinale 2004). Even coarse data about known species interactions that are relatively easy to identify (e.g., feeding links, pollinator links, and habitat or nest-site provisioning) can provide useful information about multivariate changes in community structure across environmental gradients or in response to a human impact (e.g., Brose et al. 2004; Harper-Smith et al. in review).

6.2 WHAT IS KNOWN

Biodiversity loss is a practical issue for conservation biologists and a problem for which solutions are needed in a short time frame. If food webs are to provide part of the solution, conservation biologists need a practical means of collecting food web data. Currently, putting together a food web is a time consuming (and therefore expensive) task. Insects form the vast bulk of biodiversity and most of the recent high quality and highly resolved webs have been multiyear studies (e.g., Lewis et al. 2002; Memmott et al. 1994; Muller et al. 1999; Rott and Godfray 2000). The time and resources needed for this approach are simply not available to conservation biologists working at the coalface. Bioblitz days and the Rapid Assessment Program were developed to provide a rapid, qualitative census of the biological resources of a natural area. While these represent a departure from the painstaking methods of food web construction, similar rapid techniques may need to be developed for conservation biologists who need species interaction data and who do not have the time/resources for a conventional approach.

A "quick and dirty" approach to characterizing interaction webs contrasts with the traditional structural food web approach of gathering detailed inter-action data from very few (usually one) sites; for example see Williams and Martinez (2000) and Dunne et al. (2004). The more intensive and less extensive approach to compiling typical food web data is partly the result of the traditional focus by structural food web ecologists on identifying "universal" patterns of link structure across diverse ecosystems, rather than on applications for conservation. Early food web theory was criticized as being based on incomplete webs (Cohen et al. 1990; Polis 1991) and this critique led to a more exclusive priority on gathering higher quality, detailed webs that could be used to test general theory about food web structure (Cohen et al. 1993; Dunne et al. 2004). While site-specific, intensive data on food web structure are essential in this context, we propose that data on trophic links or other interactions for one community type over time, across multiple sites, or across gradients of human impacts are invaluable for comparative analyses that quantify changes in community structure and suggest mechanisms warranting more detailed empirical investigation (Schoenly and Cohen 1991; Woodward and Hildrew 2001). Data on how these interaction webs, even if incomplete, are assembled over time within a given habitat can provide key information about community resilience (e.g., Harper-Smith et al. in review; Knapp et al. 2001) and can indicate whether or not restoration is likely to be successful.

We stress that useful ecological networks are relatively easy to construct, particularly by resource agency specialists who have an intimate knowledge of the natural history of their system, provided two conditions: (1) They are explicitly acknowledged as works in progress, and the relative certainty of each link is clearly specified; and (2) they are used primarily for comparisons within one system over space, time, or other conditions (e.g., invaded or not). The first condition stresses the heuristic value of the process by considering each web to be an operating hypothesis that helps target further empirical or theoretical investigations. The latter condition facilitates comparisons by minimizing variation among webs due to link uncertainties or investigator biases such as inconsistent levels of resolution within a web (e.g., top predators identified to species level and all phytoplankton lumped into one group).

6.3   WHAT NEEDS TO BE KNOWN

New "ecoinformatic" technologies for mining, managing, and analyzing data will greatly facilitate the development of extensive ecological network databases that are relevant to conservation. Many conservation sites or large-scale monitoring projects have, or will have, extensive species lists stored in spatial mapping databases. Examples include: (1) Sierra Nevada Forest Plan for bioregional monitoring (Powell and Blackwell 2001; USFS 2001) and (2) Millennium ecosystem assessment (www.millenniumassessment.org). For each species, attributes could be added that include its potential prey and predators from among the master species list, as well as species with which it is known to strongly interact in other ways (e.g., other species that provide critical nesting or nursery habitat). Then for any specified location, a *potential* network of interactions could be derived from this "meta-matrix" of interactions using the spatial information about which species coexist, and potentially interact, in that area. Thus, any interaction webs derived from this approach (called Rapid Webs) would be based on information compiled for all species found in the larger study area rather than on observations of interactions at each particular location. An example of how a Rapid Web program would work, the case of Lower Woods, a 285-hectare ancient woodland Nature Reserve in the UK, is presented in Box 1.

While these literature-based data may be incomplete or problematic to use for testing general food web theory, the consistency in how link structure is specified across all samples would facilitate comparisons among sites and across gradients of habitat fragmentation. Selective detailed studies to quantify links at individual sites could test the validity of this "meta-matrix" approach. Any developing species interaction database should include standards for specifying levels of certainty and sources for each link, and it should be flexible to accommodate new information that becomes available. Similarly, new developments in "semantic web" technology (www.w3.org/2001/sw/) could facilitate data mining of dispersed web-based documents to compile food webs, and provide for data inference based on information from related species and systems to fill in trophic links where local data are lacking. These types of semantic web technologies for biodiversity and food-web applications are being developed by the Semantic Prototypes in Research Ecoinformatics Project (SPIRE; http://spire.umbc.edu/).

Box 1: A Rapid Web Program

Lower Woods is a 285-hectare ancient woodland nature reserve in Gloucestershire, UK, and while a species list of 2771 species exists for the site (Martin 2004) no interaction data were available. However, to date (and this is still an ongoing project) nearly 7000 trophic interactions involving species found in this woodland are listed in the scientific literature and these, together with the species list (to determine whether both predator and prey are present), can be used to construct a food web. There are problems with the web dataset, the most obvious being that the literature is not a random sample of interactions; rather it is bias towards particular groups. For example, while there are 566 potential trophic links between the 19 mammal species and their prey species, very few links can be found between the 120 species of spider and their prey. Moreover entire groups of species are missing; for example, there are apparently no aphids, earwigs, or nematodes in the wood and only 24 species of fly! However, while there are great weaknesses in the dataset, there are also great strengths. For example, the web overall contains a number of different types of networks such as predation webs, herbivory webs, decomposition webs, and pollination webs. The web also bridges the divide between freshwater and terrestrial webs and between vertebrate and invertebrate webs. To put the web together using field techniques would take decades, cost considerable amounts of money, and would have required a team of entomologists, mammalogists, botanists, mycologists, arachnologists, herpetologists, lichenologists, and ornithologists. In comparison the literature search took three months and provided a serviceable web, albeit with (known) imperfections.

7 CONCLUSION

The five questions posed in this chapter demonstrate that there is considerable scope for adding a food web approach to the more traditional research tools used by conservation biologists. Moreover, as illustrated by the Lower Woods food web constructed from the literature, the time/cost constraints of conventional food web construction may perhaps be avoided. There may also be scope for using less detailed webs from replicate habitats as there are statistical advantages of having a small amount of data from numerous replicate habitats versus the traditional food web approach leading to a lot of data from very few (usually one) habitat. For example, data from a small section of a web for 20 degraded prairies could be compared to similar data from 20 old prairies to determine how they differ. This approach would be statistically more viable than comparing a detailed web from a single degraded prairie with a detailed web from a single old prairie. And the time it takes to gather data on the 40 small webs may indeed be comparable to the time it takes for the 2 detailed webs.

One question remains to be answered though—does knowing the structure of the food web actually help conservation biologists to fix it once it is damaged? For example, the impact of habitat destruction on connectance is of limited practical use to a team of conservation biologists who have five thousand dollars to restore a patch of heathland before the end of the financial year. We obviously believe that the approach has value, and is even valuable. For example, Lyzau Forup (2003) constructed replicated plant-pollinator networks for pairs of ancient and restored British heathlands. His data will provide very practical conservation advice on choosing how to restore ancient heathlands. Thus, the standard restoration approach involves positioning restored heaths adjacent to ancient heaths in the belief that insects will move from one to the other. Lyzau Forup's networks revealed that adjacent heaths shared few species and that factors other than having an adjacent ancient heath should perhaps be given greater priority in the decision making process. Moreover, his webs allowed him to ask whether an ecosystem service, that of pollination biology, was likely to have been restored.

REFERENCES

Albert, R., and A. L. Barabasi. 2002. "Statistical Mechanics of Complex Networks." *Rev. Mod. Phys.* 74:47–97.

Allan, B. F., F. Keesing, and R. S. Ostfeld. 2003. "Effect of Forest Fragmentation on Lyme Disease Risk." *Conserv. Biol.* 17:267–272.

Bascompte, J., and R. V. Solé. 1998. "Effects of Habitat Destruction in a Prey-Predator Metapopulation Model." *J. Theor. Biol.* 195:383–393.

Bascompte, J., H. Possingham, and J. Roughgarden. 2002. "Patchy Populations in Stochastic Environments: Critical Number of Patches for Persistence." *Amer. Natur.* 159:128–137.

Bascompte, J., P. Jordano, C. J. Melian, and J. M. Olesen. 2003. "The Nested Assembly of Plant-Animal Mutualistic Networks." *PNAS* 100:9383–9387.

Bishop Museum. 2002. Hawaii Biological Survey Databases. http://hbs.bishopmuseum.org/hbsdb.html (accessed May 4, 2005).

Boettner, G. H., J. S. Elkinton, and C. J. Boettner. 2000. "Effects of a Biological Control Introduction on Three Nontarget Native Species of Saturniid Moths." *Conserv. Biol.* 14:1798–1806.

Bolker, B. M. 2003. "Combining Endogenous and Exogenous Spatial Variability in Analytical Population Models." *Theor. Pop. Biol.* 64:255–270.

Bond, W. J. 1994. "Do Mutualisms Matter? Assessing the Impact of Pollinator and Disperser Disruption on Plant Extinction." *Phil. Trans. Roy. Soc. Lond. B.* 344:83–90.

Brose, U., A. Ostling, K. Harrison, and N. D. Martinez. 2004. "Unified Spatial Scaling of Species and Their Trophic Interactions." *Nature* 428:167–171.

Chapin, F. S., E. S. Zavaleta, V. T. Eviner, R. L. Naylor, P. M. Vitousek, H. L. Reynolds, D. U. Hooper, S. Lavorel, O. E. Sala, S. E. Hobbie, M. C.

Mack, and S. Diaz. 2000. "Consequences of Changing Biodiversity." *Nature* 405:234–242.

Cohen, J. E., E. F. Briand, and C. M. Newman. 1990. *Community Food Webs: Data and Theory.* New York: Springer-Verlag.

Cohen, J. E., R. A. Beaver, S. H. Cousins, D. L. Deangelis, L. Goldwasser, K. L. Heong, R. D. Holt, A. J. Kohn, J. H. Lawton, N. Martinez, R. O'Malley, L. M. Page, B. C. Patten, S. L. Pimm, G. A. Polis, M. Rejmanek, T. W. Schoener, K. Schoenly, W. G. Sprules, J. M. Teal, R. E. Ulanowicz, P. H. Warren, H. M. Wilbur, and P. Yodzis. 1993. "Improving Food Webs." *Ecology* 74:252–258.

Connor, E. F., and E. D. McCoy. 1979. "Statistics and Biology of the Species-Area Relationship." *Amer. Natur.* 113:791–833.

Costanza, R., R. d'Arge, S. Farber, M. Grasso, B. Hannon, K. Limburg, S. Naeem, R. V. O'Neill, J. Paruelo, R. G. Raskin, P. Sutton, and M. van den Belt. 1997. "The Value of the World's Ecosystem Services and Natural Capital." *Nature* 387:253–260.

Costello, C., and S. Polasky. 2004. "Dynamic Reserve Site Selection." *Res. & Energy Econ.* 26:157–174.

Courchamp, F., J. L. Chapuis, and M. Pascal. 2003. "Mammal Invaders on Islands: Impact, Control and Control Impact." *Biol. Rev.* 78:347–383.

Crawley, M. J., and J. E. Harral. 2001. "Scale Dependence in Plant Biodiversity." *Science* 291:864–868.

Dicks, L. V., S. A. Corbet, and R. F. Pywell. 2002. "Compartmentalization in Plant-Insect Flower Visitor Webs." *J. Animal Ecol.* 71:32–43.

Drossel, B., P. G. Higgs, and A. J. McKane. 2001. "The Influence of Predator-Prey Population Dynamics on the Long-Term Evolution of Food Web Structure." *J. Theor. Biol.* 208:91–107.

Dunne, J. A., R. J. Williams, and N. D. Martinez 2002a. "Food-Web Structure and Network Theory: The Role of Connectance and Size." *PNAS* 99:12917–12922.

Dunne, J. A., R. J. Williams, and N. D. Martinez. 2002b. "Network Structure and Biodiversity Loss in Food Webs: Robustness Increases with Connectance." *Ecol. Lett.* 5:558–567.

Dunne, J. A., R. J. Williams, and N. D. Martinez. 2004. "Network Structure and Robustness of Marine Food Webs." *Marine Ecol.-Prog. Ser.* 273:291–302.

Ehrenfeld, J. G. 2000. "Defining the Limits of Restoration: The Need for Realistic Goals." *Restoration Ecol.* 8:2–9.

Funasaki, G. Y., P. Lai, L. M. Nakahara, J. W. Beardsley, and A. K. Ota. 1988. "A Review of Biological Control Introductions in Hawaii: 1890 to 1985." *Proc. Hawaiian Entomol. Soc.* 28:105–160.

Gerber, L. R., L. W. Botsford, A. Hastings, H. P. Possingham, S. D. Gaines, S. R. Palumbi, and S. Andelman. 2003. "Population Models for Marine Reserve Design: A Retrospective and Prospective Synthesis." *Ecol. App.* 13:S47–S64.

Gilbert, F., A. Gonzalez, and I. Evans-Freke. 1998. "Corridors Maintain Species Richness in the Fragmented Landscapes of a Microecosystem." *Proc. Roy. Soc. Lond. Ser. B-Biol. Sci.* 265:577–582.

Hanski, I. 1994. "A Practical Model of Metapopulation Dynamics." *J. Animal Ecol.* 63:151–162.

Harper-Smith, S., E. L. Berlow, R. A. Knapp, R. J. Williams, and N. D. Martinez. In review. "Communicating through Food Webs: Visualizing the Impacts of Introduced Predators on Alpine Lake Communities in the Sierra Nevada, CA. I." In *Food Webs*, ed. P. de Ruiter, J. C. Moore and V. Wolters. Elsevier/Academic Press.

Henneman, M. L., and J. Memmott. 2001. "Infiltration of a Hawaiian Community by Introduced Biological Control Agents." *Science* 293:1314–1316.

Holt, R. D. 2002. "Food Webs in Space: On the Interplay of Dynamic Instability and Spatial Processes." *Ecol. Res.* 17:261–273.

Hubbell, S. P. 2001. *The Unified Neutral Theory of Biodiversity and Biogeography.* Princeton, NJ: Princeton University Press.

Ives, A. R., and B. J. Cardinale. 2004. "Food-Web Interactions Govern the Resistance of Communities after Non-random Extinctions." *Nature* 429:174–177.

Jackson, J. B. C. 2001. "What was Natural in the Coastal Oceans?" *PNAS* 98:5411–5418.

Jen, E. 2003. "Stable or Robust? What's the Difference?" *Complexity* 8:12–18.

Jordano, P., J. Bascompte, and J. M. Olesen. 2003. "Invariant Properties in Coevolutionary Networks of Plant-Animal Interactions." *Ecol. Lett.* 6:69–81.

Kearns, C. A., D. W. Inouye, and N. M. Waser. 1998. "Endangered Mutualisms: The Conservation of Plant-Pollinator Interactions." *An. Rev. Ecol. & System.* 29:83–112.

Kevan, P. G., and H. G. Baker. 1983. "Insects as Flower Visitors and Pollinators." *An. Rev. Entomol.* 28:407–453.

Knapp, R. A., K. R. Matthews, and O. Sarnelle. 2001. "Resistance and Resilience of Alpine Lake Fauna to Fish Introductions." *Ecol. Monogr.* 71:642–642.

Kondoh, M. 2003. "Foraging Adaptation and the Relationship between Food-Web Complexity and Stability." *Science* 299:1388–1391.

Lambert, T. D., G. H. Adler, C. M. Riveros, L. Lopez, R. Ascanio, and J. Terborgh. 2003. "Rodents on Tropical Land-Bridge Islands." *J. Zool.* 260:179–187.

Lawton, J. H., and R. M. May, eds. 1996. *Extinction Rates.* Oxford University Press.

Lewis, O. T., J. Memmott, J. Lasalle, C. H. C. Lyal, C. Whitefoord, and H. C. J. Godfray. 2002. "Structure of a Diverse Tropical Forest Insect-Parasitoid Community." *J. Animal Ecol.* 71:855–873.

Lubchenco, J., S. R. Palumbi, S. D. Gaines, and S. Anelman, eds. "The Science of Marine Reserves." Special Issue of *Ecol. App.* 13:S1–S228.

Lundberg, P., E. Ranta, and V. Kaitala. 2000. "Species Loss Leads to Community Closure." *Ecol. Lett.* 3:465–468.

Lyzau Forup, M. L. 2003. "The Restoration of Plant-Pollinator Interactions." Ph.D. diss., University of Bristol, Bristol.

Martin, M. H. 2004. "Lower Woods Nature Reserve Guild and Species List." Gloucestershire Wildlife Trust. http://www.gloucestershirewildlifetrust.co.uk (accessed July 6, 2005).

McCann, K. S. 2000. "The Diversity-Stability Debate." *Nature* 405:228–233.

Melian, C. J., and J. Bascompte. 2002. "Food Web Structure and Habitat Loss." *Ecol. Lett.* 5:37–46.

Memmott, J., H. C. J. Godfray, and I. D. Gauld. 1994. "The Structure of a Tropical Host Parasitoid Community." *J. Animal Ecol.* 63:521–540.

Memmott, J., N. M. Waser, and M. V. Price. 2004. "Tolerance of Pollination Networks to Species Extinctions." *Proc. Roy. Soc. Lond. Ser. B-Biol. Sci.* 271:2605–2611.

Menge, B. A. 1995. "Indirect Effects in Marine Rocky Intertidal Interaction Webs—Patterns and Importance." *Ecol. Monogr.* 65:21–74.

Montoya, J. M., and R. V. Solé. 2003. "Topological Properties of Food Webs: From Real Data to Community Assembly Models." *Oikos* 102:614–622.

Morris, R. J., O. T. Lewis, and H. C. J. Godfray. 2004. "Experimental Evidence for Apparent Competition in a Tropical Forest Food Web." *Nature* 428:310–313.

Muller, C. B., I. C. T. Adriaanse, R. Belshaw, and H. C. J. Godfray. 1999. "The Structure of an Aphid-Parasitoid Community." *J. Animal Ecol.* 6:346–370.

Munro, V. M. W., and I. M. Henderson. 2002. "Nontarget Effect of Entomophagous Biocontrol: Shared Parasitism between Native Lepidopteran Parasitoids and the Biocontrol Agent *Trigonospila brevifacies* (Diptera: Tachinidae) in Forest Habitats." *Envir. Entomol.* 31:388–396.

Nakagiri, N., and K. Tanaka. 2004. "Indirect Effects of Habitat Destruction in Model Ecosystems." *Ecol. Model.* 174:103–114.

Neuhauser, C. 1998. "Habitat Destruction and Competitive Coexistence in Spatially Explicit Models with Local Interactions." *J. Theor. Biol.* 193:445–463.

Palmer, M. A., R. F. Ambrose, and N. L. Poff. 1997. "Ecological Theory and Community Restoration Ecology." *Restoration Ecol.* 5:291–300.

Pauly, D., V. Christensen, J. Dalsgaard, R. Froese, and F. Torres. 1998. "Fishing Down Marine Food Webs." *Science* 279:860–863.

Pemberton, R. W., and D. R. Strong. 2000. "Safety Data Crucial for Biological Control Insect Agents." *Science* 290:1896–1897.

Pimm, S. L., and J. H. Lawton. 1977. "Number of Trophic Levels in Ecological Communities." *Nature* 268:329–331.

Polis, G. A. 1991. "Complex Trophic Interactions in Deserts—An Empirical Critique of Food-Web Theory." *Amer. Natur.* 138:123–155.

Post, W. M., and S. L. Pimm. 1983. "Community Assembly and Food Web Stability." *Math. Biosci.* 64:169–192.

Powell, B. E., and J. Blackwell. 2001. USDA Forest Service, Sierra Nevada Forest Plan Amendment: Record of Decision. http://www.fs.fed.us/r5/snfpa/ (accessed July 6, 2005).

Rott, A. S., and H. C. J. Godfray. 2000. "The Structure of a Leafminer-Parasitoid Community." *J. Animal Ecol.* 69:274–289.

Schlapfer, F., B. Schmid, and L. Seidl. 1999. "Expert Estimates about Effects of Biodiversity on Ecosystem Processes and Services." *Oikos* 84:346–352.

Schmitz, D.C. and Simberloff, D. 1997. "Biological Invasions: A Growing Threat." *Issues in Sci. & Tech.* 13:33–40.

Schoenly, K., and J. E. Cohen. 1991. "Temporal Variation in Food Web Structure— 16 Empirical Cases." *Ecol. Monogr.* 61:267–298.

Schonrogge, K., and M. J. Crawley. 2000. "Quantitative Webs as a Means of Assessing the Impact of Alien Insects." *J. Animal Ecol.* 69:841–868.

Solé, R. V., and J. M. Montoya. 2001. "Complexity and Fragility in Ecological Networks." *Proc. Roy. Soc. Lond. Ser. B-Biol. Sci.* 268:2039–2045.

Solé, R. V., D. Alonso, and A. McKane. 2002. "Self-Organized Instability in Complex Ecosystems." *Phil. Trans. Roy. Soc. Lond. Ser. B-Biol. Sci.* 357:667–681.

Stone, L. 1995. "Biodiversity and Habitat Destruction—A Comparative Study of Model Forest and Coral-Reef Ecosystems." *PNAS* 261:381–388.

Stork, N. E., and C. H. C. Lyal. 1993. "Extinction or Co-Extinction Rates." *Nature* 366:307–307.

Strogatz, S. H. 2001. "Exploring Complex Networks." *Nature* 410:268–276.

Terborgh, J., L. Lopez, P. Nunez, M. Rao, G. Shahabuddin, G. Orihuela, M. Riveros, R. Ascanio, G. H. Adler, T. D. Lambert, and L. Balbas. 2001. "Ecological Meltdown in Predator-Free Forest Fragments." *Science* 294:1923–1926.

Terborgh, J. T., L. Lopez, J. Tello, D. Yu, and A. R. Bruni. 1997. "Transitory States in Relaxing Ecosystems of Land Bridge Islands." In *Tropical Forest Remnants: Ecology, Management, and Conservation of Fragmented Commu-*

nities, ed. W. F. Laurance and R. O. Bierregaard, 256–274. Chicago, IL: University of Chicago Press.

Thornton, I. W. B., D. Runciman, S. Cook, L. F. Lumsden, T. Partomihardjo, N. K. Schedvin, J. Yukawa, and S. A. Ward. 2002. "How Important were Stepping Stones in the Colonization of Krakatau?" *Biol. J. Linnean Soc.* 77:275–317.

Thornton, I. W. B., R. A. Zann, P. A. Rawlinson, C. R. Tidemann, A. S. Adikerana, and A. H. T. Widjoya. 1988. "Colonization of the Krakatau-Islands by Vertebrates—Equilibrium, Succession, and Possible Delayed Extinction." *PNAS* 85:515–518.

Tilman, D. 1994. "Competition and Biodiversity in Spatially Structured Habitats." *Ecology* 75:2–16.

Tilman, D., R. M. May, C. L. Lehman, and M. A. Nowak. 1994. "Habitat Destruction and the Extinction Debt." *Nature* 371:65–66.

U.S. Forest Service. 2001. Sierra Nevada Forest Plan Amendment: Record of Decision (ROD). U.S. Department of Agriculture. http://www.fs.fed.us/r5/snfpa/ (accessed July 6, 2005).

Weitz, J. S., and D. H. Rothman. 2003. "Scale-Dependence of Resource-Biodiversity Relationships." *J. Theor. Biol.* 225:205–214.

Williams, J. C., C. S. ReVelle, and S. A. Levin. 2004. "Using Mathematical Optimization Models to Design Nature Reserves." *Frontiers in Ecol. & Envir.* 2:98–105.

Williams, R. J., and N. D. Martinez. 2000. "Simple Rules Yield Complex Food Webs." *Nature* 404:180–183.

Williams, R. J., and N. D. Martinez. 2004. "Stabilization of Chaotic and Non-Permanent Food-Web Dynamics." *Europ. Phys. J. B* 38:297–303.

Woodward, G., and A. G. Hildrew. 2001. "Invasion of a Stream Food Web by a New Top Predator." *J. Animal Ecol.* 70:273–288.

Yodzis, P. 1998. "Local Trophodynamics and the Interaction of Marine Mammals and Fisheries in the Benguela Ecosystem." *J. Animal Ecol.* 67:635–658.

Yodzis, P. 2000. "Diffuse Effects in Food Webs." *Ecology* 81:261–266.

Conclusion

Challenges for the Future: Integrating Ecological Structure and Dynamics

Mercedes Pascual
Jennifer A. Dunne
Simon A. Levin

The visual tension between the orderly representation of the *tocapu* designs in the *pallai* and the apparent scattering of elements in the *pampas* reflects an important Inca's aesthetic principle: a chaos and order that is analogous to the relationship between nature and culture.

— The Colonial Andes: Tapestries
and Silverworks, 1530–1830
(Metropolitan Museum, NY, 2004).

1 INTRODUCTION

The tapestries of the Incas, particularly the large areas of those tapestries called "pampas" that include a myriad of animals and plants, provide one example of humans' understanding of the complexity of nature. Remarkably, a sense of order emerges from these apparently chaotic representations because of their interplay with the more regular patterns included in the whole. Standing in front

Ecological Networks: Linking Structure to Dynamics in Food Webs,
edited by Mercedes Pascual and Jennifer A. Dunne, Oxford University Press.

of one of these marvelous representations of the natural world, one is reminded of the enormous complexity of ecological systems and of the past and present efforts to unravel the order behind it. The presence in the pampas of human figures and hybrid deities, with more and more apparent elements of outside influences, is a reminder of the importance of uncovering the critical properties of ecological networks that confer robustness to change. A better understanding of the properties that sustain the persistence of diverse ecosystems and their functions is crucial at this time of rapid environmental change and habitat loss, when perturbations of structure are unavoidable.

We started this book with a reference to Darwin's famous "tangled web," one of the earliest scientific images of the complexity of food webs and ecological networks, and one that continues to be relevant today. The challenging subject of food-web structure and its relationship to dynamics and stability has fascinated ecologists for a long time and has stimulated heated debates (Box 3, Pascual and Dunne Chapter 1; Dunne Chapter 2). In our view, the recurrent tensions regarding evidence for the existence of generalities in network structure and for valid theoretical approaches to address the dynamics of high-dimensional nonlinear systems have played an essential role in the development of the field. Pimm et al. (1991), in a review of an earlier period of intense research on food webs, argued for the existence of generalities in structure and for their relevance to dynamics. While improved data have challenged the generality of these early patterns, new regularities are emerging as are new approaches for investigating the interplay between structure and dynamics and their relationship to different aspects of ecosystem stability. In this book, we return to the subject at a time when network research has become a fundamental focus of the study of complex systems across many fields. While considerable advances have been made in the description of network structure for many types of systems, modeling the nonlinear dynamics of large networks remains a challenge for now and the future (Strogatz 2001).

The first section of this book reviewed what we know about food-web structure in light of these advances, as well as recent extensions to ecological networks that encompass other interactions, including parasitism and mutualism. The second and third sections illustrated recent and incipient efforts to tackle the full nonlinear dynamics of large networks in connection to structure, as well as recent theoretical advances in models for the evolution, assembly, and dynamics of ecological networks. Several chapters introduced computational approaches from the theory of complex adaptive systems, such as genetic algorithms and digital organisms, to highlight their potential utility for food-web research. The final section discussed issues at the intersection of ecological network research and more applied issues, for example, the importance of a multitrophic level perspective for conservation and restoration, with particular emphasis on the impact of habitat loss and fragmentation. In spite of all of the recent and ongoing research on networks in general and food webs in particular, a new synthesis of the relationship between structure and dynamics of complex networks remains premature. For that reason, we focus here on some areas we consider key for future theory to achieve such a synthesis in ecological network research. The sections that follow touch on three basic properties of complex systems that make the study of their dynamics especially difficult: their adaptive character, the nonlinearity of their interactions, and their large number of components. We emphasize a dynamic view of whole-system robustness that is consonant with, and can emerge from, instability at smaller scales.

2   WEAK INTERACTIONS AND THE NONLINEAR DYNAMICS OF LARGE NETWORKS

One aspect of structure likely to be particularly important to the dynamics of large networks is the distribution of variable "interaction strengths" of the links in the web. The theoretical finding that interaction strength is one key property that promotes persistence in nonlinear models of food webs has attracted considerable attention. Weak interactions have been proposed as the "glue" that binds large networks together (McCann et al. 1998), with ramifications for the value of biodiversity itself, and in particular for the potential importance of the many species in nature whose low abundances and weak per-capita consumption rates might otherwise be taken as evidence of a negligible role (McCann 2000). The potentially important functional role of weak interactions resonates with criticisms that studies of food-web structure are based on overly simplified binary link networks, and there have been strong pleas for experimental measurements of interaction strength in the field (Paine 1992; Lawton 1992). Much research on interaction strength followed those criticisms and built on earlier pioneering empirical (e.g., Paine 1980) and theoretical (e.g., Yodzis 1981) work. However, in almost all cases there remains a significant gap in how field- and model-based inquiries define and measure interaction strength, which slows progress (Berlow et al. 2004). We focus here primarily on theoretical work, but acknowledge that such approaches need to be thoughtfully connected to field-based experimentation (e.g., Brose et al. ms.).

The influential theoretical work of McCann et al. (1998) relied on dynamical models of low-dimensional food webs consisting of two to three species and a basal resource. This approach made possible detailed numerical simulations examining how the strengths of particular interactions affect the degrees of temporal variability in species abundances. In particular, the amplitude of population fluctuations and the likelihood of low population abundances were considered to be determinants of extinction. The main working hypothesis, that inhibiting strong interactions within a food web would promote persistence, was based on the detailed studies of another low-dimensional system, a three-species food chain in which a basal resource is consumed by an herbivore, and the herbivore is, in turn, consumed by a top predator (Hastings and Powell 1991). Chaotic dynamics with large population fluctuations were shown to occur in this system from the interplay of two coupled oscillators. It followed that the stabilization of the individual oscillators, obtained by either reducing their amplitudes or by promoting equilibrium behavior, should, therefore, eliminate chaos or large amplitude cycles at the level of the whole system, reducing the likelihood of population extinction. This hypothesis on the inhibition of strong interactions was addressed by studying a number of different configurations for low-dimensional food webs (McCann et al. 1998). The extrapolation from these theoretical findings to properties of real food webs, particularly from a small to a large number of species, is an obvious but non-trivial step, as demonstrated by the study of large nonlinear systems of coupled oscillators, which remains a challenge across physics and biology (Strogatz 2001). In ecology, only a handful of theoretical studies have tried to tackle the role of interaction strength in the dynamics of large communities (e.g., Wilmer et al. 2002; Kokkoris et al. 1999). The stabilizing role of weak links is at least in part consistent with the earlier finding that donor-controlled predator-prey interactions have a stabilizing role in Lotka-Volterra models of communities (Pimm 1982).

One such recent study on weak interactions in large networks assembled a web recursively through a random process (Wilmer et al. 2002). It demonstrated that communities with weaker mean interaction strength support a larger number of species than those with strong mean interaction strengths. In their model, species are allowed to sequentially invade the previously established community. The strengths of their interaction links are then drawn at random from a prescribed probability distribution whose variance provides a measure of interaction strength. Local stability is then evaluated as the criterion for the addition or subtraction of the species in the assembly process. Thus, although no explicit nonlinear dynamics are considered, the long-term number of species in the persistent communities fluctuates around a stationary stochastic equilibrium, whose mean can be quite large (e.g., approximately 200). Furthermore, as the number of species increases, the communities appear more robust, in the sense that fluctuations in species numbers relative to the mean abundance decrease in amplitude. Another theoretical study considered randomly assembled communities of competing species using Lotka-Volterra equations (Kokkoris et al. 1999). Mean competition strength falls as the assembly process develops and most established interactions are weak. However, close examination of the reported distribution of interaction strengths shows that this effect is not very pronounced, with a mode at 0.5 (values range between 0 and 1) and most values falling between 0.3 and 0.6. A key observation here is that communities become refractory to further invasion once their mean interaction strength is below that of the species pool from which invaders are drawn. It is apparent from these examples that the role of interaction strength in large ecological networks remains an open subject, as only local stability close to equilibria, and simple (type I) functional responses inherent in Lotka-Volterra formulations have been considered (but see Brose et al. in review). Nevertheless, these studies produce results that may be directly compared with long-term studies of community and food-web assembly, such as have been collected from Krakatau (Memmott et al. Chapter 14; Dobson in press).

More importantly, recent work indicates the limitations of just looking at the presence of weak interactions per se. With a generalized Lotka-Volterra model, Jansen and Kokkoris (2003) proposed that the variance, and to a lesser extent the mean, of interaction coefficients can be used to understand the local stability of ecosystems. In particular, an increase in the variance (for a constant mean) tends to decrease the probability that an equilibrium is both stable and feasible (i.e., positive equilibrium population densities). They argue that connectance per se, a basic measure of complexity in May's early work, is not a critical determinant of stability. Instead, they proposed an explanation for the U-shaped relationship between connectance and stability as described in Rozdilsky and Stone (2001). Their work indicates that the presence of weak interactions cannot be evaluated without consideration of the statistical distribution of interaction coefficients.

Ultimately, the emergent properties of food-web structure are likely to go beyond the properties of the statistical distribution of interaction strengths; this is apparent in recent findings by Kokkoris et al. (2002) on the impact of correlations between coefficients and in the combined theoretical and empirical work of Neutel et al. (2002) on soil food webs. For example, Neutel et al. (2002) showed that long loops of interactions in soil food webs contain relatively many weak links and suggested that this structure is a natural consequence of the biomass pyramid of ecosystems. In the absence of weak links, long loops would be destabilizing in the mathematical sense of requiring strong intraspecific regulation to achieve local stability. One critical implication is that it is not the statistical distribution of interaction strength that matters, but instead the distribution of such weights among the links of the webs. A similar message arises from an earlier and very different type of analysis by Haydon (2000), in which the linkage between weakly and strongly self-regulated elements appears critical to the local stability of community matrices. Thus, structure involves not just the presence and absence of the links, but also the distribution of interaction strengths across these links. These features are inseparable and cannot be evaluated independently. One challenge is to identify key properties of this distribution and develop ways to characterize them in network data. Dynamical consequences for large nonlinear systems remain an open issue.

3 ALLOMETRIC SCALING: A BRIDGE TO REALISTIC STRUCTURES?

Another major challenge in the study of the structure and dynamics of large food webs is the large number of parameters typically required for modeling studies. Given that the analysis of their nonlinear behavior is typically restricted to numerical simulations, the choice of realistic ranges of parameters becomes critical. The problem is exacerbated as we move away from simple linear functional responses for consumption. One potential avenue for realistic, simplified parametrizations is given by allometric scalings describing how specific parameters vary as functions of the size of organisms. A recent study shows that consideration of allometric relationships greatly increases the apparent stability in the dynamics of a well-described estuarine food web (Emmerson and Raffaelli 2004). By contrast, plankton ecosystem models that incorporate allometry to parametrize key rates show a striking propensity for unstable equilibria and pronounced fluctuations, raising questions about their realism. On the terrestrial side, remarkable allometric scalings have been described in both physiological and community patterns, conforming to power law patterns over many orders of magnitude (e.g., Enquist et al. 1998; West et al. 1997; Williams 1997). Much elegant theory has focused on explanations of these patterns but not on the use of the patterns themselves in the formulation of dynamical models for ecological interactions and associated networks. Consideration of deviations from both steady-state conditions and power-law patterns is likely to be important in understanding the dynamical properties that characterize the response of ecosystems to perturbations.

Regularities in the size structure of pelagic marine ecosystems were first described by Sheldon et al. (1972) who examined the distribution of particulate matter in different size classes for the upper ocean. He discovered that there are roughly equal amounts of material in logarithmically equal size intervals. Similarly, the size distribution for the living component of particulate matter in these systems conforms to a power-law relationship (of slope -1 when normalized by the nominal biomass of each class; Chisholm 1992). Theoretical efforts to explain this pattern of the biomass size spectrum were closely linked to properties of trophic interactions such as predator-prey size ratios and efficiency of biomass conversion (Borgmann 1987), and to allometric scalings at the physiological level (Platt and Denman 1978). Platt and Denman (1978) considered, for example, the biomass size spectrum as continuous and described the flow of biomass up this spectrum based on the weight dependence of growth and metabolism. Although the literature on the subject is enormous, the theory is rooted for the most part in steady-state assumptions and does not address nonlinear dynamics explicitly. In one exception, Silvert and Platt (1980) discuss potential consequences of nonlinear feedbacks, including the possibility that "small oscillations in food supply can drive large resonant swings in population," with a continuous biomass flow model that incorporates predation.

Separate from theory on biomass size spectra, plankton ecologists have more recently considered the nonlinear dynamics of size-based food webs (e.g., Moloney and Field 1991). Simple food chains with a single compartment per trophic level (i.e., phytoplankton and zooplankton) were deemed insufficient to represent the dynamics of plankton ecosystems. However, additional compartments result in additional parameters. This problem was circumvented by specifying parameters based on empirically determined body-size relationships from the literature. The resulting model was applied to three different ecosystems off the west coast of southern Africa (Moloney et al. 1991). From a more theoretical perspective, Pahl-Wostl (1993) built similar food-web networks with trophic levels subdivided into size classes. One general property of the models was the propensity for unstable equilibria, with rapid and pronounced fluctuations of the size classes including chaotic behavior. Interestingly, stability properties in these models are a function of hierarchical scale. Thus, at the trophic level, the total biomass obtained by aggregating the component size classes exhibited much more regular dynamics with equilibrium behavior. The realism of the fast oscillations at the lower levels has been controversial. For particular size classes representing specific taxonomic groups, the fluctuations were considered pathological in their short time scales (Moloney and Field 1991; Amstrong, 1999). However, field data appear to support the "dynamic, rapidly fluctuating plankton community" simulated for the summer season at one of the sites (the Agulhas bank; Moloney et al. 1991). The mathematical origins of the within-chain instabilities have been analyzed by Armstrong (1999), using particular symmetric structures of the food web in which functional forms and parameters within classes of a given trophic level were assumed to be equal. This work further examined and proposed modifications to the structure of the basic model that dampen the instabilities and are consistent with the observed ecosystem patterns of biomass size spectra.

These examples from plankton ecology provide only a glimpse at the fascinating potential for merging allometry and food-web dynamics. The model used by Martinez et al. (Chapter 6), based on the bioenergetic framework of Yodzis and Innes (1992), provides a natural framework within which to incorporate allometric scalings and consider the dynamical consequences in large networks. This type of formulation can specifically incorporate taxonomic diversity while using a currency of relevance to biogeochemistry, ecosystem science, and evolutionary studies (see also Gillooly et al. Chapter 8). The resulting theoretical predictions are also relevant to the recent characterizations of ecosystems that consider both trophic structure and biomass distribution within a single descriptive framework (e.g., Cohen et al. 2003).

4 VARIATIONS ON REPRESENTATIONS OF STRUCTURE

The graphical representation of food webs as networks of nodes and links (e.g., fig. 1, Pascual and Dunne Chapter 1) pervades our perception of "structure" as a tangible and fixed object of study. This perception is supported by an extensive body of work that revealed network-structure regularities in early collections of food-web data (e.g., Cohen et al. 1990a; Pimm et al. 1991) as well as in more recent and better-sampled data sets (see review by Dunne Chapter 2). Additional types and aspects of ecological interactions may greatly complicate our view of structure and open avenues for novel characterizations.

For example, the addition of parasitic links will require extensions of the analysis of networks to account for different types of links explicitly. This is fairly straightforward within a network structure framework (e.g., Huxham et al. 1996; Memmott et al. 2000), but it is more challenging and thus far less explored within a dynamical framework. The limiting step for assessing parasitic interactions within ecological networks is still the lack of detailed data, but new datasets continue to be compiled (e.g., Dobson et al. Chapter 4; Lafferty et al. in press; Thompson et al. 2005). Data for coastal salt marsh ecosystems (Lafferty et al. in press) reveals that the parasitic links are not distributed randomly upon the predator-prey part of the network but exhibit a higher degree of clustering (Warren et al. in prep.). This pattern is well captured by an extension of the structural niche model (Williams and Martinez 2000) developed by Warren et al. (in prep.). The analysis of data on other types of interactions, for example between plants and their mutualists such as pollinators and seed dispersers, has revealed patterns of nestedness clearly distinguishable from random patterns (Bascompte and Jordano Chapter 5). As with parasites, implications for dynamics remain to be examined. For example, do mutualistic interactions of this type have similar implications to those described for cooperative links in catalytic networks (Jain and Krishna 2001)? The relevance of mutualistic interactions for conservation is highlighted by Bascompte and Jordano (Chapter 5) and Memmott et al. (Chapter 14), with a suite of important open questions regarding the impacts of habitat destruction.

Another source of open questions on how we represent and analyze structure is the recognition that interaction strength varies not just as a function of the two species directly involved in the interaction, but with the abundance of other species in the network (see Bolker et al. 2003 for a review). One important example is phenotypic plasticity: changes in behavioral or physiological traits of a prey species that are driven by the presence of a predator, that in turn affect its own ability to forage (e.g., Peacor and Werner 2001). Because indirect interactions of this sort are not mediated by density but by traits, appropriate functional forms to represent them in typical equation-based models of food webs are not well understood. One promising theoretical approach is the use of individual-based models to derive functional forms at the population level. Peacor et al. (Chapter 10) describe a computational framework based on digital organisms that could be used for this purpose. As they demonstrate, a simple formulation of this "adaptive" approach reveals two novel mechanisms for a stabilizing role of phenotypic plasticity on dynamics. Interestingly, the model was not formulated to test specific a priori hypotheses, but instead served as a tool to explore the emergence of persistence mechanisms from individual-based assumptions. Experiments and other models typically formulated at the population level have also demonstrated the influence of phenotypic plasticity on dynamics (e.g., Ives and Dobson 1987; Kondoh 2003), but so far there is no consensus on its stabilizing role. Regardless of the modeling approach, the challenge of large systems remains.

With regard to structure, the modifications that result from phenotypic plasticity contrast with the lethal outcomes of predation typically represented in graphs of food webs. We can, therefore, think of structure as composed of not only the direct links between nodes but also the additional links that modulate their strength. Plastic interactions in both data and models may require a fundamental reexamination of structure.

5 NETWORKS AND FLOWS

Properties of structure will also change in networks that more accurately represent the flow of energy in the system. The importance of currency, specifically energy vs. population numbers, has been raised by Gilooly et al. (Chapter 8). Martinez et al. (Chapter 6) also discuss some of the pathways that should be added to their "bioenergetic" food web model for further realism, including detrital pathways. These extensions would close the gap between food-web models in community ecology that describe interactions between species and those in ecosystem science that incorporate detrital compartments and nutrient pools.

The view of food webs as flow networks goes back to the early "trophic-dynamic viewpoint" of Lindeman (1942) and is found in more recent food-web studies that use energy as the basic currency (Ulanowicz 1986; Christian and Luczkovich 1999; Heymans et al. 2002). The characterization of cycles is explicitly the core focus in flow networks, but for large systems requires enormous computational time, as it increases more than exponentially with the number of nodes. Another interesting element of structure related to cycles that is basic within graph theory but is only beginning to be applied to food webs (Allesina 2004; Allessina et al. in press) is that of strongly connected components (SCCs) (Skiena 1990; Read and Wilson 1998). SCCs apply only to directed networks and are defined as subsets of vertices that are connected through cycles: two species or vertices belong to the same SCC if we can find paths that connect them in both directions. In this way, SCCs define subsets of species connected through cycles, but their identification is much faster than that of cycles because computational time grows only linearly with system size (Tarjan 1972; Allessina et al. in press). These subsets are intriguing in that they provide one type of ecological compartment associated with flows (see also Krause et al. 2003) whose dynamical consequences in ecological networks are thus far unknown. A recent study of generic directed networks in the physics literature (Variano et al. 2004) raises the possibility that SCCs influence local stability and other aspects of robustness. Consideration of other types of subgraphs, such as those composed of species with the same number of links to all other species in the subgraph (Melián and Bascompte 2004), is also warranted.

6   FROM DETERMINISTIC TO STOCHASTIC MODELS OF NETWORKS

The high-dimensional food-web model of Martinez et al. (Chapter 6) achieves persistence for a large number of species for the full nonlinear dynamics of the system. This model is a hybrid in which the structure is determined by a static stochastic model that draws the links between species using different distribution rules (i.e., niche, cascade, random), while the dynamics are specified with a set of ordinary differential equations based on the bioenergetic model of Yodzis and Innes (1992). Numerical simulations show that the niche model, which best reproduces a set of structural properties of real food webs (Williams and Martinez 2000; Stouffer et al. in press), also confers higher dynamical persistence. Thus, structural properties of real food webs appear critical to dynamical properties of nonlinear food webs (see also review by Jordán and Scheuring 2004). Using an opposite approach, McKane and Drossel (Chapter 9) develop a dynamic and stochastic model of food-web evolution and assembly to produce ecological networks (McKane and Drossel Chapter 9; Drossel et al. 2001). In their "webworld" model, ecological time scales are considered to be much faster than evolutionary ones, with population dynamics treated deterministically allowed to reach an attractor between periods of evolutionary stasis. The two approaches are complementary and provide useful frameworks for more detailed investigations of the connection between structure and dynamics. In particular, the evolutionary-assembly approach can inform our understanding of how the relative temporal scales of ecology and evolution influence relationships uncovered with the structure-dynamics approach. It can also help to define relevant regions of parameter space, and allows investigation of a variety of interesting questions, including: Do evolutionary trajectories produce transient communities with longer persistence times as structure develops? Do communities become more robust to perturbations along this trajectory? Is robustness at the level of the system as a whole obtained by variability at the level of component species, as discussed below?

More extensive numerical explorations of parameter space are possible for dynamical systems that are still high-dimensional but have fewer species than the model of Martinez et al. (Chapter 6). Fussman and Herber (2002) provide an interesting example using parallel computation to address the likelihood of chaos in a deterministic model, similar to that of Martinez et al., as complexity, including species richness, increases. They show that chaotic dynamics become less likely, occurring in a more restricted region of parameter space, while the frequency of equilibria also decreases. These findings indirectly imply that cyclic behavior should become more common. It is unclear whether such conclusions extend to larger systems, but the question is clearly important for the relevance of the many studies that consider stability properties and linear approximations close to equilibria (see review by Dunne et al. 2005).

An earlier study using a hybrid model (Cohen et al. 1990b; Chen and Cohen 2001) provides one of the few examples of analytical results on the stability of large nonlinear networks. Lotka-Volterra equations were built upon a network of links produced by the cascade model. Their mathematical analysis tackles the global qualitative stability of the system, related only to the signs and not the strength of interactions. They demonstrated the existence of a phase transition that divides stable from unstable networks as a function of three stochastic parameters determining the type of predator-prey interactions. The possibility of extending this type of analysis to other nonlinear models of ecological networks opens many challenging, and interesting mathematical problems for the future. Analytical approaches would provide powerful tools to establish general results in spite of high dimensionality.

Much more work remains to be done to develop a general dynamical theory that explains within one framework not only structural properties of ecological networks (Dunne Chapter 2; Cartozo et al. Chapter 3) but also macroscopic community patterns. A number of macroscopic patterns have an important place in the description of ecological communities; examples include the well-known species-area curves and species-rank abundance curves. In a theory where these patterns emerge in a dynamic context, we can also hope to understand the implications of those patterns for instability. Stochastic assembly models are perhaps the best candidates to build such a theory. They are a powerful tool to capture the stochastic nature of immigration/colonization and extinction via species interactions, and have a venerable tradition in ecology (e.g., Post and Pimm 1982; Law and Blackford 1992). Hubbell's neutral theory of biodiversity (Hubbell 2001) uses this type of formulation, but does not incorporate multiple trophic levels and other aspects of trophic and non-trophic interaction networks. The stochastic assembly model by Solé et al. (2002) does encompass in an abstracted representation predator-prey interactions. In this model, structure emerges and continuously changes as the result of innovation (colonization) and extinction (Solé et al. 2002; McKane et al. 2000; Solé et al. 2000). Macroscopic (statistical) patterns at the level of the whole community, including species-area and rank abundance curves, emerge from processes at the level of interacting populations and individuals. Another emerging pattern, relevant to food-web structure, is given by the following relationship between species richness S and connectance C which sets a limit to diversity:

$$S \sim C^{-1+\varepsilon}, \tag{1}$$

where $\varepsilon^{-1} = \ln(1-\mu)/\mu$ and μ denotes the immigration probability of new species (Solé et al. 2002; McKane et al. 2000). Note that eq. (1) encompasses a wide array of curves depending on the specific value of the parameter μ. For a small immigration parameter, ε is close to zero and eq. (1) becomes the hyperbolic relationship previously proposed by May (1973) from the local stability analysis of a community of interacting species near equilibrium. The history of this earlier relationship and empirical evidence supporting considerable deviations from it, are reviewed in detail elsewhere (Dunne Chapter 2). Here, we note that eq. (1) is equivalent to the following expression relating the number of links to the number of species

$$L \sim S^{2-\frac{1}{1-\epsilon}} , \qquad (2)$$

where the definition of connectance $C = L/S^2$ has been used. Thus, depending again on the strength of immigration, eq. (1) is consistent with the power-law dependence of links with species observed in real food webs, including in the limit of high immigration, the hypothesis of constant connectance (see fig. 1 in Chapter 2). More importantly, eq. (1) sets a dynamic limit to diversity: communities in the model are set at this boundary by the dynamic tension between innovation (immigration, speciation) and species interactions. Thus, at the macroscopic level of the whole community, this curve functions as an "attractor" and macroscopic quantities such as species richness reach a stationary state, while at the microscopic level instability is rampant, with recurrent species extinctions and unpredictable population fluctuations (Solé et al. 2002). The idea that the stability of systems at aggregated, higher levels of organization are sustained by instability at lower levels has already been recognized as important in ecology (Tilman 1999; May 2001). We return to it below in the context of ecosystem robustness.

Assembly models are also a natural framework for incorporating evolution, an important area for the future. A number of formulations that couple evolutionary and ecological dynamics in a network context are beginning to emerge (see McKane and Drossel Chapter 9; Martinez Chapter 12 for a brief summary). Although it is still too early, a major goal of future theory should be to establish generalizations, which are closely connected to empirical network patterns from the results of multiple models. Abstract formulations (e.g., Solé et al. 2002) can play an important role in synthesizing and interpreting aspects of the behavior of more detailed models (e.g., McKane and Drossel Chapter 9; Peacor et al. Chapter 10). Comparisons should not be limited to theory within ecology, as similar types of questions on the emergence of complex structures, the existence of critical boundaries, and the associated dynamics of large systems are being asked in a variety of other fields with models of evolving networks.

One remarkable example concerns the evolution of an idealized chemical system with both "catalytic" and "inhibitory" interactions in a model that couples the dynamics of "species' populations" (at fast time scales) and of links between them (at slow time scales) (Jain and Krishna 2001). A selection process results in the extinction of populations with small abundances as the result of interactions between "species." Innovation occurs through the immigration of new types to the system. The model demonstrates the emergence of cooperation as a small autocatalytic set appears by chance but inevitably, providing the seed for the rapid growth of a complex network dominated by cooperative interactions (Jain and Krishna 2001). The differential equations used for the population dynamics are not immediately transposable to ecological interactions, but analogies for systems that include mutualistic as well as antagonistic interactions are intriguing. Questions on possible functional similarities between autocalytic sets and other types of cycles in ecological networks that include energy flows also arise. Furthermore, one interesting property seen in the dynamics of this model is the intermittent collapse of the system, with the rapid loss of a "keystone" species that "plays an important organizational role in the network despite its low population" (Jain and Krishna 2001). The system loses a large number of species and reorganizes again, with periods of relative stasis in diversity punctuated by intermittent catastrophic events. Thus, self-organization in this system leads to a complex structure poised in a state of apparent stability that is nevertheless susceptible to unpredictable and intrinsic catastrophic events.

Instability boundaries take many forms in models of network evolution and dynamics, including concepts of self-organized criticality (SOC) and phase transitions (see Solé and Goodwin 2000 for examples); they are described in the literature by different images including the well-known "edge of chaos" (Kauffman 1993). In an ecological framework, Solé et al. (2002) argue that SOC is too restrictive in its assumptions to fully encompass the type of critical boundaries relevant to ecological systems. They propose the more general concept of *self-organized instability* to describe a critical boundary in diversity generated by the tension between immigration and extinctions in the self-organization of communities. The questions of whether ecological networks function near instability boundaries and what their implications are for species' extinctions, population fluctuations, and responses to specific perturbations, can only be addressed by further theory in order to develop a general understanding of which processes underlie different types of boundaries, and which properties are key for differentiating empirically among these types. In a similar but much simpler context, Pascual and Guichard (2005) review the conditions leading to different types of criticality in spatial ecological systems with disturbance, including predation, disease, and abiotic perturbations. The temporal and spatial scales of recovery and disturbance appear key to the type of critical behavior observed, with clear implications for responses to perturbations and the occurrence of large shifts in system state. Are there similar organizational principles for types of "self-organized instabilities" in ecological networks? If so, what patterns and properties of specific processes should be measured to empirically identify a particular type in nature? The relative temporal and spatial scales of the different evolutionary and ecological processes are likely to play an important role.

7　SOME SPATIAL CHALLENGES

The spatial dimension is largely missing from studies of large networks (see review by Brose et al. in press), although it is a key component of the most serious current threats to ecological systems. Memmott et al. (Chapter 14) outline a series of critical questions on the effects of habitat destruction and fragmentation, and emphasize the importance of a multi-trophic perspective to pressing problems in conservation and restoration. Solé and Montoya (Chapter 13) illustrate the importance of a multi-trophic perspective with a network model that addresses habitat destruction. A spatial scaling theory of species diversity and their trophic interactions has been proposed by Brose et al. (2004). Their theory accurately predicts how trophic links scale with area and species richness across scales from microcosms to metacommunities. However, the issue of space is missing from most dynamical models of complex food webs.

Lotka-Volterra equations and their descendants used so far to address deterministic dynamics of ecological networks with many species are "mean-field" models. In these models, space is not treated explicitly because individuals are assumed to be well mixed and to interact at mean densities as a result of fast movement or long dispersal distances relative to the rate of interactions and their spatial extent, respectively. Local interactions among individuals are well known to be capable of dramatically affecting the dynamics of ecological models (e.g., Durrett and Levin 1994) and can also play a role in "adaptive" or plastic responses of prey (Peacor et al. Chapter 10). Clearly, however, tracking individuals for a large number of species would quickly become computationally prohibitive, and generate results difficult if not impossible to interpret. Some recent findings on individual-based models for a single predator and its prey indicate that simpler, low-dimensional models in which space is represented implicitly apply at the population level (Pascual et al. 2001; Pascual et al. 2002). Thus, although individual predators and prey interact locally in a spatial lattice, the aggregated dynamics of "global" abundances at the level of the whole lattice are well approximated with a simple "modified mean-field" model. Specifically, the predator-prey dynamics at the population level can be modeled as if space did not matter by modifying the functional forms of the mean-field equations. Additional parameters are introduced in the functional forms describing how the per-capita predation rate and prey growth rate vary with population abundances. The resulting equations exhibit dynamics that are more stable than their mean-field counterparts: permanent cycles become transient oscillations that approach equilibria, and transient cycles exhibit smaller amplitudes and faster convergence to equilibria. Although this type of scaling approach, from the local (or individual) to the global (or population) scale, is still under investigation for the case of only two interacting species, its potential application to the study of food webs should be considered.

In the same spirit, higher-dimensional numerical models for a large number of species' trophic interactions also do not treat space explicitly, but use different nonlinear functional responses to implicitly represent spatial and other non-trophic effects. Changing the form of the response (for example, from Type II to Type III) can be used to reflect the existence of prey refuges, or behavior modification of prey in the presence of predators (Martinez et al. Chapter 6). Dynamics can change dramatically with shifts in the functional response (e.g.,

Williams and Martinez 2004), yet another indication that spatial effects can be critical.

8   ROBUSTNESS AND SCALE

As will be clear from the discussion so far and from the papers in this volume, fundamental to the consideration of networks is the relation between structure and dynamics. How do the microscopic interactions among individuals lead, over ecological and evolutionary time, to the range of topological structures that are observed at higher levels of organization, and what are the implications for stability and robustness of macroscopic features of webs? Key features or robust systems relate to their heterogeneity and functional diversity, as well as levels of redundancy and modularity (Levin 1999). We have already seen, for example, that weak links can make food webs more robust by introducing a degree of modularity that limits the spread of catastrophic disruptions. But food webs are not systems that have been designed for robustness; their features are emergent. Is there any reason then to expect food webs to be maximally robust to disturbances at any scale? Such questions are rich areas for research.

Indeed, even when designing a web as a robust structure, issues of scale assume importance. What makes a system robust on short time scales may be at the cost of increased fragility on longer time scales; this concept of "robust, yet fragile" is indeed a central theme in recent advances in control theory (Carlson and Doyle 2002). Similarly, as already mentioned, robustness at higher levels of organization or broader spatial scales may be in spite of the fact, or indeed possibly because of the fact, that there is a lack of robustness at lower levels or at smaller scales. That very lack of robustness at the lower levels provides the capacity for the network to "adapt" in response to perturbations. This is not to imply that adaptive capacity is an evolutionary response to change; the focus of selection, of course, is at levels well below that of the whole food web. Indeed, food webs, by their diffuse nature, are clearly not evolutionary units. This, however, only makes more interesting a comparison of emergent properties with what would be expected were these designed systems.

9   IN CLOSING

In the concluding chapter of his inspiring book *SYNC*, Strogatz (2003) brings forward both the current excitement and the limitations of the studies of complex systems. He writes, "I hope I've given you a sense of how thrilling it is to be a scientist right now. It feels like the dawn of a new era. After centuries of studying nature by teasing it into smaller and smaller pieces, we're starting to ask how to put the pieces back together again." But after referring to a number of problems in a variety of fields on the emergence of order and self-organization, he also writes "we're still waiting for a major breakthrough in understanding, and it could be a long time in coming. I think we may be missing the conceptual equivalent of calculus, a way of seeing the consequences of myriad interactions that define a complex system." He refers here to breakthroughs in the mathematical analysis of qualitative dynamical properties, but more broadly perhaps to the emergence of a general theory of complex systems.

The dynamics of ecological networks are one particularly important challenge whose relevance and urgency cannot be overemphasized. From a theoretical perspective, we believe that general principles can result from the synthesis of many parallel efforts in the types of models discussed here. Given the enormous difficulties and time involved in empirical studies, more consideration should be given to the types of data required to assess specific hypotheses on stability, robustness, and the consequences of perturbationss.

REFERENCES

Allesina, S. 2004. "Ecological Flow Networks: Topological and Functional Features." Ph.D. diss., Environmental Sciences Department, University of Parma.

Allesina, S., A. Bodini, and C. Bondavalli. 2005. "Ecological Subsystems via Graph Theory: The Role of Strongly Connected Components." *Oikos* 110(1): 164–176.

Armstrong, R. A. 1999. "Stable Model Structures for Representing Biogeochemical Diversity and Size Spectra in Plankton Communities." *J. Plankton Res.* 21:445–464.

Berlow, E. L., A.-M. Neutel, J. E. Cohen, P. De Ruiter, B. Ebenman, M. Emmerson, J. W. Fox, V. A. A. Jansen, J. I. Jones, G. D. Kokkoris, D. O. Logofet, A. J. McKane, J. Montoya, and O. L. Petchey. 2004. "Interaction Strengths in Food Webs: Issues and Opportunities." *J. Animal Ecol.* 73 585–598.

Bolker, B., M. Holyoak, V. Krivan, L. Rowe, and O. Schmitz. 2003. "Connecting Theoretical and Empirical Studies of Trait-Mediated Interactions." *Ecology* 84:1101–1114.

Borgmann, U. 1987. "Models on the Slope of, and Biomass Flow Up, the Biomass Size Spectrum." *Can. J. Fish. Aquat. Sci.* 44:136–140.

Brose, U., E. L. Berlow, and N. D. Martinez. 2005 (in review). "Simple Explanations for Variable Keystone Effects in Complex Ecological Networks." *Nature.*

Brose, U., A. Ostling, K. Harrison, and N. D. Martinez. In press. "Unified Spatial Scaling of Species and Their Trophic Interactions." *Nature* 428:167–171.

Brose, U., M. Pavao-Zuckerman, A. Eklof, J. Bengtsson, M. P. Berg, S. H. Cousins, C. Mulder, H. A. Verhoef, and V. Wolters. In press. "Spatial Aspects of Food Webs." In *Dynamic Food Webs: Multispecies Assemblages, Ecosystem Development, and Environmental Change*, ed. J. Moore, P. de Ruiter, and V. Wolters. New York: Academic Press.

Carlson, J. M., and J. Doyle. 2002. "Complexity and Robustness." *PNAS* 99(1): 2538–2545.

Chen, X., and J. E. Cohen. 2001. "Global Stability, Local Stability and Permanence in Model Food Webs." *J. Theor. Biol.* 212:223–235.

Chisholm, S. W. 1992. "Phytoplankton Size." In *Primary Productivity and Biogeochemical Cycles in the Sea*, ed. P. G. Falkowski and A. D. Woodhead, 213–237. New York: Plenum Press.

Cohen, J. E., F. Briand, and C. M. Newman. 1990a. *Community Food Webs: Data and Theory*. New York: Springer-Verlag.

Cohen, J. E., T. Jonsson, and S. R. Carpenter. 2003. "Ecological Community Description using the Food Web, Species Abundance, and Body Size." *PNAS* 100(4):1781–1786.

Cohen, J. E., T. Luczak, C. M. Newman, and Z.-M. Zhou. 1990b. "Stochastic Structure and Nonlinear Dynamics of Food Webs: Qualitative Stability in a Lotka-Volterra Cascade Model." *Proc. Roy. Soc. Lond. B.* 240:607–627.

Christian, R. R., and J. J. Luczkovich. 1999. "Organizing and Understanding a Winter's Seagrass Foodweb Network through Effective Trophic Levels." *Ecol. Model* 117:99–124.

Dobson, A. P., D. M. Lodge, J. Alder, G. Cumming, J. E. Keymer, H. A. Mooney, J. A. Rusak, O. E. Sala, D. H. Wall, R. Winfree, V. Wolters, and M. A. Xenopoulos. 2005 (in preparation). "Habitat Loss, Trophic Collapse and the Decline of Ecosystem Services." *Ecology*.

Drossel, B., P. G. Higgs, and A. J. McKane. 2001. "The Influence of Predator-Prey Population Dynamics on the Long-Term Evolution of Food Web Structure." *J. Theor. Biol.* 208:91–107.

Dunne, J. A., U. Brose, R. J. Williams, and N. D. Martinez. 2005. "Modeling Food-Web Dynamics: Complexity-Stability Implications." In *Aquatic Food Webs: An Ecosystem Approach*, ed. A. Belgrano, U. Scharler, J. A. Dunne, and R. E. Ulanowicz. New York: Oxford University Press.

Durrett, R. and S. A. Levin. 1994. "On the Importance of Being Discrete (and Spatial)." *Theor. Pop. Biol.* 46:363–394.

Emmerson, M. C., and D. Rafaelli. 2004. "Predator-Prey Body Size, Interaction Strength and the Stability of a Real Food Web." *J. Animal Ecol.* 73:399–409.

Enquist, B. J., J. H. Brown, and G. B. West. 1998. "Allometric Scaling of Plant Energetics and Population Density." *Nature* 395:163–166.

Fussmann, G. F., and G. Heber. 2002. "Food Web Complexity and Chaotic Population Dynamics." *Ecol. Lett.* 5:394–401.

Harte, J., E. Conlisk, A. Ostling, J. L. Green, and A. B. Smith. 2005 (in press). "A Theory of Spatial-Abundance and Species-Abundance Distributions in Ecological Communities at Multiple Spatial Scales." *Ecol. Monogr.*

Hastings, A., and T. Powell. 1991. "Chaos in a Three-Species Food Chain." *Ecology* 72:896–903.

Haydon, D. T. 2000. "Maximally Stable Model Ecosystems can be Highly Connected." *Ecology* 81(9):2631–2636.

Heymans, J. J., R. E. Ulanowicz, and C. Bondavalli. 2002. "Network Analysis of the South Florida Everglades Graminoid Marshes and Comparison with Nearby Cypress Ecosystems." *Ecol. Model* 149:5–23.

Hubbell, S. P. 2001. *The Unified Neutral Theory of Biodiversity and Biogeography.* Princeton, NJ: Princeton University Press.

Huxham, M., S. Beany, and D. Raffaelli. 1996. "Do Parasites Reduce the Chances of Triangulation in a Real Food Web?" *Oikos* 76:284–300.

Ives, T., and A. Dobson. 1987. "Antipredator Behavior and the Population Dynamics of Simple Predator-Prey Systems." *Amer. Natur.* 130:431–447.

Jain, S., and S. Krishna. 2001. "A Model for the Emergence of Cooperation, Interdependence, and Structure in Evolving Networks." *PNAS* 98(2):543–547.

Jansen, V. A. A., and G. D. Kokkoris. 2003. "Complexity and Stability Revisited." *Ecol. Lett.* 6:498–502.

Jordán, F., and I. Scheuring. 2004. "Network Ecology: Topological Constraints on Ecosystem Dynamics." *Phys. Life Rev.* 1:139–229.

Kauffman, S. A. 1993. *The Origins of Order: Self-Organization and Selection in Evolution.* New York: Oxford University Press.

Kokkoris, G. D., A. Y. Troumbis, and J. H. Lawton. 1999. "Patterns of Species Interaction Strength in Assembled Theoretical Competition Communities." *Ecol. Lett.* 2:70–74.

Kokkoris, G. D., V. A. A. Jansen, M. Loreau, and A. Y. Troumbis. 2002. "Variability in Interaction Strength and Implications for Biodiversity." *J. Animal Ecol.* 71:362–371.

Kondoh, M. 2003. "Foraging Adaptation and the Relationship between Food-Web Complexity and Stability." *Science* 299:1388–1391.

Krause, A. E., K. A. Frank, D. M. Mason, R. E. Ulanowicz, and W. W. Taylor. 2003. "Compartments Revealed in Food-Web Structure." *Nature* 426:282–285.

Law, R., and J. C. Blackford. 1992. "Self-Assembling Food Webs: A Global Viewpoint of Coexistence of Species in Lotka-Volterra Communities." *Ecology* 73:567–578.

Lawton, J. H. 1992. "Feeble Links in Food Webs." *Nature* 355:19–20.

Levin, S. A. 1999. *Fragile Dominion: Complexity and the Commons*. Reading, MA: Perseus Books Group.

Lindeman, R. L. 1942. "The Trophic-Dynamic Aspect of Ecology." *Ecology* 23: 399–418.

May, R. M. 1973. *Stability and Complexity in Model Ecosystems*. Princeton, NJ: Princeton University Press.

May, R. M. 2001. *Stability and Complexity in Model Ecosystems*. Landmarks in Biology. Princeton, NJ: Princeton University Press.

McCann, K. S. 2000. "The Diversity-Stability Debate." *Nature* 405:228–233.

McCann, K., A. Hastings, and G. Huxel. 1998. "Weak Trophic Interactions and the Balance of Nature." *Nature* 395:794–798.

McKane, A., D. Alonso, and R. V. Solé. 2000. "A Mean Field Stochastic Theory for Species-Rich Assembled Communities." *Phys. Rev. E* 62:8466–8484.

Melián, C. J., and J. Bascompte. 2004. "Food Web Cohesion." *Ecology* 85:352–358.

Memmott, J., N. D. Martinez, and J. E. Cohen. 2000. "Predators, Parasitoids and Pathogens: Species Richness, Trophic Generality and Body Sizes in a Natural Food Web." *J Animal Ecol.* 69:1–15.

Moloney, C. L., and J. G. Field. 1991. "The Size-Based Dynamics of Plankton Food Webs. I. A Simulation Model of Carbon and Nitrogen Flows." *J. Plankton Resh.* 13:1003–1038.

Moloney, C. L., J. G. Field, and M. I. Lucas. 1991. "The Size-Based Dynamics of Plankton Food Webs. II Simulations of Three Contrasting Southern Benguela Food Webs." *J. Plankton Resh.* 13(5):1039–1092.

Neutel, A. M., J. A. P. Heesterbeek, and P. C. de Ruiter. 2002. "Stability in Real Food Webs: Weak Links in Long Loops." *Science* 296:1120–1123.

Paine, R. T. 1980. "Food Webs, Linkage Interaction Strength, and Community Infrastructure." *J. Animal Ecol.* 49:667–685.

Paine, R. T. 1992. "Food-Web Analysis through Field Measurement of Per Capita Interaction Strength." *Nature* 355:73–75.

Pahl-Wostl, C. 1993. "The Influence of a Hierarchy in Time Scales on the Dynamics of, and the Coexistence within, Ensembles of Predator-Prey Pairs." *Theor. Pop. Biol.* 43(2):184–216.

Pascual, M., and F. Guichard. 2005. "Criticality and Disturbance in Spatial Ecological Systems." *Trends Ecol. & Evol.* 20(2):88–95.

Pascual, M., P. Mazzega, and S. A. Levin. 2001. "Oscillatory Dynamics and Spatial Scale in Ecological Systems: The Role of Noise and Unresolved Pattern." *Ecology* 82(8):2357–2369.

Pascual, M., M. Roy, and A. Franc. 2002. "Simple Models for Ecological Systems with Complex Spatial Patterns." *Ecol. Lett.* 5:412–419.

Peacor, S. D., and E. E. Werner. 2001. "The Contribution of Trait-Mediated Indirect Effects to the Net Effects of a Predator." *PNAS* 98(7):3904–3908.

Pimm, S. L. 1982. *Food Webs*. Chapman and Hall. Reprinted in 2002 as a 2d ed. by University of Chicago Press.

Pimm, S. L., J. H. Lawton, and J. E. Cohen. 1991. "Food Web Patterns and Their Consequences." *Nature* 350:669–674.

Platt, T., and K. Denman. 1978. "The Structure of Pelagic Marine Communities." *Rapp. P. Reun. Cons. Int. Explor. Mer* 173:60–65.

Post, W. M., and S. L. Pimm. 1983. "Community Assembly and Food Web Stability." *Math. Biosci.* 64:162–192.

Read, R. C., and R. J. Wilson. 1998. *An Atlas of Graphs*. Oxford University Press.

Rozdilsky, I. D., and L. S. Stone. 2001. "Complexity can Enhance Stability in Competitive Systems." *Ecol. Lett.* 4:397–400.

Sheldon, R. W., A. Prakash, and W. H. Sutcliffe. 1972. "The Size Distribution of Particles in the Ocean." *Limnology and Oceanography* 17:327.

Silvert, W., and T. Platt. 1980. "Dynamic Energy-Flow Model of the Particle Size Distribution in Pelagic Ecosystems." In *Evolution and Ecology of Zooplankton Communities*, ed. W. C. Kerfoot, 754–763. Hanover, NH: The University Press of New England.

Skiena, S. 1990. "Strong and Weak Connectivity." In *Implementing Discrete Mathematics: Combinatorics and Graph Theory with Mathematica*, ed. S. Skiena, 172–174. Addison-Wesley.

Solé, R. V., and B. Goodwin. 2000. *Signs of Life: How Complexity Pervades Biology*. New York: Basic Books.

Solé, R. V., D. Alonso, and A. McKane. 2000. "Scaling on a Network Model of a Multispecies Ecosystem." *Physica A* 286:337–344.

Solé, R. V., D. Alonso, and A. McKane. 2002. "Self-Organized Instability in Complex Ecosystems." *Phil. Trans. Roy. Soc. Lond. B* 357:667–668.

Stouffer, D. B., J. Camacho, R. Guimera, C. A. Ng, and L. A. N. Amaral. 2005. "Quantitative Patterns in the Structure of Model and Empirical Food Webs." *Ecology* 86:1301–1311.

Strogatz, S. H. 2001. "Exploring Complex Networks." *Nature* 410:268–275.

Strogatz, S. H. 2003. *Sync: The Emerging Science of Spontaneous Order*. New York: Hyperion.

Tarjan, R. 1972. "Depth First Search and Linear Graph Algorithms." *SIAM J. Comp.* 1:146–160.

Thompson, R. M., K. N. Mouritsen, and R. Poulin. 2005. "Importance of Parasites and Their Life Cycle Characteristics in Determining the Structure of a Large Marine Food Web." *J. Animal Ecol.* 74(1):77–85.

Tilman, D. 1999. "The Ecological Consequences of Changes in Biodiversity: A Search for General Principles." *Ecology* 80:231–251.

Ulanowicz, R. E. 1986. *Growth and Development: Ecosystem Phenomenology.* New York: Springer Verlag.

Variano, E. A., J. H. McCoy, and H. Lipson. 2004. "Networks, Dynamics and Modularity." *Phys. Rev. Lett.* 92(18):188701-1–188701-4.

Warren, C., M. Pascual, K. Lafferty, and A. Kuris. 2005 (in preparation). "Parasitic Networks on Food Webs: Clustering, Nestedness and the Inverted Niche Model." Unpublished manuscript.

West, G. B., J. H. Brown, and B. J. Enquist. 1997. "A General Model for the Origin of Allometric Scaling Laws in Biology." *Science* 276:122–126.

Williams, N. 1997. "Fractal Geometry Gets the Measure of Life's Scales." *Science* (commentary) 276:34–35

Williams, R. J., and N. D. Martinez. 2000. "Simple Rules Yield Complex Food Webs." *Nature* 404:180–183.

Williams, R. J., and N. D. Martinez. 2004. "Stabilization of Chaotic and Non-Permanent Food-Web Dynamics." *European Phys. J. B* 38:297–303.

Willmer, C. C., S. Sinha, and M. Brede. 2002. "Examing the Effects of Species Richness on Community Stability: An Assembly Model Approach." *Oikos* 99(2):363–367.

Yodzis, P. 1981. "The Stability of Real Ecosystems." *Nature* 289:674–676.

Yodzis, P., and S. Innes. 1992. "Body-Size and Consumer-Resource Dynamics." *Amer. Natur.* 139:1151–1173.

Index